TISSUE ENGINEERING: ROLES, MATERIALS AND APPLICATIONS

TISSUE ENGINEERING: ROLES, MATERIALS AND APPLICATIONS

STEVEN J. BARNES AND LAWRENCE P. HARRIS
EDITORS

Nova Science Publishers, Inc.
New York

For permission to use material from this book please contact us:
Telephone 631-231-7269; Fax 631-231-8175
Web Site: http://www.novapublishers.com

LIBRARY OF CONGRESS CATALOGING-IN-PUBLICATION DATA

Tissue engineering : roles, materials, and applications / Steven J. Barnes and Lawrence P. Harris, editors.
p. ; cm.
Includes bibliographical references and index.
ISBN 978-1-60456-293-4 (hardcover)
1. Tissue engineering. I. Barnes, Steven J. II. Harris, Lawrence P.
[DNLM: 1. Tissue Engineering. 2. Biocompatible Materials. QT 37 T6169 2008]
R857.T55T57 2008
610.28--dc22

2007050808

Published by Nova Science Publishers, Inc. New York

CONTENTS

PREFACE

Tissue engineering is the use of a combination of cells, engineering and materials methods, and suitable biochemical and physio-chemical factors to improve or replace biological functions. While most definitions of tissue engineering cover a broad range of applications, in practice the term is closely associated with applications that repair or replace portions of or whole tissues (i.e., bone, cartilage, blood vessels, bladder, etc.). Often, the tissues involved require certain mechanical and structural properties for proper function. The term has also been applied to efforts to perform specific biochemical functions using cells within an artificially-created support system (e.g. an artificial pancreas, or a bioartificial liver). The term regenerative medicine is often used synonymously with tissue engineering, although those involved in regenerative medicine place more emphasis on the use of stem cells to produce tissues. This new book presents important research in the field.

Chapter 1 - Posttraumatic peripheral nerve defects repair continues to be a major challenge in restorative medicine and microsurgery. Although progress has been made in surgical techniques over the last 30 years, functional recovery after severe lesion of a major nerve trunk, especially for large nerve gaps is often incomplete and often unsatisfactory with long-time functional deficits. Treatment of lesions causing complete transection of peripheral nerve is primarily aimed at recreating continuity of the nerve to allow axonal regrowth into the nerve stump distal to the lesion. To date, the standard for bridging peripheral nerve defects involves the use of autografts. But disadvantages with this method include harvesting the autografts from a second operative site, with consequent loss of end-organ innervation, prolonged surgical procedures, and shortage of autogenous nerve graft material for extensive nerve injuries. This dictates research towards the development of biomimetic contructs as alternatives. With the tissue engineering rapidly progressing, tissue engineered bioartificial nerves have been served as new alternatives, which have the advantages of unlimited availability and avoidance of donor-site morbidity. Based on current concepts, this review summarizes various strategy of nerve tissue engineering, including different extracellular matrics, scaffolds, growth factors and seed cell sources that have been shown to promote peripheral nerve regeneration. Also, the importance of using appropriate animal models and criteria for evaluating tissue engineered nerve constructs suitability for bridging peripheral nerve gaps is highlighted.

Chapter 2 - Tissue engineering has emerged as a response to the problems associated with the substitution of tissues lost due to diseases or trauma. Nowadays, xenografts, allografts or autografts are commonly used to replace the damaged tissue. However, these options involve numerous drawbacks such as rejection, chronic inflammation and severe organ donor shortages. Tissue engineering keeps the promise to overcome these limitations by creating biological substitutes capable of replacing the injured tissue.

Cells, temporary 3D constructs and biochemical signals that trigger tissue regeneration cascades are the three major components for engineering tissues. The adequate combination of these components leads to the successful repair of damaged tissues. In addition to these three main elements, the use and development of bioreactors has come out as an essential tool for the study and achievement of engineered tissues under *in vitro* conditions while mimicking the *in vivo* environment.

Thus, tissue engineering is an interdisciplinary field that applies the principles of engineering and life sciences to develop biological substitutes that restore, maintain, or improve the tissue functions.

This chapter discusses the challenge of creating engineered bone tissues by combining biological and engineering skills. In particular, it will review the issue of stem cell biology, the fabrication methods and artificial materials used for 3D constructs, the principal growth factors used in bone applications and the key role played by bioreactors. More importantly, it will highlight the interdependence of all these parameters and the need for interdisciplinary research for a successful approach.

Chapter 3 - Cardiac diseases are leading causes of morbidity and mortality in men and women worldwide. Extensive damage to the heart is presently treatable by surgical reconstruction, implantable devices, and ultimately transplantation. These current treatments are "mechanical" and do not cure heart disease. The interdisciplinary field of tissue engineering has evolved to offer the hope of biologic, natural solutions for tissue regeneration and repair. While engineering an entire heart is an ambitious goal, the use of growth factors, cell therapy, and biomaterial scaffolds offers the promise of rebuilding the injured heart from its component parts. Of the many biomaterials that have been tested for such applications, collagen-based scaffolds are considered as one of the most attractive options. Collagens are highly biocompatible, biodegradable and possess weak antigenecity; and the physical properties of their gels can be controlled by regulating cross-linking reactions. In addition to these highly desirable characteristics, the heart's extracellular matrix is comprised mostly of collagens, making collagen a highly suitable choice for cardiac engineering purposes. This review discusses current progress made in the use of collagen biomaterials in the development of vascular grafts and heart valves, and in cell delivery and myocardial tissue engineering for the treatment of cardiac disease.

Chapter 4 - Electrospinning is a specialized fiber fabrication technique that uses the potential difference of a static electric field as the driving force to propel a fine, charged stream of polymer solution or melt from a capillary orifice to a grounded collection device. The fibers produced can range from 5 microns to 20 nanometers. Since polymer solution electrospinning is more widely used, the focus of this chapter will be entirely on this method.

Chapter 5 - Recently, the development of biomaterials for liver tissue engineering has focused on the design of biomimetic glycopolymers that can perform specific hepatocyte behaviors and direct liver regeneration through molecular recognition. Molecular recognition of glycopolymers by hepatocytes has been achieved by chemical modification of biomaterials with galactose moiety as a specific ligand that can incur specific interactions with asialoglycoprotein receptors (ASGPR) on the hepatocytes. The biomimetic glycopolymers potentially mimic many roles of extracellular matrix (ECM) in liver tissues, such as attachment, proliferation, migration, differentiation and survival of hepatocytes. This review discusses mechanism of specific interaction between galactose moieties of the galactose-carrying polymers and ASGPR of hepatocytes. Liver tissue engineering typically employs synthetic three-dimensional ECMs to induce long-term and stable differentiated hepatocyte functions. Therefore, the synthetic ECMs can be fabricated from two classes of biomaterials: natural polymers and synthetic ones. The criteria to design biomimetic glycopolymers such as galactose density, topology of ECM, deposited natural ECM, coculture system and cell source are addressed. The potential mechanisms on hepatocellular signaling by the ECM are also summarized.

Chapter 6 - Almost three decades have passed since tissue engineering started for the repair of diseased skin tissues of patients. Numerous studies have focused on the cells that are required for tissue engineering. Indeed, it is impossible to regenerate a tissue without cells. It should be, however, stressed that scaffold also plays a key role in tissue engineering, especially when tissue engineering aims at regenerating large, three-dimensional tissues. In this case, not only the cells in the tissue, but also the associated extracellular matrix has been diseased or lost. An important function of the scaffold is to temporarily substitute the lost extracellular matrix that had assisted the proliferation, differentiation, and functions of the cells that had presented in the pre-diseased tissue.

The purpose of this review article is to demonstrate how critical a scaffold is for the success of preclinical and clinical applications of tissue engineering. The resorbable biomaterial that has been most often selected in the tissue engineering studies described in this article is an L-lactide(LA)-ε-caprolactone(CL) copolymer with a molar LA/CL ratio of 50/50. This copolymer exhibits a bioabsorption kinetics ranging between that of poly(glycolide) (PGA) and that of poly(L-lactide) (PLLA). If a porous scaffold fabricated from this copolymer is to be used for *in situ* regeneration of large tissues, the scaffold should be reinforced with PGA or PLLA fabrics. The reinforced scaffold has proven very suitable, in particular, for the *in situ* cardiovascular tissue engineering using bone marrow cells.

Chapter 7 - In the current review, the authors aim to give an overview of different rapid prototyping (RP) methods which have been applied so far for the production of polymeric scaffolds for tissue engineering applications. After a general introduction on tissue engineering, scaffolds and biomaterials, both the well established RP techniques and the novel more recent developments in the field will be discussed. Comparison between RP and the classical techniques for producing scaffolds, each with their specific advantages and drawbacks will enable the reader to gain insight in the power of RP as an elegant scaffold production technique. Finally, an overview of some commonly used techniques for characterising the materials developed will be given.

Chapter 8 - To date, there is no ideal method for treating the segmental bone defect. Autograft bone is considered to be an ideal bone graft material because of its good biological properties and compatible mechanical strength, however, several disadvantages accompany with it. Current bone tissue engineering is using *in vitro* expanded mesenchymal stem cells and synthetic scaffolding materials to produce synthetic bone substitutes comparable to autologous bone. Such an approach is theoretically possible to repair damaged bone tissue with tissue engineered bone in clinics, instead of autologous bone; however, partially due to the nutrition problem it is difficult to produce large living tissue-engineered bone in practice at the moment. Increasing data is showing that osteoinductive synthetic materials can attract mesenchymal stem cells in non-osseous tissues and form ectopic bone. *In vivo* bone tissue engineering, by using osteoinductive synthetic materials to produce bone in soft tissues, would be another approach to produce synthetic "autologous" bone for bone repair. *In vivo* bone tissue engineering means to use the body as bioreactor to construct bone graft with intrinsic osteoinductive biomaterials in the non-osseous or osseous sites in absence of any exogenous growth factors or cells, then transplant the graft to repairing bone defect. Three parts are involved in this chapter. Part I, Osteoinductive materials for *in vivo* bone tissue engineering. The physicochemical properties of materials which are important for osteoinduction will be highlighted. Part II, Biological mechanisms of constructing *in vivo* tissue engineered bone. Histological characteristics of calcium phosphate-induced bone formation will be compared to natural bone regarding biosafety. The material-dependence and animal –dependence of calcium phosphate induced-bone formation will be discussed. Part III, Techniques of constructing and the feasibility of applying in vivo tissue engineered bone. Although limited,current data have already raised the possibility to construct large bone graft with good blood supply and certain mechanical strength with *in vivo* bone tissue engineering technique.

Chapter 9 - Blends between gellan (GL) and gelatin (GE) with various compositions (GL/GE 0/100, 25/75, 50/50, 75/25, 100/0 wt/wt) were produced, as candidate materials for biomedical applications. GL/GE films obtained by casting were subjected to a chemical treatment (CT) in an acidic coagulation bath at a pH promoting ionic interactions between GE and GL. A subsequent dehydrothermal treatment (DHT) was performed, allowing the formation of amide and/or ester bonds between the macromolecules. The effects of blend composition and the type of treatment on the physicochemical properties of the samples were evaluated by scanning electron microscopy, infrared analysis and release tests. Interactions between GE and GL by H-bonding were found to increase with decreasing the GL amount and, for each composition, by applying CT+DHT. Infrared spectra of samples after CT+DHT displayed a new adsorption band which was indicative of ester bonds formation. Release tests evidenced that resistance of GL/GE blends to water dissolution greatly increased after CT+DHT. Mechanical properties of GL/GE samples, determined through stress-strain tests, displayed significant variations as a function of composition. GL/GE two-dimensional scaffolds were fabricated by micro-molding (μML), a technique based on the use of a poly-(dimethylsiloxane) (PDMS) mould to create patterns with micro-scale resolution on cast films. The influence of blend composition on the adhesion of primary endothelial cells was studied: GL/GE blends with a high GE content (50 and 75 wt. %) were found to support cells adhesion and proliferation which made them promising candidates for uses in the field of tissue engineering. Coating of GL/GE samples with gelatine enhanced endothelial cell response only for blends containing a high (75 wt%) GE content. Cells tests using endothelial cells were preliminary to future investigations on the potentiality of GL/GE blends as constituent materials for vascular grafts.

Chapter 10 - Scaffolds intended for use in tissue engineering applications often invoke a foreign body response upon implantation leading to undesirable effects, such as encapsulation or rejection of the scaffold. Therefore to develop their full potential, biodegradable tissue engineering scaffold surfaces must be modified to overcome this foreign body response. Current research has suggested that negatively charged and hydrophilic surfaces avoid this response by inducing apoptosis in the responsible immune cells. One approach which has shown potential in functionalising biodegradable polymer surfaces with a negatively charged surface is oxygen plasma treatment. It has the advantages of controllability, versatility, and simultaneous sterilisation of the implant. Examples from the authors studies into the use of oxygen plasma treatment of the commonly used polymer poly(lactic-*co*-glycolic acid) (PLGA) will illustrate a number of important issues. In these studies PLGA was thinly and uniformly coated on glass coverslips to mimic the internal pore surfaces of tissue engineering scaffolds produced by thermally induced phase separation (TIPS). It was found that oxygen plasma, at pressures within the limits of current testing ($4x10^{-4}$ to 10^{-4} mbar; 10-60 s), significantly reduces the advancing and receding water contact angle. Contact angles obtained from samples washed once in ethanol after treatment consistently yielded contact angle values up to 10° higher than unwashed samples, supporting theories that oxygen plasma treatment cleaves PLGA chains, liberating soluble oligomeric products. XPS data indicates an increase in atomic oxygen content after plasma treatment by 1% in washed samples and by 2% in unwashed samples. AFM was used to visualise the extent of etching resulting from the oxygen plasma treatment, and surface roughness was found to change from < 1 nm (untreated sample) to 14 ± 3 nm (10 s oxygen plasma treatment at 150 mTorr). The findings of this study highlight the dual effect of oxygen plasma treatment and cautions the use of materials subjected to oxygen plasma in biological assays that have not been washed prior to their use.

Chapter 11 - The current review will address alternative methods for the tissue engineering of blood vessels and heart valves. This is in contrast with most current reports on cardiovascular tissue engineering (CVTE), which are generally limited to one specific paradigm or certain aspects thereof. The paradigm in question described the creating of autologous tissue, by in vitro seeding of mainly synthetic scaffolds followed by a phase of *in vitro* maturation, finally resulting in an implantable construct suitable for only one specific patient. Although this approach is very promising it has suffered from several draw-backs, such as costs, instigating several groups to search for alternative approaches. Mainly aiming at the creation if living tissue by *in vivo* approaches and thereby reducing the costs and risks associated to the *ex vivo* treatment of cells, these approaches have already proven the ability to create living blood vessels and blood vessels by exploiting and guiding naturally occurring processes, such as foreign body reaction and wound healing. The techniques to obtain such living prostheses can be based on intraperitoneal implantation and seeding, as was shown by the group of Campbell for blood vessels [1] and by the authors group for heart valves [2]. In both cases the methods have been developed and evaluated *in vivo*. A second alternative are the smart matrices. These scaffolds are designed to enhance *in situ* recellularisation [3,4] but have some implications in the *in vitro* tissue engineering as well.

Chapter 12 - The use of adult stem cells for cell-based tissue engineering and regeneration represents a promising approach for repairing damaged tissue. This emerging field of regenerative medicine will require a reliable source of stem cells in addition to biomaterial scaffolds and growth factors. Recent studies have shown that stem cells can be isolated from a wide variety of tissues including bone marrow, muscle, and adipose tissue. The cellular component of this regenerative approach will play a key role in bringing tissue-engineered constructs from the laboratory bench to the clinical bedside. An ideal source of stem cells still remains undefined, however, and may differ depending upon the required application.

Muscle tissue engineering approaches could revolutionize current therapies for irreversible skeletal muscle damage and muscular dystrophies, and may significantly improve the quality of life of millions of patients. Skeletal muscle tissue engineering approaches rely mainly on myoblasts and biomaterials; however, myoblasts lack expansion capability. Cells from other sources, including adult stem cells, are being investigated to overcome this problem. Several cell sources have been proposed for creation of engineered skeletal tissue, but the literature to date unfortunately does not assist in the identification of an ideal cell source. The authors describe here the advantages and limitations of candidate cell sources for skeletal muscle tissue engineering.

As with bone marrow, adipose tissue is a mesoderm-derived organ that contains a stromal population including microvascular endothelial cells, smooth muscle cells, and stem cells. This population shares many characteristics with its bone marrow counterpart including extensive proliferative potential and an ability to undergo multilinear differentiation. Adipose tissue also has the ability to dynamically expand and shrink throughout the life of an adult. This capacity is mediated by the presence of vascular and non-vascular cells that provide a pool of stem and progenitor cells with unique regenerative capacities. Human adipose tissue-derived adult stem cells (ASCs) therefore meet many of the requirements of "ideal" cells for stem cell therapy and tissue engineering. ASCs have been used in cell-based tissue engineering of hard tissues such as bone and cartilage, but their capacity to assist in regeneration of damaged muscle has not yet been investigated in detail.

Here the authors review recent progress in the muscle tissue engineering field and describe one of their experiments as an example of successful skeletal muscle tissue engineering. They engineered skeletal muscle tissue using a combination of ASCs and injectable PLGA spheres. These findings suggest that mixtures of ASCs and injectable scaffolds may provide excellent tools for muscle tissue engineering and that such combinations may find applications in clinical settings requiring muscle tissue regeneration therapy.

In: Tissue Engineering: Roles, Materials and Applications ISBN: 978-1-60456-293-4
Editors: S. J. Barnes, L. P. Harris, pp. 1-17

Chapter 1

PERIPHERAL NERVE TISSUE ENGINEERING

Sai-yun Hou*[*1,2,] and Jia-kai Zhu[1]

[1] Department of Orthopedic and Microsurgery, First Affiliated Hospital, SunYat-Sen University, GuangZhou, China

[2] Department of Orthopedic Surgery, Stanford University School of Medicine, Stanford, California, USA

ABSTRACT

Posttraumatic peripheral nerve defects repair continues to be a major challenge in restorative medicine and microsurgery. Although progress has been made in surgical techniques over the last 30 years, functional recovery after severe lesion of a major nerve trunk, especially for large nerve gaps is often incomplete and often unsatisfactory with long-time functional deficits. Treatment of lesions causing complete transection of peripheral nerve is primarily aimed at recreating continuity of the nerve to allow axonal regrowth into the nerve stump distal to the lesion. To date, the standard for bridging peripheral nerve defects involves the use of autografts. But disadvantages with this method include harvesting the autografts from a second operative site, with consequent loss of end-organ innervation, prolonged surgical procedures, and shortage of autogenous nerve graft material for extensive nerve injuries. This dictates research towards the development of biomimetic contructs as alternatives. With the tissue engineering rapidly progressing, tissue engineered bioartificial nerves have been served as new alternatives, which have the advantages of unlimited availability and avoidance of donor-site morbidity. Based on current concepts, this review summarizes various strategy of nerve tissue engineering, including different extracellular matrics, scaffolds, growth factors and seed cell sources that have been shown to promote peripheral nerve regeneration. Also, the importance of using appropriate animal models and criteria for evaluating tissue engineered nerve constructs suitability for bridging peripheral nerve gaps is highlighted.

Keywords: Schwann cells, stem cells, peripheral nerve regeneration, growth factors, scaffords, nerve tissue engineering.

---

* Correspondence concerning this article should be addressed to: Dr. Saiyun Hou, PM&R (117), 3801 Miranda Ave, Palo Alto, CA 94304, USA E-mail: housaiyun@hotmail.com

1. Introduction

Peripheral nerve injury is a common problem. There were in excess of 50,000 peripheral nerve repair procedures performed in the United States in 1995 [1]. Up to 300,000 cases present each year in Europe alone and despite surgical repair, disability often persists for the lifetime of many patients [2]. Widely accepted method by most surgeons is bridging the defect with an autologous donor nerve, which is the gold standard of repair of a peripheral nerve defect. This is associated with several disadvantages, including harvesting the autografts from a second operative site, with consequent loss of end-organ innervation, prolonged surgical procedures, an extra incision for removal of a healthy sensory nerve ultimately resulting in a sensory deficit and shortage of autogenous nerve graft material for extensive nerve injuries. Finally, donor graft material is limited, particularly for managing extensive lesions (e.g. brachial plexus), which require several lengths of nerve graft. This dictates research towards the development of biomimetic contructs as alternatives. Fortunately, recent advances in neuroscience, cell culure, genetic techniques, and biomaterials provide optimism for new treatments for nerve injuries. In the search for alternatives to nerve grafts what is required is a substrate mimicking a nerve graft by providing a scaford substituting for the Schwann cells (SCs) basal laminae present in a graft. In addition, a substitute for the growth factors which are normally provided by the SCs in the graft is needed [3]. This can be done either by impregnating or providing the nerve conduit with available factors, or by supplying cells producing such factors. The cells can be pre-cultured or acutely dissociated SCs or gene-modified cells engineered to produce factors. An ideal alternative to nerve grafts should be tissue-engineered to provide these various components. To fully understand current approaches to nerve engineered constructs, several critical comments should be considered: scaffold for mechanical support, support cells, growth factors, and extra-cellular matrix.

2. Scaffords

A scaffold to provide a construct for the cells to grow is a crucial component for tissue engineering. These “nerve guides” or “nerve guidance channels” serve to direct axons sprouting from the proximal nerve end, provide a conduit for the diffusion of growth factors secreted by the injured nerve ends, and reduce the infiltration of scar tissue. During the nineteenth century, physicians began to focus considerable effort on nerve restoration[4]. Alternative materials, including fat sheaths, gauze, bone, and metal tubes were used in an attempt to repair damaged nerves. Unfortunately, most of techniques resulted in poor functional recovery[5]. It was not until the 1960s that Millesi pioneered microsurgical techniques to accurately align nerve fascicles in the direct resection of nerve ends, with improved functional outcomes[6]. Past research in this area has focused either on existing natural or synthetic materials; however, nonc of the materials studied to date have matched or exceeded the performance of the nerve autograft. As a result, researchers are now focusing on the combination of materials, cells and desired biomolecules to create new composite materials that can actively stimulate nerve regeneration. In addition, methods to minimize the immune response to nonautologous tissue could provide a source of natural material for nerve repair. Now some scaffolds in use made of many kinds of materials are presenting.

Autologous Tissue Grafts

Autologous nerve tissue are more likely to be biocompatible than artifical materials, are less toxic, and provide a support structure to promote cell adhesion and migration. As a result, autologous nerve grafts are still gold standard for repair of peripheral nerve. Generally, nerve autografts are typically derived from one of several cutaneous nerves, such as the sural or saphenous nerve, with an available length up to 40 cm and a cable diameter of 2-3 cm[7]. In addition to autologous nerve grafts, there are other natural tissues, such as autologous vein[8,9], muscle[10,11], tendon[12], epineurial sheaths[13], muscle-vein combined grafts[14,15]. These autologous tissues have also been combined with growth factors or coated with collagens for further stimulation of axonal outgrowth and SCs migration and proliferation. Currently, Raimondo et al. confirmed that a vein filled with skeletal muscle as a suitable means of sustaining SCs migration and proliferation[16]. Tos et al. were of the opinion that a piece of fresh muscle inside a vein would expand the vessel, that the muscle basal lamina would help to support axonal outgrowth[17]. Such conbined vein-muscle nerve conduits were used to bridge 1-2cm gaps in rat sciatic nerves. Although they show the good biocompatible, the shortage of these materials limit the use of them in the clinic, because these tissue must be removed from the patients.

Acellular Grafts and Nonautologous

Nerve allografts or xenografts (donor nerves from cadavers and animals, respectively) are considered a good alternative for nerve conduits, because they poss the natural nerve structure like human nerve. Also they have the advantages that supplies can be large and their use does not require harvest from the patients. However, these tissues possess some risk of disease transmission and must either be used in conjunction with immunosuppressants or must be processed to remove immunogenic components. Fortunately, recent work has demonstrated that low-dose immunosuppression with FK506 has yielded improved regeneration with the use of allografts [18,19]. Acellular grafts, although non-immunogenic, tend to result in a significant degree of inflammation when implanted, mainly because the decellularization fails to completely extract cellular remnants or leaves damaged extracellular matrics. As a result, many efforts are being made to improve to process intact nonautologous tissue, making it less immunogenic for clinical use. So far three different methods have been explored, such as thermal technique, rediation and chemical processes.

Thermal decellularization which is the most common decellularization technique involves repeated freeze-thaw cycles to kill and fragment the cells [20-22]. Radiation treatments destroy cells in tissues and produce relatively little damage to the matrix structure, but they also fail to extract all cellular components [23]. Checmical treatments are the more effective in the complete removal of cell debris [24,25]. Hu et al demonstrated that the successful use of chemically extracted allogenic nerve grafts implanted with cultured autologous bone marrow stromal cells (MSCs) to bridge 40mm, a long distance peripheral nerve gap in the rhesus monkey points to the potential promise for clinical use in humans [26]. Further study showed that chemically extracted acellular nerve allografts have good cellular clearance, same basal lamina tubes and three-dimensional structure as the normal intact nerve, providing an effective pathway or conduit for regenerating axons. So the use of acellular tissues for clinical nerve repair may become a viable option in the future.

Synthetic Materials

Synthetic materials are attractive because their chemical and physical properties (degradation rate, porosity, mechanical strength) can be specifically optimized for a particular application. However, the biocompatibility of synthetic materials has a challenge because of the inflammation issue. Permanent materials are less desirable because they pose a higher risk for infection, are more likely to provoke a chronic inflammatory response, and have the potential to compress the nerve over time. Thus, a nerve guide that degrades as the nerve regenerates is preferred. Additionally, guidance channels should be pliable, but should maintain their shape and resist collapse during implantation and over the time course for regeneration. Several biodegradable synthetic materials have been demonstrated to support nerve regeneration. Polyesters such as polylactic acid, polyglycolic acid, and poly(lactic-co-glycolic acid) (PLGA) were early choices for investigation because of their availability and FDA approval [27]. To date, these materials have been processed into foams and seeded with SCs to improve their regenerative potential [28,29]. Recent approaches, using a biodegradable PLGA tube inserted with plexiform PLGA filaments, seeded with the differentiated SCs has shown excellent structural preservation as normal nerve tissue and good adhesion for axon to migrate [30]. (Figure 1, Figure 2)

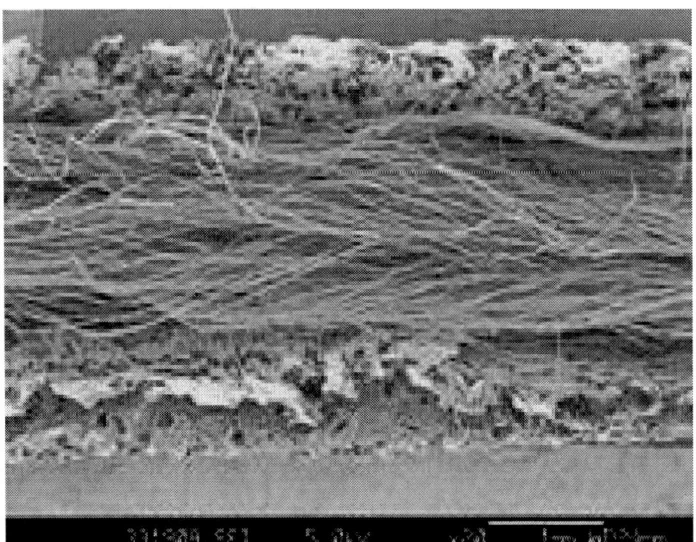

Figure 1. Scanning electron microscopy (SEM) of a PLGA conduit inserted with plexiform PLGA filaments. (20X).

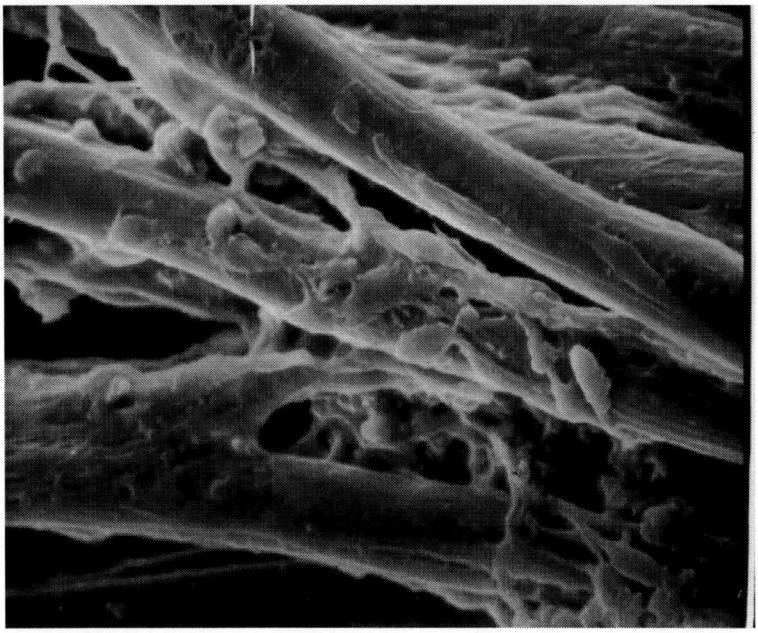

Figure 2. SEM of SCs differentiated by MSCs on the ECM-gel-coated PLGA filaments for 3 d, exhibiting tight attachment between cell and scaffolds and cell-bridge formation. (1000X).

In addition to poly(esters), biodegradable poly(urethane) (31), poly(organo phosphazene) (32), methacrylate-based hydrogels (33), and poly(3-hydroxybutyrate) (34) have shown a capacity for guiding regeneration. However, approval of PLGA for clinical use has encouraged its use over other materials. Physical properties, such as conduit dimensions, wall porosity, surface texture, and inherent electrical properties, also dramatically influence the effects of nerve regeneration.

There are several key properties that all scaffolds should possess: (1) they must be readily formed into a conduit having a desired diameter and wall thickness, (2) they should possess good biocompatibility, without toxic, inflammation and immuological rejection, (3) they should be biodegradable. The speed of degradation for given scaffolds must match the speed of axonal regeneration. Finally they can be completely absorbed, (4) they should have good structural compatibility. They possess the strength and plasticity to protect the 3-D structure of scaffolds from collapse. Permeable conduits benefit the exchange of nutrient substance and excreta, (5) they should be simple to implant using microsurgical techniques, and (6) they must be sterilizable. No characteristics of physics, chemistry and biology happen after sterilization. To date, no engineered graft conduit has been found as effective as an autologous nerve graft. The ideal scaffold is still unknown. A combination of natural and symthetic materials will be future direction, which can combine the both benefits to design as suitable scaffolds as autologous nerve graft.

3. Support Cells

Schwann Cells

Alternatives to autologous nerve grafts have been studied since last century. The reason for failture of nerve regeneration is the lack of viable SCs. The SCs play a vital role in maintaining the peripheral nervous system. In the normal nerve, SCs seem to be quiescent. However, after an injury occurs in the peripheral nervous system, there tends to be disruption in the conductance of neuronal signals due to axonal damage. Such an injury is accompanied by an inflammation response by SCs. The SCs help to phagocytize the damaged end of the axon and then forms the regeneration tube of the axon that is connected to the cell body. While forming this tube, SCs proliferate rapidly and provide the axon with a path to grow along. This trail is formed by the SCs (Bungner's Band) that are still present in the distal stump [35]. SCs are also able to attract injured neurons by secreting neurotrophic factors such as nerve growth factor (NGF) to aid axon elongation. The SCs also promote elongation by secreting F-Spondin, a developmental protein of the floor plate, after peripheral nerve injury [36]. As a result, SCs play an important role in peripheral nerve regeneration probably through several mechanisms. These include the formation of excellular matrix (ECM) [37,38], synethesis of cell adhension molecules (CAM) [39], and the release of growth factors and their receptors. Many studies have demonstrated excellent axonal regeneration by placing SCs within a scaffold [40-42]. Hadlock et al. demonstrated that rolled SCs monolayer grafts implanted in a 7 mm gap in rat sciatic nerve increased functional regeneration after 10.5 weeks compared to acellular controls [43]. Xu et al. have shown that SCs, when implanted with Matrigel, can aid axonal regeneration across spinal cord transections in adult rats at levels above that of Matrigel alone [44]. Kierstead et al. extended these studies to show that SCs transplantation in combination with demyelination enhanced axonal growth in rat spinal cord injuries [45].

Stem Cells

Although SCs play an important role in axonal regeneration, there still are some problems to limit its use in peripheral nerve tissue engineering. The culture of sufficient numbers of autologous SCs is time consuming, limiting the usefulness of this technique as delay compromises axonal regeneration [46]. Use of allogeneic SCs is involved in immunological rejection [47]. So new seed cells need to replace SCs. Therefore, considerable efforts have focused on inducing other cells to turn into SCs, so as to expand their sources for the need of nerve tissue engineering and cell therapy.

With the growing applicability of stem cells in all avenues of medicine, increased opportunities become available to allow the use of a variety of cell lines for nerve differentiation such as embryonic stem cells, neural stem cells, bone marrow stromal cells,somatic pluripotent stem cells, adipose-derived stem cells, olfactory ensheathing cells(OEC), boundary cap cells, and hair follicle stem cells[48,49]. There are several key properties that all seed cells should possess:(1) convenience to draw the materials; (2) strong proliferation in vitro; (3) easy adaptation to biomaterial and host environment; (4) complete replacement of the function of SCs.

Neural stem cells have also been effective at restoring motor function in the setting of chronically denervated peripheral nerves [50]. But the source of donor tissue and expansion difficulty of such cells could pose a significant barrier[51].

Adipose-derived adult stem cells are a population of mesenchymal stem cells extracted from discarded adipose tissue. Many studies have reported the differentiation of adipose-derived stem cells into chondrocytes, myocytes, osteoblasts, and, most recently, neural progenitor cells. Kingham et al demonstrated that adipose tissue contains a pool of regenerative stem cells which can be differentiated to SCs phenotype and may be of benefit for treatment of peripheral nerve injuries [52].

Boundary cap cells can generate neurons as well as peripheral glia during embryonic development recently, and the boundary cap was shown to contain multipotent stem cells [53]. Aquino et al have explored the competence of boundary neural crest stem cells (bNCSCs) to differentiate into mature functional SCs in vitro and in vivo. Furthermore, bNCSCs failed to differentiate into SCs in vitro when cultured in a defined media and in vivo when grafted into adult rat sciatic nerves. However, in the presence of neuregulins, during long-term cultures, the majority of bNCSCs differentiated into SCs [54]. Hair follicle stem cells, which may be easier to access, have been transdifferentiated into SCs and indicate promise in enhancing rate of restoration of nerve function [55].

Olfactory ensheathing cells (OEC's) have been described as the progenitor cells of the peripheral nervous system [56]. The use of these cells in spinal cord regeneration has been extensively reviewed [57]. But for the peripheral nerve repair, the effect of OECs was seldom reported. Verdu et al investigated that 15mm sciatic nerve resection was repaired by bridging the stumps with a silicone tube prefilled with a laminin gel. Successful axonal regeneration was gotten, when purified OECs were added to the gel filling the tube. Therefore, OECs help repair severe peripheral nerve injuries, in addition to their ability to promote axonal regeneration within the central nervous system [58].

Presently, MSCs have been the highlight of many tissue engineering studies because of their multipotential properties. They contribute to the regeneration of tissues such as bone, cartilage, fat, and muscle [59-62]. Recent data suggest that MSCs can be induced to differentiate into Schwann-like cells that exhibiting characteristics similar to SCs in vitro [30,63] and in vivo [64-66], which can promote nerve regeneration. MSCs are relatively easy to isolate from small aspirates of bone marrow that can be obtained under local anesthesia, and they can be expanded easily in vitro. MSCs could provide a vast source of autologous cells to prevent immunological rejection and the need of a second surgical step. As a result, MSCs are great potential seed cells for peripheral nerve tissue engineering.

To date, no seed cell has been found as effective as an autologous SCs. It is still unclear how a stimulation process might direct a cell towards a given phenotype and morphological shape, and whether phenotypical and morphological expression really equates with functional ability of SCs. How long can these cells survive and can these cells remain viable for transportation and implantation?

4. Growth Factors

Growth factors play critical roles in neuronal survival after nerve injury[67-70]. Protecting neurons from death after injury will increase the potential for axon regeneration. The local presence of growth factors plays an important role in controlling survival, migration, proliferation, and differentiation of the various cell types involved in nerve regeneration. Some of these factors include nerve growth factors (NGF), neurotrophin 3 (NT-3), brain-derived neurotrophic factor (BDNF), insulin-like growth factor (IGF-1, IGF-2), platelet-derived growth factors (PDGF), fibroblast growth factor (FGF), and ciliary neurotrophic factor(CNTF), glial cell line-derived growth factor (GDNF). These factors stimulate peripheral nerve migration directly or indirectly.

NGF is vital to the development and regeneration of the nervous system. NGF is expressed at low levels in healthy peripheral nerve and is upregulated in the distal stump upon injury [71]. NGF enhances axon growth of DRG neurons induced by laminin, whereas neurotrophins or laminin alone induce limited neurite growth from either CNS or peripheral nervous system (PNS) neurons during development [72,73].

Application of a low dose of BDNF to the injury site only promoted axon regeneration in chronic axotomy-induced lesion, whereas it had no effect on acute nerve injury; however, high doses of BDNF inhibited regeneration in both injury models [74]

NT-3, like BDNF, promotes motor neuron survival and outgrowth as well as sensory axon growth[75-77]. In vivo, NT-3 plays a vital role in aiding the regeneration of peripheral nerves [78,79]. But Young et al. demonstrated that there were no long-lasting improvements of anatomical or functional recovery in NGF- or NT3-treated animals 12 weeks following sciatic nerve transection [80]. NT-4/5 has been required for the successful growth of axons in peripheral nerves. English et al. investigated that when grafts from wild-type or NT-4/5(-/-) mice were treated acutely at the time of surgical repair either with exogenous BDNF or NT-4/5, the lengths of axon profiles measured in the grafts from NT-4/5 knockout mice were significantly longer than those measured in grafts from untreated wild-type[81].

CNTF promotes motor neuron survival, outgrowth, and sprouting [82]. The application of exogenous CNTF has been associated with increased levels of regeneration following injury in peripheral nerve [83]. Hu et al. demonstrated that LV-CNTF-transduced SCs enhanced axonal regeneration to an even greater extent than that of LV-GFP controls. Many regenerated axons were myelinated [84].

Glial cell line-derived neurotrophic factor (GDNF) is a potent motoneuron factor. It prevents motoneuron degeneration in mice and rats after axotomy[85]. GDNF also has been correlated with improved peripheral nerve regeneration [86]. Li et al demonstrated that when bridging a rat sciatic nerve defect with a conduit filled with GDNF-transfected SCs, nerve regeneration was better than that of the control [87]. Further, Iwase et al. demonstrated that GDNF utilizes neural cell adhesion molecule (NCAM) signaling pathways to regulate SC function prior to myelination and at early stages of myelin formation [88].

aFGF and bFGF have been associated with enhanced regeneration following injuries in the peripheral nerve [89]. FGF-2 and its high-affinity receptors are expressed in the intact peripheral nervous system and regulated following nerve injury. After exogenous application, FGF-2 promotes neuronal survival and neurite outgrowth in vitro and in vivo. It has been demonstrated that administration of FGF-2 protects neurons from injury-induced cell death and promotes axonal regrowth [90]. Jungnickel J et al. used transgenic mice over-expressing FGF-2 (TgFGF-2) to demonstrate that endogenously synthesized FGF-2 influences early peripheral nerve regeneration by regulating SCs proliferation, axonal regrowth, and remyelination [91]. Within the group of physiologically relevant trophic factors, the FGF-2 system seems to be crucially involved in the peripheral nerve development and regeneration.

Insulin-like growth factor I (IGF-I) promotes survival of motor and sensory neurons after injury, as well as enhances survival and proliferation of SCs [92,93]. IGF-I also promotes axonal regeneration and myelination, and appears that locally applied IGF-I binds to its receptor and binding protein and is transported to the cell body in a retrograde fashion[94].

Gene Delivery

Therapies with relevant growth factors received increasing attention in recent years. However, growth factor therapy is a difficult task because of the high biological activity, pleiotrophic effects, and short biological half-life of these protein drugs. The delivery of growth factors to support regeneration has several intrinsic challenges, including the toxicities and poor stability associated with many bioactive factors. Growth factors can be delivered through a variety of mechanisms. Traditionally, growth factors were delivered exogenously through the application of tubes or injections. The disadvantage of delivery still remains, because the concentration of growth factors can not be adjusted. Polymer matrics can overcome this challenge. Synthetic and naturally derived polymers are widely used in controlled release devices for protein delivery. These devices are designed such that bioactive factors are released in a spatially and temporally controlled manner. One example of a polymer matrix delivery device is the nerve guidance channel. By incorporating growth factors into the conduit wall, the nerve guidance channel itself becomes a delivery device. Ethylene vinyl acetate [95] and fibronectin mats [96] are examples of polymeric nerve guidance channel materials that have been successfully modified to deliver growth factors to regenerating peripheral nerve.

Biodegradable, polymeric microspheres have been considered for the delivery of neuronal growth factors in the context of both the implantation in the brain and the combination with conduit for peripheral nerve repair[97]. They are suitable for a variety of delivery applications and are commonly used to deliver molecules to the CNS. The problem with microsphere delivery of growth factors is that degradation can be variable. Growth factor levels may require alternative delivery times and sequences. In an attempt to incorporate these issues in growth factor delivery, focus has turned to plasmid or viral vectors as modes of controlled gene expression of growth factors.

A number of viral and nonviral gene delivery techniques are available. Viral vectors have been investigated in a variety of tissue systems and include methods based on retrovirus, herpes virus, adenovirus, and adeno-associated virus. Herpes virus [98] and adenovirus [99] are more common in neural application. Viral gene transfer is able to promote high levels of gene expression from the vector [100]. Target cells for somatic gene therapy of the injured PNS are mainly SCs. Gravvanis et al. performed viral transduction of SC ex vivo in order to induce expression of sialyl-transferase-X(STX) in these cells. Treatment with STX transduced SC resulted in improved motor recovery and better morphological tissue regeneration[101]. Furthermore, over-expression of GDNF was achieved by retroviral transduction of SC with this factor ex vivo. A silicone tube filled with the GDNF over-expression cells was introduced into a 10-mm defect in the sciatic nerve of adult rats, such treatment resulted in increased regeneration regarding functional and morphological criteria[102]. Haastert et al. established non-viral methods enabling somatic gene transfer to repair large defects in the PNS. SC were transfected in vitro with non-vial plasmid DNA in order to induce over-expression of fibroblast growth factors-2(FGF-2) isoforms in these cells. Silicone tubes filled with FGF-2 over-expressing SC were then transplanted between the proximal and distal stumps of transected sciatic nerves of adult rats to bridge a 15 mm gap. They found differential promoting effects of different FGF-2 isoforms on regeneration across long gaps[103]. Nonetheless, some level of risk is associated with administering a therapy based on a viral agent. Before they can be used in the clinic, viral vectors must be proven to be safe and any inflammatory response must be minimized [104,105]

5. Extracellular Matrix

Extracellular matrix (ECM) is also an important part of a tissue engineered construct. ECM molecules, such as laminin, collagen, and fibronectin, have been shown to play a significant role in axonal development and repair in the body. Silicone tubes filled with laminin, fibronectin, and collagen show improved regeneration over a 10 mm rat sciatic nerve gap compared to empty silicone controls [106]. Collagen filaments have also been used to guide regenerating axons across 20–30 mm defects in rats [107]. Further studies have shown that oriented fibers of collagen within gels, aligned using magnetic fields, provide an improved template for neurite extension compared to randomly oriented collagen fibers [108]. ECM gel coated on PLGA filaments to adhere differentiated SC have been used to bridge 10 mm nerve defects in rats, with results close to that of the nerve autograft [30] (figures 1,2). Chitosan, a polysaccharide derivative of chitin, possesses antitumor and antibacterial properties and has been shown to promote migration and proliferation of SCs, thereby gaining increased interest to promote nerve regeneration [109].

6. Animal Models and Evaluation of Regeneration

In addition to above critical components for designing the ideal nerve engineered contructs for peripheral nerve regeneration, the animal models, methods of analysis that determine success, and the criteria used to define success are also important. For rat model, two factors should be involved, a gap greater than 15mm, and controls involving autografts. Further, a large animal such as primate should used to test the nerve regeneration for tissue engineered construct, with gaps greater than 40mm[26].

Assessment of nerve regeneration is frequently done by histomorphometrical evaluation, for example, by counting myelinated axons. However, the number of myelinated axons does not generally correlate with functional parameters such as fluorogold retrolabeling, nerve conduction studies, muscle contraction force under nerve stimulation, or mobility of limbs as quantifiable by video gait or walking track analysis[110]. It should be realized that different testing procedures are generally poorly correlated so that multiple methods should be used to evaluate the effect of any intervention on nerve regeneration. Perfect assessment of nerve regeneration should include both negative controls and positive controls.

7. Conclusion

Peripheral nerve regeneration is complicated. No approach has so far allowed an effective clinical therapy. Although significant advances have been made in alternatives to autologous grafts, including use of bioengineered conduits, induction of stem cell differentiation and manipulation of growth factor expression through genetic enigeering, the problem still remain to be overcome for clinical application. For example, most studies on regeneration have focused on how to promote axon elongation in the CNS. However, the exact molecular mechanism involved in successful peripheral axon regeneration is still not clear. The identification of the pathways that trigger dedifferentiation and proliferation of SCs after injury is still unclear and requires future experimentation. Synthetic nerve conduits have not been a widely used replacement for autografts, particularly in long defects. The mechanism of stem cell differentiating into SC and methods controlled of differentiation are not completely understood. Defining the appropriate expression sequence, dosing, and duration for growth factors is critical. Non-viral strategies for ex vivo gene therapy will probably be part of future therapeutic approaches of tissue engineered nerve constructs in reconstruction of severely injured peripheral nerves. The future progress will be greatly made through the combined efforts of scientists and engineers from a variety of disciplines, and other developments in different fields of tissue engineering.

References

[1] National Center for Health Statistics based on Classification of Diseases, 9th Revision, Clinical Modification for the following categories: ICD-9 CM Code: 04.3, 04.5, 04.6, 04.7.

[2] Wiberg M, Terenghi G. Will it be possible to produce peripheral nerves? *Surg. Technol. Int.* 2003: 11, 303–310.

[3] Lundborg G: A 15-year perspective of peripheral nerve surgery: Evolving neuroscientific concepts and clinical significance. *J. Hand Surg.* [Am] 2000;25:391-414.

[4] Ferrara G. Nuova selva di chirurgia divisiain tre parti. *Venice: S Combi;* 1608.

[5] Glück T. Ueber neuroplastik auf dem wege der trans-plantation. *Arch. Klin. Chir.* 1880; 25: 606–16.

[6] Millesi H, Ganglberger J, Berger A. Erfharungen mit der mikrochirurgie peripherer nerven. *Chirurgica Plastica.* 1967; 34: 47.

[7] Seddon HJ. 1975. Surgical Disordersof the Peripheral Nerves. New York: Churchill Livingston 40. Sunderland S. 1991. *Nerve Injuries and Their Repair.* London: Churchill Livingston.

[8] Quarles RH. Myelin sheaths: glycoproteins involved in their formation, maintenance and degeneration. *Cell Mol. Life Sci.* 2002;59:1851—71.

[9] Risitano G, Cavallaro G, Merrino T, Coppolino S, Ruggeri F. Clinical results and thoughts on sensory nerve repair by autologous vein graft in emergency hand reconstruction. *Chir. Main.* 2002,21:194–97

[10] Battiston B, Tos P, Cushway TR, Geuna S. Nerve repair by means of vein filled with muscle grafts I. Clinical results. *Microsurgery.* 2000,20:32–36

[11] Meek MF, Varejao AS, Geuna S. Muscle grafts and alternatives for nerve repair. *J. Oral Maxillofac. Surg.* 2002, 60:1095–96

[12] Brandt J, Dahlin LB, Lundborg G. Autologous tendons used as grafts for bridging peripheral nerve defects. *J. Hand Surg. [Br.]* 1999, 24:284–90

[13] Karacaoglu E, Yuksel F, Peker F, Guler MM. Nerve regeneration through an epineurial sheath: its functional aspect compared with nerve and vein grafts. *Microsurgery.* 2001, 21:196–201

[14] Geuna S, Tos P, Battiston B, Guglielmone R, Giacobini-Robecchi MG. A stereological study of long-term regeneration of rat severed sciatic nerve repaired by means of muscle-vein-combined grafts. *Ital. J. Anat. Embryol.* 2000,105:65–73 49.

[15] Pagnotta A, Tos P, Fornaro M, Gigante A, Geuna S, et al. Neurotrophins and their receptors in early axonal regeneration along muscle-vein-combined grafts. *Microsurgery.* 2002,22:300–3

[16] Raimondo S, Nicolino S, Tos P, Battiston B, Giacobini- Robecchi MG, Perroteau I, Geuna S. Schwann cell behavior after nerve repair by means of tissue-engineered muscle-vein combined guides. *J. Comp. Neurol.* 2005; 489:249–59.

[17] Tos P, Battiston B, Geuna S, et al. Tissue specificity in rat peripheral nerve regeneration through combined skeletal muscle and vein conduit grafts. *Microsurgery.* 2000;20:65-71.

[18] Sobol JB, Lowe III JB, Yang RK, Sen SK, Hunter DA, Mackinnon SE. Effects of delaying FK506 administration on neuroregeneration in a rodent model. *J. Reconstr. Microsurg.* 2003; 19: 113–8. 21.

[19] Myckatyn T, MacKinnon SE, McDonald JW. Stem cell transplantation and other novel techniques for promoting recovery from spinal cord injury. *Transpl. Immunol.* 2004;12: 343–58.

[20] Ide C, Tohyama K, Tajima K, Endoh K, Sano K, et al. Long acellular nerve transplants for allogeneic grafting and the effects of basic fibroblast growth factor on the growth of regenerating axons in dogs: a preliminary report. *Exp. Neurol.* 1998, 154:99–112 53.

[21] Frerichs O, Fansa H, Schicht C, Wolf G, Schneider W, et al. Reconstruction of peripheral nerves using acellular nerve grafts with implanted cultured Schwann cells. *Microsurgery.* 2002, 22:311–15

[22] Gulati AK, Cole GP. Immunogenicity and regenerative potential of acellular nerve allografts to repair peripheral nerve in rats and rabbits. *Acta Neurochir.* (Wien) 1994,126:158–64

[23] Marmor L. The repair of peripheral nerves by irradiated homografts. *Clin. Orthop.* 1964, 34:161–69

[24] Sondell M, Lundborg G, Kanje M. Regeneration of the rat sciatic nerve into allografts made acellular through chemical extraction. *Brain Res.* 1998,795:44–54 56.

[25] Sondell M, Lundborg G, Kanje M. Vascular endothelial growth factor stimulates Schwann cell invasion and neovascularization of acellular nerve grafts. *Brain Res.* 1999, 846:219–28

[26] Hu J, Zhu Q, Liu X, Xu Y, Zhu J. Repair of extended peripheral nerve lesions in rhesus monkeys using acellular allogenic nerve grafts implanted with autologous mesenchymal stem cells. *Experimental Neurology,* 2007, 204: 658-666

[27] Hudson R, Evans GRD, Schmidt C. Engineering strategies for peripheral nerve repair. *Clin. Plast. Surg.* 1999; 4: 617– 28. (Reprinted in Orthop Clin N Am. 2000; 3 1: 485–97.)

[28] Evans GR, Brandt K, Niederbichler AD, Chauvin P, Herrman S, et al. Clinical long-term in vivo evaluation of poly(Llactic acid) porous conduits for peripheral nerve regeneration. *J. Biomater. Sci. Polym. Ed.* 2000, 1:869–78 110.

[29] Evans GR, Brandt K, Katz S, Chauvin P, Otto L, et al. 2002. Bioactive poly(L-lactic acid) conduits seeded with Schwann cells for peripheral nerve regeneration. *Biomaterials.* 23:841–48.

[30] Hou SY, Zhang HY, Quan DP, Liu XL, Zhu JK. Tissue-engineered peripheral nerve grafting by differentiated bone marrow stromal cells. *Neuroscience.* 2006;140,110-110

[31] Soldani G, Varelli G, Minnocci A, Dario P. Manufacturing and microscopical characterisation of polyurethane nerve guidance channel featuring a highly smooth internal surface. *Biomaterials.* 1998, 19: 1919–24

[32] Nicoli Aldini N, Fini M, Rocca M, Giavaresi G, Giardino R. Guided regeneration with resorbable conduits in experimental peripheral nerve injuries. *Int. Orthop.* 2000, 24:121–25

[33] Dalton PD, Flynn L, Shoichet MS..Manufacture of poly(2-hydroxyethyl methacrylate-co-methyl methacrylate) hydrogel tubes for use as nerve guidance channels. *Biomaterials.* 2002, 23:3843–51

[34] Young RC, Wiberg M, Terenghi G. Poly-3-hydroxybutyrate (PHB): a resorbable conduit for long-gap repair in peripheral nerves. *Br. J. Plast. Surg.* 2002, 55:235–40

[35] Kanav Bhatheja, Jeffrey Field Schwann cells: Origins and role in axonal maintenance and regeneration. *The International Journal of Biochemistry and Cell Biology.* 2006,38:1995–1999

[36] Burstyn-Cohen, T., Frumkin, A., Xu, Y. T., Scherer, S. S., and Klar,A. Accumulation of F-spondin in injured peripheral nervepromotes the outgrowth of sensory axons. *Journal of Neuroscience.* 1998,18(21), 8875–8885.

[37] Tonge,DA. Golding,JP.Edbladh M.et al.Effect of excellular matrix on axonal autogrowth from peripheral nerve of adult animal in vitro. *Exp neuro,* 1997,146:81.

[38] Sterne GD,Brown RA,Green CJ,et al.Neurotrophin-3 delivered locally via FN mats enhances peripheral nerve excellular matix ligands. *Dev. boil,* 1997,185(2):215.

[39] Martin R. Formation and main tenance of myelin sheath in the peripheral nerve :the roles cell adhesion molecules and gap junction connexin 32. *Microsc. Res. Tech,* 1998, 41(5):403-415

[40] Bryan DJ, Wang KK, Chakalis-Haley DP. Effect of Schwann cells in the enhancement of peripheral-nerve regeneration. *J. Reconstr. Microsurg.* 1996;12(7):436-439

[41] Frerichs O, Fansa H, Schicht C, et al. Reconstruction of peripheral nerves using acellular nerve grafts with implanted cultured Schwann cells. Reconstruction of peripheral nerves using acellular nerve grafts with implanted cultured Schwann cells. *Microsurgery.* 2002;22(7):311-5.

[42] Rodriguez FJ, Verdu E, Ceballos D et al. Nerve guides seeded with autologous schwann cells improve nerve regeneration. *Exp. Neurol.* 2000 Feb;161(2):571-84.

[43] Hadlock TA, Sundback CA, Hunter DA, Vacanti JP, Cheney ML. A new artificial nerve graft containing rolled Schwann cell monolayers. *Microsurgery.* 2001,21:96–101

[44] Xu XM, Chen A, Guenard V, Kleitman N, Bunge MB. Bridging Schwann cell transplants promote axonal regeneration from both the rostral and caudal stumps of transected adult rat spinal cord. *J. Neurocytol.* 1997,26:1–16

[45] Keirstead HS, Morgan SV, Wilby MJ, Fawcett JW. Enhanced axonal regeneration following combined demyelination plus Schwann cell transplantation therapy in the injured adult spinal cord. *Exp. Neurol.* 1999,159:225–36

[46] Sulaiman OAR, Gordon T. Effects of short- and long-term Schwann cell denervation on peripheral nerve regeneration, myelination and size. *Glia.* 2000; 32:234–246.

[47] Guenard V, Kleitman N, Morrissey TK, Bunge RP, Aebischer P. Syngenic Schwann cells derived from adult nerves seeded in semipermeable guidance channels enhance peripheral nerve regeneration. *J. Neurosci.* 1992, 12:3310–3320.

[48] Dezawa M. Future views and challenges to the peripheral nerve regeneration by cell based therapy. *Rinsho Shinkeigaku.* 2005; 45: 877–9.

[49] Chalfoun CT, Wirth GA, Evans GR. Tissue engineered nerve constructs: where do we stand? *J. Cell. Mol. Med.* 2006,10:309-317.

[50] Heine W, Conant K, Griffin JW, Hoke A. Transplanted neural stem cells promote axonal regeneration through chronically denervated peripheral nerves. *Exp. Neurol.* 2004; 189: 231–40.

[51] Morrison SJ, White PM, Zock C, Anderson DJ.Prospective identification, isolation by flow cytometry, and in vivo self-renewal of multipotent mammalian neural crest stem cells. *Cell.* 1999; 96: 737–49.

[52] Kingham PJ, Kalbermatten DF, Mahay D, et al. Adipose-derived stem cells differentiate into a Schwann cell phenotype and promote neurite outgrowth in vitro. *Exp. Neurol.* 2007 Oct;207(2):267-74.

[53] Maro, G.S., Vermeren, M., Voiculescu, O., Melton, L., Cohen, J., Charnay, P., Topilko, P. Neural crest boundary cap cells constitute a source of neuronal and glial cells of the PNS. *Nat. Neurosci.* 2004, 7 (9): 930-938

[54] Aquino JB, Hjerling-Leffler J, Koltzenburg M, et al. In vitro and in vivo differentiation of boundary cap neural crest stem cells into mature Schwann cells. *Exp. Neurol.* 2006 Apr;198(2):438-49

[55] Amoh Y, Li L, Campillo R, Kawahara K, Katsuoka K, Penman S, Hoffman RM. Implanted hair follicle stem cells form Schwann cells that support repair of severed peripheral nerves. *Proc. Natl. Acad. Sci. USA.* 2005; 102: 17734–8.

[56] Tohill M, Terenghi G. Stem cell plasticity and therapy for injuries of the peripheral nervous system. *Biotechnol. Appl. Biochem.* 2004; 40: 17–24.

[57] Raisman G. Repair of spinal cord injury by transplantation of olfactory ensheathing cells. *C. R. Biol.* 2007, 330(6-7):557-60. *Epub.* 2007 May 9.

[58] Verdu E, Navarro X, Gudino-Cabrera G, et al. Olfactory bulb ensheathing cells enhance peripheral nerve regeneration. *Neuroreport,* 1999 Apr 6;10(5):1097-101.

[59] Friedenstein AJ, Gorskaja JF, Kulagina NN. Fibroblast precursors in normal and irradiated mouse hematopoietic organs. *Exp. Hematol.* 1976, 4:267–274.

[60] Prockop DJ. Marrow stromal cells as stem cells for nonhematopoietic tissues. *Science.* 1997, 276:71–74.

[61] Pittenger MF, Mosca JD, Moorman MA, Simonetti DW, Craig S, Marshak DR Multilineage potential of adult human mesenchymal stem cells. *Science.* 1999, 284:143–147.

[62] Ferrari G, Cusella-De Angelis G, Coletta M, Paolucci E, Stornaiuolo A, Cossu G, Mavilio F. Muscle regeneration by bone marrow derived myogenic progenitors. *Science.* 1998, 279:1528–1530.

[63] Dezawa M, Takahashi I, Esaki M, Takano M, Sawada H Sciatic nerve regeneration in rats induced by transplantation of in vitro differentiated bone-marrow stromal cells. *Eur. J. Neurosci.* 2001,14:1771–1776.

[64] Chen X, Wang X, Chen G, et al. Study of in vivo differentiation of rat bone marrow stromal cells into Schwann cell-like cells. *Microsurgery.* 2006;26:111-115.

[65] Cuevas P, Carceller F, Dujovny M, et al. Peripheral nerve regeneration by bone marrow stromal cells. *Neurol. Res.* 2002;24:634-638

[66] Zhang P, He XJ, Liu KY, et al. Bone marrow stromal cells differentiated into functional Schwann cells in injured rats sciatic nerve. *Artificial cells, Blood Substitutes,and Biotechnology.* 2004;32: 509-518

[67] Koliatsos VE, Clatterbuck RE,Winslow JW, Cayouette MH, Price DL. Evidence that brain-derived neurotrophic factor is a trophic factor for motor neurons in vivo. *Neuron.* 1993,10:359–67

[68] Miyata Y, Kashihara Y, Homma S, Kuno M. Effects of nerve growth factor on the survival and synaptic function of Ia sensory neurons axotomized in neonatal rats. *J. Neurosci.* 1986,6:2012–18

[69] Wiese S, Metzger F, Holtmann B, Sendtner M. The role of p75NTR in modulating neurotrophin survival effects in developing motoneurons. *Eur. J. Neurosci.* 1999,11:1668–76

[70] Yan Q, Elliott J, Snider WD. Brain-derived neurotrophic factor rescues spinal motor neurons from axotomy-induced cell death. *Nature.* 1992. 360:753–55.

[71] Heumann R, Lindholm D, Bandtlow C, Meyer M, Radeke MJ, et al. Differential regulation of mRNA encoding nerve growth factor and its receptor in rat sciatic nerve during development, degeneration, and regeneration: role of macrophages. *Proc. Natl. Acad. Sci. USA.* 1987,84:8735–39

[72] Lentz SI, Knudson CM, Korsmeyer SJ, Snider WD. Neurotrophins support the development of diverse sensory axon morphologies. *J. Neurosci.* 1999,19:1038–48

[73] Liu RY, Schmid RS, Snider WD, Maness PF. NGF enhances sensory axon growth induced by laminin but not by the L1 cell adhesion molecule. *Mol. Cell. Neurosci.* 2002,20:2–12

[74] Boyd JG, Gordon T. A dose-dependent facilitation and inhibition of peripheral nerveregeneration by brain-derived neurotrophic factor. *Eur. J. Neurosci.* 2002,15:613–26

[75] Henderson CE, Camu W, Mettling C, Gouin A, Poulsen K, et al. Neurotrophins promote motor neuron survival and are present in embryonic limb bud. *Nature.* 1993, 363:266–70 188.

[76] Braun S, Croizat B, Lagrange MC,Warter JM, Poindron P. Neurotrophins increase motoneurons' ability to innervate skeletal muscle fibers in rat spinal cord–human muscle cocultures. *J. Neurol. Sci.* 1996. 136:17–23

[77] Dijkhuizen PA, HermensWT, TeunisMA, Verhaagen J. Adenoviral vectordirected expression of neurotrophin-3 in rat dorsal root ganglion explants results in a robust neurite outgrowth response. *J. Neurobiol.* 1997,33:172–84

[78] Bloch J, Fine EG, Bouche N, Zurn AD, Aebischer P. Nerve growth factorand neurotrophin-3-releasing guidance channels promote regeneration of the transected rat dorsal root. *Exp. Neurol.* 2001,172:425–32

[79] Sterne GD, Brown RA, Green CJ, Terenghi G. Neurotrophin-3 delivered locally via fibronectin mats enhances peripheral nerve regeneration. *Eur. J. Neurosci.* 1997, 9:1388–96

[80] Young C, Miller E, Nicklous DM, et al. Nerve growth factor and neurotrophin-3 affect functional recovery following peripheral nerve injury differently. *Restor. Neurol. Neurosci.* 2001;18(4):167-75..

[81] Englsih AW, Meador W, Carrasco DI. Neurotrophin-4/5 is required for the early growth of regenerating axons in peripheral nerves. *Eur. J. Neurosci.* 2005, 21:1624-34.

[82] Siegel SG, Patton B, English AW. Ciliary neurotrophic factor is required for motoneuron sprouting. *Exp. Neurol.* 2000,166:205–12

[83] Newman JP, Verity AN, Hawatmeh S, Fee WE Jr, Terris DJ. Ciliary neurotrophic factors enhances peripheral nerve regeneration. *Arch. Otolaryngol. Head Neck Surg.* 1996. 122:399–403

[84] Hu Y, Leaver SG, Plant GW, et al. Lentiviral-mediated transfer of CNTF to schwann cells within reconstructed peripheral nerve grafts enhances adult retinal ganglion cell survival and axonal regeneration. *Mol. Ther.* 2005;11(6):906-15

[85] Sakamoto T, Kawazoe Y, Shen JS. Adenoviral gene transfer of GDNF, BDNF and TGF beta 2, but not CNTF, cardiotrophin-1 or IGF1, protects injured adult motoneurons after facial nerve avulsion. *J. Neurosci. Res.* 2003;72:54_64.

[86] Fine EG, Decosterd I, Papaloizos M, Zurn AD, Aebischer P. GDNF and NGF released by synthetic guidance channels support sciatic nerve regeneration across a long gap. *Eur. J. Neurosci.* 2002;15:589–601

[87] Li Q, Ping P, Jiang H, et al. Nerve conduit filled with GDNF gene-modified schwann cells enhances regeneration of the peripheral nerve. *Microsurgery.* 2006;26(2):116-21

[88] Iwase T, Jung CG, Bae H, et al. Glial cell line-derived neurotrophic factor-induced signaling in Schwann cells. *Neurochem.* 2005 Sep;94(6):1488-99

[89] Cordeiro PG, Seckel BR, Lipton SA, D'Amore PA, Wagner J, et al. Acidic fibroblast growth factor enhances peripheral nerve regeneration in vivo. Plast. Reconstr. Surg. 1989. 83:1013–19; *Discussion.* 1020–11

[90] Grothe C, Haastert K, Jungnickel J. Physiological function and putative therapeutic impact of the FGF-2 system in peripheral nerve regeneration--lessons from in vivo studies in mice and rats. *Brain Res. Rev.* 2006 Aug;51(2):293-9

[91] Jungnickel J, Haase K, Konitzer J, et al. Faster nerve regeneration after sciatic nerve injury in mice over-expressing basic fibroblast growth factor. *J. Neurobiol.* 2006 Aug;66(9):940-8

[92] Kanje M, Skottner A, Sjoberg J, Lundborg G. Insulin-like growth factor (IGF-I) stimulates regeneration of the rat sciatic nerve. *Brain Res.* 1989;486:396—8.

[93] Delancy CL, Russel JW, Cheng HJI, Feldman EI. Insulin-like growth factor-I and over-expression bcl-xL prevent glucosemediated apoptosis in Schwann cells. *J. Neuropathol. Exp. Neurol.* 2001;60:147—60.

[94] Syroid DE, Zorick TS, Arbet-Engles C, et al. A role for insulinlike growth factor I in the regulation of Schwann cell survival. *J. Neurosci.* 1999;19:2059—68.

[95] Fine EG, Decosterd I, Papaloizos M, Zurn AD, Aebischer P. GDNF and NGF released by synthetic guidance channels support sciatic nerve regeneration across a long gap. *Eur. J. Neurosci.* 2002. 15:589–601

[96] Sterne GD, Brown RA, Green CJ, Terenghi G. Neurotrophin-3 delivered locally via fibronectin mats enhances peripheral nerve regeneration. *Eur. J. Neurosci.* 1997. 9:1388–96

[97] Pean JM, Menei P, Morel O, Montero-Menei CN, Benolt JP. Intraseptal implantation of NGF-releasing microspheres promote the survival of axotomized cholinergic neurons. *Biomaterials.* 2000, 21:2097-2101.

[98] Yamada M, Natsume A, Mata M, Oligino T, Goss J, et al. Herpes simplex virus vector-mediated expression of Bcl- 2 protects spinal motor neurons from degeneration following root avulsion. *Exp. Neurol.* 2001.168:225–30

[99] Dijkhuizen PA, HermensWT,TeunisMA, Verhaagen J. Adenoviral vectordirected expression of neurotrophin-3 in rat dorsal root ganglion explants results in a robust neurite outgrowth response. *J. Neurobiol.* 1997. 33:172–84

[100] Berry M, Barrett L, Seymour L, Baird A, Logan A. Gene therapy for central nervous system repair. *Curr. Opin. Mol. Ther.* 2001. 3:338–49

[101] Gravvanis AI, Lavdas A, Papalois AE, et al. Effect of genetically modified Schwann cells with increased motility in end-to-side nerve grafting. *Microsurgery,* 2005,25:423-432.

[102] Li Q, Ping P, Jiang H, et al. Nerve conduit filled with GDNF gene-modified Schwann cells enhances regeneration of the peripheral nerve. *Microsurgery.* 2006,26:116-121.

[103] Haasert K, Lipokatic E, Fischer M, et al. Differentially promoted peripheral nerve regeneration by grafted Schwann cells over-expressing different FGF-2 isoforms. *Neurobiol. Dis,* 2006, 21: 138-153.

[104] Hermens WT, Verhaagen J. Viral vectors, tools for gene transfer in the nervous system. *Prog. Neurobiol.* 1998,55:399–432

[105] Smith GM, Romero MI. Adenoviral-mediated gene transfer to enhance neuronal survival, growth, and regeneration. *J. Neurosci. Res.* 1999. 55:147–57

[106] Chen YS, Hsieh CL, Tsai CC, Chen TH, Cheng WC, et al. Peripheral nerve regeneration using silicone rubber chambers filled with collagen, laminin and fibronectin. *Biomaterials.* 2000. 21:1541–47

[107] Yoshii S, Oka M, Shima M, Taniguchi A, Akagi M. 30 mm regeneration of rat sciatic nerve along collagen filaments. *Brain Res.* 2002. 949:202–8

[108] Dubey N, Letourneau PC, Tranquillo RT. Guided neurite elongation and Schwann cell invasion into magnetically aligned collagen in simulated peripheral nerve regeneration. *Exp. Neurol.* 1999. 158:338–50

[109] Yuan Y, Zhang P, Yang Y, Wang X, Gu X. The interaction of Schwann cells with chitosan membranes and fibers in vitro. *Biomaterials.* 2004; 25: 4273–8.

[110] Ahmed MR. Peripheral nerve regeneration in cell adhesive peptide incorporated collagen tubes in rat sciatic nerve early and better functional regain. *J. Peripher. Nerv. Syst.* 2005,10:390-91

In: Tissue Engineering: Roles, Materials and Applications ISBN: 978-1-60456-293-4
Editors: S. J. Barnes, L. P. Harris, pp. 19-66

Chapter 2

THE CHALLENGE OF COMBINING CELLS, SYNTHETIC MATERIALS AND GROWTH FACTORS TO ENGINEER BONE TISSUE

Melba Navarro and Alexandra Michiardi
Institute for Bioengineering of Catalonia (IBEC), CIBER-BBN, Barcelona, Spain

ABSTRACT

Tissue engineering has emerged as a response to the problems associated with the substitution of tissues lost due to diseases or trauma. Nowadays, xenografts, allografts or autografts are commonly used to replace the damaged tissue. However, these options involve numerous drawbacks such as rejection, chronic inflammation and severe organ donor shortages. Tissue engineering keeps the promise to overcome these limitations by creating biological substitutes capable of replacing the injured tissue.

Cells, temporary 3D constructs and biochemical signals that trigger tissue regeneration cascades are the three major components for engineering tissues. The adequate combination of these components leads to the successful repair of damaged tissues. In addition to these three main elements, the use and development of bioreactors has come out as an essential tool for the study and achievement of engineered tissues under *in vitro* conditions while mimicking the *in vivo* environment.

Thus, tissue engineering is an interdisciplinary field that applies the principles of engineering and life sciences to develop biological substitutes that restore, maintain, or improve the tissue functions.

This chapter discusses the challenge of creating engineered bone tissues by combining biological and engineering skills. In particular, it will review the issue of stem cell biology, the fabrication methods and artificial materials used for 3D constructs, the principal growth factors used in bone applications and the key role played by bioreactors. More importantly, it will highlight the interdependence of all these parameters and the need for interdisciplinary research for a successful approach.

INTRODUCTION

Bone degenerative problems including bone fractures, osteoporosis and bone retrieval due to cancer and other pathologies affecting the musculoskeletal system represent half of all chronic diseases in people over 50 in developed countries. This trend is expected to double in the next 15 years due to the high aging rate in the population. Thus, there is a real need to develop new systems to remedy this situation.

Tissue engineering has emerged as a potential alternative to currently used techniques such as allografts, autografts and xenografts which present several limitations, namely, donor site scarcity, rejection, diseases transfer, harvesting costs and postoperative morbidity [1-3]. Tissue engineering and regenerative medicine are recent research fields that explore how to repair and regenerate tissues and organs using the biological cell signalling pathways and their components such as growth factors and peptide sequences among others, in combination with synthetic scaffolds and stem cells [4].

There are three main approaches described for bone tissue engineering (illustrated in figure 1), which are: i) guidance of tissue regeneration using engineered 3D porous scaffolds; ii) injection of cells in the implantation place; and/or iii) combination of 3D scaffolds with cells to guide bone regeneration. The first method involves implanting a scaffold at the site of interest and allowing cell ingrowth from surrounding tissues to colonize the scaffold. The second method also known as cell therapy is less invasive. It involves the use of cells optionally genetically manipulated prior to injection or infusion. In the third method, 3D porous scaffolds designed to mimic extracellular matrices are loaded with cells and located in the implantation site.

In practice, organ-specific cells are often seeded into the scaffold *ex vivo* prior to implantation, and with time the cells synthesize a new extracellular matrix as the scaffold degrades simultaneously and eventually produce a new properly functioning tissue [5].

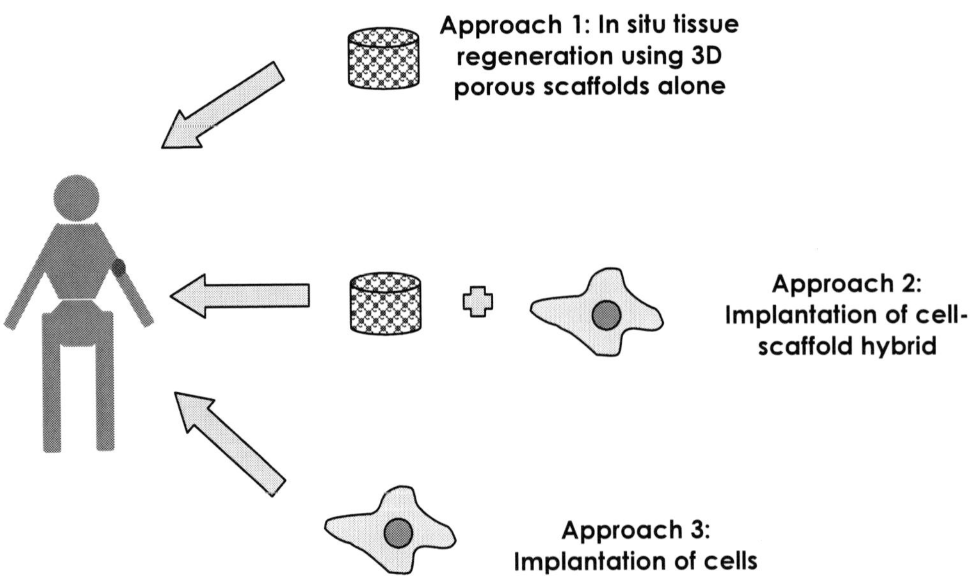

Figure 1. Main approaches to tissue engineering.

Tissue engineering is constituted of three major components (figure 2): 1) cells building the new tissue, 2) 3D porous scaffold mimicking the extracellular matrix and guiding tissue regeneration, and 3) specific signalling systems that trigger the healing cascade through the activation of genes responsible for the secretion of products responsible for tissue differentiation. These three components must be properly combined for a successful repair and regeneration of tissues.

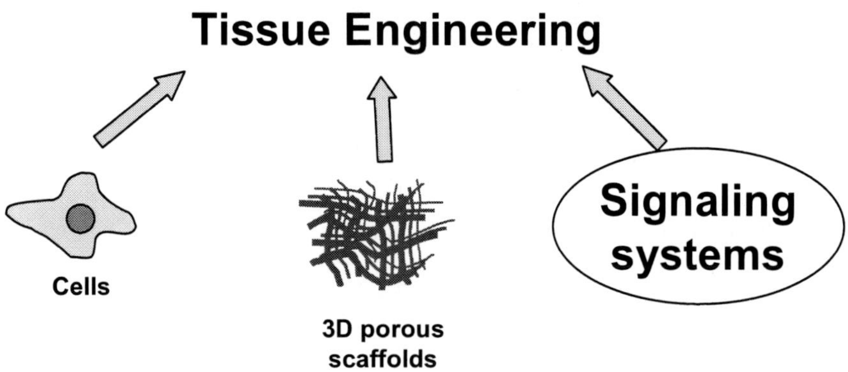

Figure 2. The three major components of tissue engineering.

Furthermore, the use and development of bioreactors has emerged as an essential tool for the study and attainment of tissues under *in vitro* conditions while mimicking physiological *in vivo* conditions.

Therefore, tissue engineering is a multidisciplinary and interdisciplinary field that requires the combination of knowledge and expertises of chemists, engineers, biologists and physicists among others in order to overcome the multiple demands and challenges inherent to the development of tissue regeneration systems.

More insight about each of these three major components as well as the approaches and strategies used to achieve bone repair and regeneration will be discussed in this chapter.

Cell Biology in Tissue Engineering

The shortage of tissues and organs for transplantation has led to the rapid development of tissue engineering as an alternative to overcome this problem. Some of the tissues found in the body possess cells capable of initiating regeneration or repair after injury. Cell regeneration potential depends on the cell type, nature and location of the injury, the size of the defect to be repaired, and patient's age among other aspects. Nevertheless, most human tissues do not regenerate spontaneously. This is the reason why fabrication of biotissues that mimic the natural ones is so promising.

In vivo bone healing is generally considered to be biologically optimal since the majority of defects in this tissue, experience spontaneous healing with minimal treatment. However, it is known that from the total amount of fractures occurring every year in the United States, 5-10% require further treatment to achieve full healing because of either interposition of soft tissue, improper fracture fixation, bone loss, metabolic disturbances, impairment of blood supply or infection [6]. Moreover, in certain clinical conditions, large pieces of bone must be retrieved to treat benign and malign tumours, osteomyelitis and also bone deficiencies and abnormal loss in the maxillofacial area. In these cases, the use of bone substitutes and systems to enhance tissue regeneration is absolutely required.

Autografts, which are grafts made of tissue obtained from the patient himself, are the gold standard in these situations, because they avoid most problems related to transfection and rejection. However, autologous bone sources are limited and the harvesting process is usually painful and involves risks of infection.

Synthetic substrates as well as the use of allografts that are grafts made of tissue from other donor (usually post-mortem), and xenografts which are grafts of tissue proceeding from other species, have been used in the last years to overcome the drawbacks of autografts. Nevertheless, these three alternatives also present important limitations. Both allografts and xenografts imply the risk of immunological reactions and infections, and in the case of allografts, availability of donors is also a main concern. At present, as an alternative to the before mentioned grafts, synthetic substrates have only been used successfully in small defects in direct contact with bone tissue.

Thus, all these limitations and drawbacks related to the attainment and harvesting of autologous bone have impelled the development of novel alternative techniques including the use of growth factors and stem cells.

Independent of the therapeutic approach to regenerate the diseased or injured tissue, cell sourcing is the first issue to consider in the development of bioengineered tissues. Nowadays, a number of cell sources are being investigated; including adult stem cells, adult differentiated cells, embryonic and fetal stem cells, cells generated by nuclear transportation, and *ex vivo* manipulated cells [7].

The Role of Stem Cells

Stem cells are characterized by three main features i) they are capable of self-renewal ii) they have the ability to give rise to different cell lineages and iii) they are capable of *in vivo* functional regeneration of the tissues to which they give rise [8]. Stem cells biology offers the potential to grow tissues by following a developmental pathway, which means that they have the capability of mimicking the developmental process followed during the formation of an embryo, where cells from a fertilized egg are able to differentiate and form a complex multicellular organism. This ability to differentiate into a variety of cell types is known as potency. Stems cells potency varies depending on their source; cells that are able to produce only one type of cell are unipotent cells, whereas cells able to produce a wide variety of cells are pluripotent. Totipotent cells are those able to differentiate in any cell lineage comprised in the body (stems cells hierarchy is illustrated in figure 3). This potency ability of stem cells has raised great interest and is considered as a promising resource for tissue engineering applications and transplantation. Besides, the use of stem cells in regenerative medicine has raised great interest given the insights they offer into the understanding of tissue repair and regeneration processes.

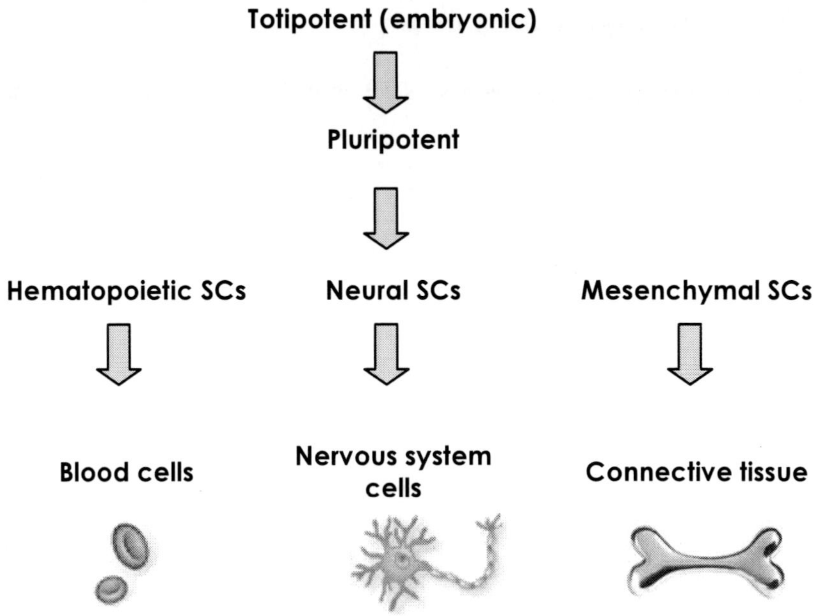

Figure 3. Stem cells hierarchy.

Stem cells can be retrieved from both embryonic and adult tissues. Totipotency of embryonic stem cells makes them the only cell source capable to give rise to all sorts of adult cell lineages. Embryonic stem cells can only be collected at very early stages of embryogenesis. Sources of embryonic stem cells are supernumerary embryos produced by in vitro fecundation or embryos created via the nuclear transfer technique. Both the collection and usage of this type of cells deals with important ethical issues. For instance, it must be emphasized that using human embryonic progenitor cells is banned in many countries, although a modification of the so-called bioethical law allows importation of human embryonic cell lines for research purposes.

Adult progenitor or stem cells have a more limited potential than embryonic ones. Indeed, most of them are unipotent, and at present only those collected from bone marrow, brain, and adipose tissue seem to be pluripotent cells [9]. These cells have been isolated from a wide variety of tissues and although they lack tissue-specific characteristics, they can differentiate into specialized cells with a phenotype different from that of the precursor when they are stimulated with the appropriate signals. One of the most important points in the utilization of adult stem cells is that their collection does not involve ethical restrictions.

Although it has been demonstrated that adult stem cells are available from different sources such as bone marrow, brain and adipose tissue among others, up to now, only those hematopoietic stem cells which give rise to all the blood cell types including myeloid (monocytes and macrophages, neutrophils, basophils, eosinophils, erythrocytes, megakaryocytes/platelets, dendritic cells) and lymphoid lineages such as T-cells, B-cells and NK-cells, and mesenchymal stem cells collected from the bone marrow have been used for therapeutic applications.

The Promise of Mesenchymal Stem Cells

The term mesenchymal stem cells (MSCs) is commonly used to refer to connective tissue cells in adult tissues such as (myo)fibroblasts, bone, cartilage, fat, tendon, muscles, and nerve tissue. These cells represent a sub-group of precursor cells that sticks to the stem cell features (self-renewal and the ability to give rise to different cell types). Indeed, their ability to differentiate in mature cells and several lineages and generate new tissues has been shown by several authors [10]. MSCs can be isolated from several sources, including bone marrow, fat, umbilical cord blood, and also peripheral blood [11-13] and may differentiate into several tissue cells, including osteoblasts, chondroblasts, myoblasts and adipocytes. Recently, mice models have provided evidence that such cells may differentiate into cardiac myocytes and even neurones.

Interest in the use of mesenchymal stem cells is mainly focused on their abilities to produce various functional tissues. Furthermore, another property known as induction of immune tolerance to grafts or even immunosuppression has been recently attributed to these cells, which makes then even more appealing for tissue engineering applications.

Among the different MSCs sources, bone marrow is the most currently used. Bone marrow is a natural reservoir of skeletal mesenchymal stem cells. These cells contained in the stromal compartment of bone marrow were first identified by Friedenstein and Petrakova [14] who isolated bone-forming progenitor cells from rat marrow. These cells represent a minimal fraction (0.001-0.01%) of the total population of nucleated cells in marrow [10].

From the bone tissue engineering point of view, mesenchymal stem cells have a number of advantages: i) their isolation is easy as it relies primarily on the ability of these cells to adhere to tissue-culture plastic [15]; ii) they have a high proliferative potential [16-18]; iii) the default pathway of MSCs is the osteogenic pathway; iv) bone formation is not correlated to the number of cell passages as long as the human stem cells retain their proliferative potential [15;19]; and v) freezing conditions do not affect the osteogenic potential of MSCs, a condition that greatly facilitates their storage [17].

Besides bone marrow, there are other cell sources with osteogenic potential, namely, periosteum [20-22], trabecular bone [23-25] adipose tissue [26-28], synovium [29], skeletal muscle [30], lung [31] and deciduous teeth [32]. Nonetheless, further studies in clinically relevant animal models are needed to validate the osteogenic potential of stem cells isolated from these alternative sources.

Cell-Based Therapies for Bone Regeneration

In general, therapies based on the use of cells for achieving bone repair include both the use of i) cells without gene modifications and ii) cells genetically engineered, as shown in figure 4. The first approach involves the implantation of stem cells directly in the place of interest in order to stimulate bone repair. In this case, the choice of the adequate scaffold is crucial. Scaffolds must possess osteoinductive (that induce the formation of new bone) and osteoconductive (that stimulates bone ingrowth) properties to facilitate bone formation. Thus, the cells/scaffold combination must trigger the osteogenic (bone formation) cascade to enhance the regeneration of the bone tissue.

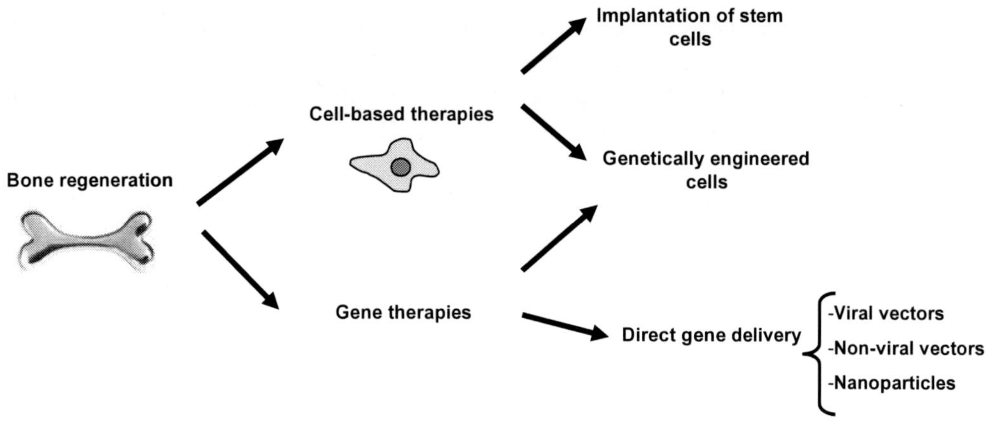

Figure 4. Cell and gene based therapies for bone regeneration.

Generally, both mesenchymal stem cells and muscle-derived stem cells are the most used for bone tissue engineering applications. Mesenchymal stem cells play a key role in cell-based therapies due to their inherent ability to differentiate in several lineages. Osteogenic activation can be triggered in the presence of β-glycerol-phosphate, ascorbic acid-2-phosphate, dexamethasone and fetal bovine serum. When cultured in monolayer in the presence of these supplements the cells acquire an osteoblastic morphology with upregulation of alkaline phosphatase activity and deposition of a calcium-rich mineralized extracellular matrix.

As already mentioned, an alternative source for adult stem cells is the muscle tissue. It has been demonstrated that a population of muscle-derived stem cells (MDSCs) isolated from mouse skeletal muscle produced alkaline phosphatase in a dose dependent manner when stimulated with recombinant human bone morphogenetic proteins 2 and 4 (rhBMP-2 and rhBMP-4). These cells were also able to differentiate toward the osteogenic lineage and, consequently, improved bone healing in calvarial defects in mice [33-35].

The second approach is based in the use of genetically engineered cells for bone regeneration. In the cell-based gene therapy cells are genetically engineered *ex vivo* to express a certain transgene that is then implanted *in vivo*. Cell-based gene delivery is becoming a primary approach in bone tissue engineering. It takes advantage of the ability of adult stem cells to express a transgene that can both affect the cells themselves and recruit host stem cells to achieve bone formation [36].

Gene-Based Therapies for Bone Regeneration

Therapies that involve gene modifications to enhance bone formation consist of two different approaches: a) direct *in vivo* delivery of genes into the injured place, and b) *ex vivo* transduction and subsequent transplantation of cells expressing the osteoinductive factor (figure 4). The selection of the gene delivery approach depends on several factors, including the particular gene of interest, indication targeted, desired duration of gene expression, and delivery vector.

MSCs can also be genetically engineered to express a desired gene. In the case of bone tissue engineering cells are genetically modified to express an osteogenic gene, which induces them to form bone *in vivo*. The efficiency of using stem cell-based gene therapy for bone formation has been demonstrated in many studies [26;37-41].

It is important to mention that if stem cell-based gene therapy is used, the number of MSCs necessary for implantation can be reduced. Each engineered cell will produce the specific therapeutic protein, leading to the recruitment of additional MSCs from host tissues and subsequent differentiation into bone cells.

The effectivity of the cell-gene-based techniques using engineered MSCs to generate bone tissue has been evaluated using micro–computed tomography and molecular analyses of gene and protein expression [36;42].

Most gene therapy studies in the field of bone tissue engineering focus on the Bone Morphogenetic Proteins (BMP) family (that will be described in detail further in this chapter). However, some studies have also involved the use of other genes that induce osteogenic differentiation.

The delivery of genes into MSCs can be done by different methods; most of them involve the use of viral vectors (modified viruses for gene delivery), such as adenovirus and retrovirus. These vectors are potentially hazardous because of the risks of insertional mutagenesis or potential evoking of the immune system. Until recently, there was no nonviral alternative for these vectors because all conventional nonviral vectors failed in efficiently transfecting (introducing genes into cells) MSCs. In this sense, the electroporation method, which is based in the increase in permeability of the cell membrane due to an externally applied electrical field has been suggested as an alternative to viral vectors. It has been successfully modified and used to transfect bone marrow–derived MSCs without the need of a viral vector [43]. To demonstrate the efficacy of the electroporation strategy, MSCs were nucleofected with BMP-2 and BMP-9 plasmids and then implanted ectopically. Bone formation was evident after 4 weeks, proving that *ex vivo* nonviral gene delivery into adult bone marrow–derived stem cells is a viable alternative to viral methods of transduction [38]. Nevertheless, further studies should be conducted to determine more deeply the ability of electroporated cells to regenerate a critical-size defect in an animal model.

Another promising method for transfecting cells without the usage of viral vectors is the use of polymer nanoparticles for the entrapment and delivery of genetic material. This issue will be discussed later in this chapter.

In order to enhance bone formation effectively, genes involved in the osteogenesis process must be known. Genetically engineered MSCs can also be used to find novel candidate therapeutic genes for bone repair. To this end, different studies based on gene arrays followed by a clustering analysis are used to generate a database of genes that play a major role in the bone formation process that is induced by genetically modified MSCs.

Scaffolds: Synthetic 3D Structures for Bone Tissue Engineering

Bone is a complex tissue with multiple cell phenotypes, distinct tissue types, high vascularisation and plays a very demanding mechanical role. Therefore, the attainment of systems that substitute original bone is an extremely difficult task.

Bone transplantation is one of the most successful methods to restore bone functionality in bone defects and injuries. In particular, trabecular bone autografts are the gold standard in bone transplantation. The high porosity of trabecular bone allows the tissue ingrowth and vascularisation of the graft within few weeks and new bone formation within months. Thus, the ideal bone substitute should resemble trabecular bone's architecture, biochemistry and mechanical properties [44].

Cells count on a natural scaffold which consists of various macromolecules and is known as extracellular matrix (ECM). The ECM is a combination of specific proteins that are secreted by cells to form a complex network. This network is a supporting framework where cells grow and differentiate giving the form and shape of the desired tissue. The ECM provides the proper environment within which migratory cells can move and interact with one another and stationary cells are anchored. Cell-matrix interactions are paramount to initiate and mediate biological events that control cell growth, migration, differentiation, survival, tissue organization, and matrix remodelling among others. ECM also plays a structural role. For instance, in the case of connective tissues, ECM is the "stress bearing" element. Most of the mechanical stresses undergone by this type of tissue are borne by the ECM. The function of the tissue engineering scaffold is to act as a synthetic ECM mimicking its properties for support of cell growth and tissue development.

Scaffolds are short -term matrices for bone growth and provide a specific environment and architecture for tissue development. The material composition as well as structural characteristics such as porosity, interconnectivity and pore size and distribution among others are crucial for the success of tissue engineering approaches based on the usage of acellular constructs [45;46].

Temporary synthetic 3D porous constructs that stimulate cells attachment, in-growth, proliferation and differentiation into functional osteoblastic cells are currently being explored as provisional support for the formation of the new tissue. Materials used in the elaboration of these scaffolds are placed within the context of the third generation of biomaterials, which deals with the development of new materials able to stimulate specific cellular responses at the molecular level [47].

Ideal scaffolds that mimic autologous bone grafts must possess three key properties: osteoconductivity, osteoinductivity and osteogenicity. Bone substitute materials usually possess two of these key characteristics.

Besides these three properties, porous structures aimed to tissue engineering applications need to fulfil the following criteria [48;49]:

- The material must be biocompatible and its degradation by-products non-cytotoxic.
- The scaffold must be biodegradable and should resorb at the same rate as the tissue is repaired.
- The scaffold must possess a highly interconnected porous network, formed by a combination of macro and micro pores that enable a proper tissue ingrowth, vascularization and nutrient delivery.
- The mechanical properties of the scaffold must be appropriate to regenerate bone tissue in load bearing sites. Moreover, the material must keep its structural integrity during the first stages of the new bone formation.

Other wanted properties are an adaptable shape to the defect size and geometry as also radioopaque qualities to allow radiographic differentiation between the implanted material and the new formed bone [50].

Thus, biomaterials used as tissue engineering scaffolds must combine both biodegradability and bioactivity properties among other aspects in order to success. Surface functionalization with peptide sequences that mimic the extracellular matrix components, delivery of biochemical factors, and control of cell behaviour through mechanotransduction in order to trigger specific cell responses are some of the approaches that are currently under study in order to complete the three key properties and add osteoinductive characteristics to the materials [51-53].

Scaffolds' porosity must include a wide variety of pore sizes. Microporosity with pores less than 10 μm is needed for capillary ingrowth and to promote protein adhesion and consequently cell-matrix interactions such as cell adhesion and proliferation. Macroporosity with pore sizes ranging between 150 and 900 μm allows for nutrient supply and waste removal of cells grown on the scaffold [54;55]. In spite of the controversy about the optimum pore size, in general, an average pore size around 100-350 μm is considered adequate [56]. Pore diameters over 100μm are necessary for cell ingrowth and diameters over 200μm are required for osteoconduction.

Various materials have been used to produce scaffolds that meet the requirements mentioned above. The most important are natural and synthetic biodegradable polymers as well as ceramics.

Polymeric Scaffolds

Both natural and synthetic polymers have been used in the development of new 3D scaffolds for bone tissue engineering. Natural polymers can be of both plant and animal origin. These polymers usually offer the advantage of biological recognition; however, they may also include pathogenic impurities and in general offer low reproducibility.

In contrast, synthetic polymers offer high reproducibility and the possibility of large-scale production. In particular synthetic biodegradable polymers have raised special attention because they enable a better control of their physico-chemical properties and also because they have been successfully used in clinical applications. Polylactic acid (PLA), polyglycolic acid (PGA) and polycaprolactone (PCL) are the most studied polymers for bone tissue engineering purposes.

Commonly Used Biodegradable Polymers

Polylactic Acid

One of the most widely used synthetic polymeric materials is polylactic acid (PLA). PLA, which structure is shown in figure 5, is a biocompatible, thermoplastic, resorbable aliphatic polyester from the poly(α-hydroxyacid) family, it is FDA (Food and Drug Administration) approved, and has been used clinically as sutures, bone fracture fixation devices and as drug release systems. PLA is produced by the ring-opening polymerization of lactide; a cyclic diester. The polymerization requires heat and a metallic or an organometallic catalyst. Stannous octoate is the most commonly used catalyst because it is also FDA approved.

PLA is a chiral molecule, and thus exists as two stereoisomers or enantiomers: L-lactide and D-lactide

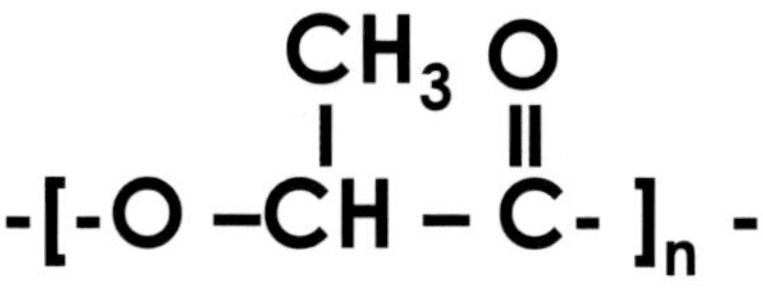

Figure 5. Polylactic acid (PLA) structure.

PLA can be polymerized into four morphologically distinct forms: poly-DD-LA, poly-LL-LA (also known as PLLA), poly-DL-LA (also known as PDLA) and poly-meso-LA. PLLA is a semicrystalline polymer, with a glass transition temperature (Tg) ranging between 50-65°C, and a melting temperature (Tm) between 170-190°C. PDLA is an amorphous polymer with a Tg ranging between 50-60°C. As with most of the polymers, crystallinity, Tg and molecular weight are factors that modulate PLA's mechanical and degradation properties. PLLA is the most frequently employed of the polylactic polymers because it yields the L-lactide acid upon hydrolytic degradation which is the naturally occurring steroisomer of lactic acid [57].

PLA degrades by random bulk cleavage of its ester bonds. This process is auto-catalyzed by the ends of the carboxylic chains that are produced during the ester hydrolysis. The autocatalytic cleavage of the PLA ester bonds has important effects on the degradation behaviour of PLA implants. The cleavage of the ester bonds in PLA releases lactic acid which tends to acidify the milieu of the PLA implant and may be toxic to the surrounding tissues [58]. This is one of the main drawbacks of PLA implants and its application as a tissue engineering material.

Degradation by-products (lactic acid) are finally eliminated as CO_2 and H_2O [59]. First, polylactic acid degrades into lactic acid which in turn gives pyruvic acid. In the presence of sufficient oxygen, pyruvic acid is converted into carbon dioxide, CO_2, and acetyl-coenzime A. This coenzyme is the main input to the citric acid or tricarboxilic acid cycle; a series of chemical reactions of central importance to all living cells. Within the citric acid cycle, acetyl-coenzyme A reacts with oxoaloacetate to produce citrate which is the first and the last product of the cycle. A schematic description of the process is shown in figure 6.

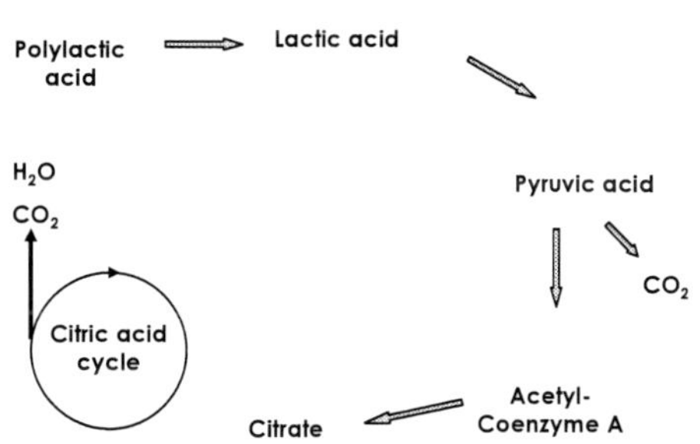

Figure 6. Schematic description of the metabolic degradation of polylactic acid.

Polyglycolic Acid

Polyglycolic acid (PGA) is the simplest linear aliphatic polyester. PGA was used to develop the first totally synthetic absorbable suture that has been patented as DEXON® since the 1960s by Davis and Geck [60;61]. Glycolide monomer is synthesized from the dimerization of glycolic acid. As in the case of PLA, PGA is obtained by ring-opening polymerization of the cyclic diesters catalized by antimony, zinc or lead. PGA is highly crystalline (45-55%) with a high melting point ranging between 220 and 225°C and a glass transition temperature between 35 and 40°C [61]. The chemical structures of PLA and PGA are similar except that the PLA has a methyl pendant group (see figures 5 and 7). This structural difference confers PGA a higher hydrophilicity in comparison to PLA. This fact contributes to significant differences in their degradation kinetics; PGA degrades much faster than PLA. While PGA degrades in few weeks [62;63], PLA can remain stable for over 1 year or more depending on its degree of crystallinity [64].

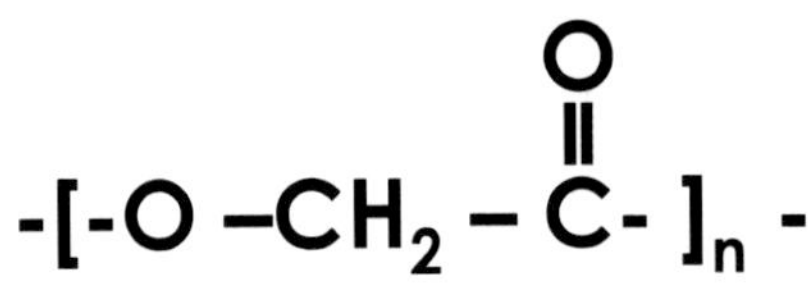

Figure 7. Polyglycolic acid (PGA) structure.

As in the case of PLA, PGA degradation begins with random hydrolysis in an aqueous environment. *In vivo*, however, enzymes are thought to enhance the initial degradation. The hydrolytic degradation *in vivo* may take place via non-specific esterases and carboxyl peptidases that produce glycolic acid monomers which are converted enzymatically into glycine, which can be used in protein synthesis, or into pyruvate that enters the tricarboxylic acid (TCA) cycle yielding energy, water and carbon dioxide. Besides, glycolic acid is partially excreted in urine.

Glycolide has been copolymerized with other monomers to reduce its stiffness and modulate its degradation rate. The most common copolymer is the poly(lactic-co-glycolic) acid.

Polycaprolactone

Poly(ε-caprolactone) (PCL): PCL is an aliphatic polyester that has also been widely studied as degradable biomaterial (figure 8). The ring opening polymerization of ε-caprolactone yields a semicrystalline polymer with a low melting point of 59-64°C and a exceptionally low glass-transition temperature around -60°C. Thus, this polymer is always in a rubbery state at room temperature. PCL can also be degraded by a hydrolytic mechanism under physiologic conditions. The degradation by-products of this polymer are metabolized intracellularly following a pathway similar to that of the poly(α-hydroxy acids). The homopolymer has a degradation time of the order of two years. Thus, it possesses a significant slower degradation rate in comparison with PLA and PGA. Different polymer blends have been investigated to combine the advantages of various polymers. For instance, PCL's degradation rate can be tailored by combining it with other polymers, such as DL-lactide which increases PCL degradation rate, or glycolide that reduces stiffness compared to pure PCL and is being sold as a suture [65].

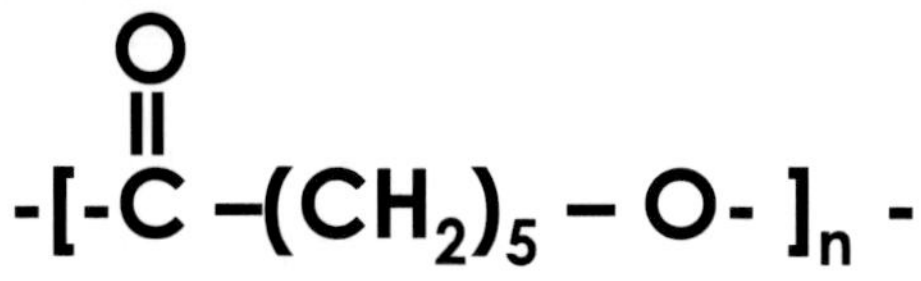

Figure 8. Poly(ε-caprolactone) (PCL) structure.

In general, hydrolysis of these biodegradable polymers depends on several aspects such as their molecular weight, surface area/weight radio, porosity and monomer concentration, geometric isomerism and conformation and crystallinity. The pH of the aqueous medium is also crucial as highly acidic or basic media strongly affects the chemical structure of these polymers accelerating there degradation rate [63;66;67].

One of the strategies to obtain scaffolds that are both biodegradable and bioactive is the combination of organic and inorganic phases in a composite material. Biodegradable composite scaffolds combining biodegradability and bioactivity offer unique advantages in the bone tissue engineering field. The incorporation of an inorganic phase into the bioabsorbable polymer matrix modifies the mechanical behaviour of the porous structure [68], enhances the scaffold bioactivity and also modifies the degradation pattern of the polymer[69], [49]. Furthermore, the incorporation of a second inorganic phase into the polymeric matrix may buffer poly(α-hydroxy acids)'s acidic degradation by-products.

Elaboration Techniques

Numerous elaboration techniques and approaches to the development of 3D scaffolds that combine biodegradability and bioactivity are currently under study. Each elaboration technique provides specific and different structural characteristics to the final scaffold. Therefore, the choice of the technique depends on the requirements of the final application. Some of the most promising techniques for the elaboration of such scaffolds are gel casting, solvent casting and particulate leaching, laminated object manufacturing (LOM), phase separation, gas saturation, fibre bonding and membrane lamination among others [70]. Scaffolds' micro and macrostructure depends strongly on the processing technique.

Solvent Casting and Particulate Leaching

The solvent casting and particulate leaching method was developed by Mikos et al [71] amongst others for polylactic and polyglycolic acid polymers, and several authors have used the method to manufacture composite scaffolds [72-74]. It consists of dissolving a polymer in a solvent and then adding particles of a leachable porogen: salt particles, glucose, paraffin spheres, etc. The mixture forms a thick paste which is left to dry in air or under vacuum until the solvent has evaporated completely. The porogen is leached out afterwards leaving behind a network of interconnected pores. In the case of composites, the second phase is added with the porogen and remains within the structure after the porogen is leached out. Additionally, a thermal treatment can be used to modulate the crystallinity of the polymer by melting the polymer and controlling the cooling rate. The advantages of the solvent casting method are that it is a simple and fairly reproducible method which does not require sophisticated instruments. The disadvantages include thickness limitations intrinsic to the particulate leaching process, limited mechanical properties, and some authors question the homogeneity and interconnection of the pores in the scaffolds, as well as the presence of residual porogen and solvent [75].

Phase Separation

Thermally induced phase separation was first applied to PLA scaffolds by Schugens et al [76], several authors have applied this technique to composite scaffolds [48;77;78]. It consists of inducing a solid-liquid or liquid-liquid phase separation. This is done by dissolving the polymer in a solvent and quenching the solution at a certain temperature. The rapid cooling process, known as quenching induces a phase separation into a polymer-rich phase and a polymer-poor phase. The solvent must then be removed from the phase separated solutions either by freeze-drying, or by solvent extraction. The solvent leaves behind a microstructural foam. The main advantage of the phase separation method is that pore morphology and orientation can be tailored by altering the thermodynamic and kinetic parameters of the processing. Its disadvantages include the use of potentially toxic solvents and a high degree of anisotropy of the porosity. The latter may actually be beneficial for certain biomedical applications such as nerve regeneration [77;79]. Figure 9 shows the morphology of polylactic scaffolds elaborated using two different techniques, solvent casting followed by particulate leaching (figure 9a) and the phase separation method (figure 9b).

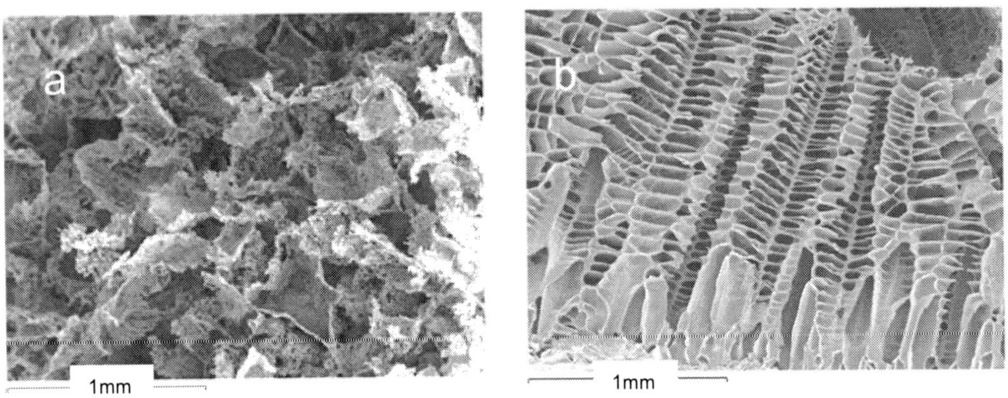

Figure 9. Polylactic acid scaffolds elaborated by the a) solvent casting and b) phase separation techniques.

Gas Foaming

The gas foaming process is used to fabricate highly porous foams without the use of organic solvents [88-90]. Organic solvents may leave residues behind which can have toxic effects *in vitro* and may cause inflammation *in vivo*. The process consists of saturating the polymer mix with gas at high temperatures and pressures. Then, a thermodynamic instability is created by quickly decreasing the temperature and pressure which stimulates the nucleation and growth of pores of gas within the polymer. Gas-foaming yields high porosities (up to 93%) and varying the temperature, pressure, and rates of parameter reductions can modulate pore sizes. The main disadvantage is due to the poor interconnectivity of the porosity, and the fact that surfaces are mostly nonporous.

Fibre Bonding

The fibre bonding method was first developed by Cima et al [80] who produced scaffolds made of polyglycolic (PGA) acid polymer. They took advantage of the fact that PGA was available as sutures and thus in the shape of long fibres. Mikos et al [71] improved the structural stability of the constructs developing a fibre bonding technique in which the PGA fibres joined at their cross-linking points by "sintering" above their melting point temperature. The main advantage of this technique is the very high surface area/volume ratio which makes them ideal for tissue engineering applications.

Electrospinning

The electrospinning is a simple and cost-effective process that uses an electric field to control the formation and deposition of polymer fibres on a substrate. Sheets and cylindrical shapes can be fabricated with this technique. This method can produce fibres with diameters between several microns down to 100nm or even less. This is of great interest in the development of tissue engineering scaffolds since these fibres can mimic ECM's proteins structure. Electrospinning is able to produce both random and oriented networks which are also a requirement when mimicking natural tissues.

Rapid Prototyping

Rapid prototyping, also known as solid free-form fabrication allow the production of complex scaffolds from 3D databases. All rapid prototyping techniques are based on the same principle of building 3D models layer by layer.

Their main advantage is the possibility of creating intricate 3D shapes and guaranteeing pore interconnectivity. One of the most consolidated techniques, and probably the oldest, is 3D printing. In this case, a thin layer of polymer powder is spread over a platform and a liquid binder is printed onto the layer. The following layer is created by lowering the platform and applying another layer of polymer powder. Thus, the previous layer is joined to the present layer by the binder creating a three-dimensional shape. The operation parameters can be controlled exactly via a computer-assisted design program and the process takes place at room temperature. Some other techniques include the following:

- Stereo-lithography: photo-polymerization by a low-power highly focused UV laser.
- Selective laser sintering: laser sintering of polymer powders.
- Fuse deposition Modelling: extrusion of the polymer, the polymers is usually standard and medical grade, elastomer, polycarbonate, polyphenilsulfone.
- Solid ground curing: similar to stereolithography: but entire layers are cured at time thanks to the use of photomasks.
- Ink-jet printing: 3D binding of polymer powders by a binderfluid; usually needs post process infiltration.

The main limitation of these techniques is their resolution limit (50-300 μm depending on the technique) which makes it difficult to design scaffolds with fine microstructures [81].

Ceramic Scaffolds

Some ceramics and glasses have been used for the elaboration of porous scaffolds. As in the case of polymers, a high degree of macro, micro and nanoporosities is needed. In addition, the degradation rate of the biomaterial and the regeneration rate of the tissue must be equilibrated in order to enable a proper bone turn-over process [82].

Ceramics

Calcium phosphate ceramics have been studied over many years and are currently used for clinical applications. In particular hydroxyapatite ($Ca_{10}(PO_4)_6(OH)_2$, HA) and tricalcium phosphate (especially β-TCP) that have similarities with the chemistry and crystallography of the mineral phase of bone. These calcium phosphates are known as biocompatible, bioactive and osteoconductive and are usually used both in compact and porous forms as well as granules. The filling of bone defects with these materials aims at stimulating a faster and stable healing process. Especially, hydroxyapatite (HA) has been and is used extensively as a substitute in bone grafts, because the crystalline phase of natural bone is basically HA. Moreover, HA as well as other calcium phosphate ceramics possesses bioactive properties and show excellent abilities to bond to bone.

HA possesses a low degradation rate, which is a hurdle for tissue engineering applications. To overcome this problem, HA is usually combined with β-TCP that possesses a faster degradation rate. This combination of materials is also known as biphasic calcium phosphates [83].

Other types of ceramics used in bone repair include porous calcium metaphosphate ($[Ca(PO_3)_2]n$) blocks [69] and natural coral.

Though these artificial calcium phosphate ceramics can possess osteoconductive and, osteogenic properties, they clearly have no osteoinductive effects on regenerating bone [84]. Their combination with bone morphogenetic proteins (BMP), mesenchymal stem cells from bone marrow, and angiogenesis inductive factors may solve this drawback [85]. Other strategy involves the use of chitosan foams covered with calcium phosphate slurries to obtain scaffolds with the advantages of the osteoinductive properties of chitosan together with the osteoconductive properties of calcium phosphates.

Current research is focused in the development of nanocrystalline structures resembling bone mineral phase crystals, organic-inorganic composites, the use of fibres and microspheres for the elaboration of porous structures, 3D amorphous calcium phosphate scaffolds, HA , tuned porosity microstructures [86] and hierarchically organized structures [87].

Calcium phosphate cements foams have been used as templates for *in situ* regeneration with very promising results. Calcium phosphate cements based on β-TCP have been foamed using albumen and other foaming agents showing positive *in vivo* results. Scaffolds with porosities around 75% and spherical pores ranging between 100-500 μm can be obtained using this technique [88;89].

Glasses and Glass-Ceramics

Biological glasses both bioactive and bioabsorbable, as well as some calcium phosphate glass-ceramics have been also used for the fabrication of 3D porous scaffolds for bone tissue engineering.

Bioactive glasses such as the well known Bioglass were first developed by Hench et al [90]. They discovered that certain glass compositions had the capacity to form a tight bone bonding. Bioactive glasses based on SiO_4 as network former are able to create a calcium deficient carbonated phosphate surface layer that allows the formation of the chemical bond to bone. The formation of this surface layer and posterior binding to bone is known as bioactivity. This bioactivity that is also present in some calcium phosphate ceramics such as HA is induced by interfacial and cell mediated reactions in the aqueous medium [91].

The formation of a tight bond to the bone tissue guarantees a good fixation between the scaffold and the surrounding tissue. Bioactive glasses have demonstrated to support enzyme activity, vascularization, osteoblast adhesion, growth and differentiation. In addition, they have also shown to induce the differentiation of mesenchymal cells into osteoblasts, which is of great importance in cellular tissue engineering applications [48;92;93].

The degradation by-products of these bioactive SiO_4 based glasses have been analysed. According to the obtained results, dissolution products from bioactive glasses upregulate the gene expression that control osteogenesis and the production of growth factors [91]. Besides, it has been shown that the presence of silicon in ceramic materials and glasses play a key role in bone mineralization and gene activation. These findings are of great impact for bone tissue engineering since they can stimulate materials' osteoinductivity.

Glasses bioactivity and osteoconductivity together with their osteogenic properties make them promising materials for bone tissue engineering applications.

Phosphate glasses also represent an interesting option as materials for tissue engineering applications given that their chemistry can be modulated in order to obtain glasses chemically similar to the bone mineral phase. Furthermore they have demonstrated to present a controlled solubility [94].In fact, their solubility can be modulated with the addition of highly charged and small radius modifying ions [95]. Additionaly, previous studies have revealed the biocompatibility of these glasses. *In vitro* cultures have shown that these formulations elicit no cytotoxicity [96]. In particular glasses in the system P_2O_5-CaO-Na_2O-TiO_2 have shown to accelerate the early differentiation of osteoblastic cells *in vitro* [96]

Different methods for the elaboration of these 3D vitreous or ceramic constructs have been proposed such as the introduction of diverse kinds of porosifiers, foaming agents and emulsifiers, the solid free form fabrication technique, the impregnation of a skeleton of porous polyurethane foam with ceramic emulsions, sol-gel and gel-casting techniques among others [89;97-101]. Among this variety of processes for fabrication of porous materials, the replication technique (also called the polymersponge method) produces porous ceramic structures that are most similar to those of spongy bone [102].

The replication method consists in impregnating a body of porous polymer foam with slurry containing the ceramic or glass powder, water and some optional additives. The polymer foam is the responsible of the porosity, interconnectivity degree and pores size. Once the foam has been impregnated with the slurry, it is exposed to a heat treatment that allows the ceramic densification and elimination of the organic foam. HA and bioactive glass scaffolds have been fabricated by means of the replication method [103;104].

Incorporation of porosifiers implies the use of organic compounds that can be pyrolitically removed leaving behind pores. The size and shape of the pores depends on the nature of the porosifier agent. Ceramic granules or particles are mixed with the porosifier agent and pressed forming the green body that will be subsequently heated in order to eliminate the organic matter and to sinter the ceramic material.

The sol-gel process involves the preparation of a mixture of distilled water, some alkoxide precursors and salts, and a catalyst for hydrolysis. Once hydrolysis is completed, aliquots of sol are foamed with the addition of surfactants and vigorous agitation. In this case, porosity strongly depends on the concentration of surfactant. After the foaming process is finished the gelation process is also completed and the constructs proceed to the aging and drying processes to remove solvents, and finally the thermal stabilisation process takes place to allow partial densification of the matrix [105].

Incorporation of foaming agents such as hydrogen peroxide (H_2O_2) or albumin has been also studied. In this case, a calcium phosphate ceramic powder or glass particles slurry is formed together with the foaming agent solution. Afterwards, the foamed mixture is heated up to remove the foaming agent and sinter the ceramic or glass particles. Figure 10 displays the fracture surface of two porous glass-ceramic materials foamed with albumin (figure 10a) and hydrogen peroxide (figure 10b). There is a clear difference in pores amount morphology, size and distribution depending on the elaboration method.

In addition, new strategies to generate highly complex three-dimensional (3D) ceramic-based scaffolds arise from RP technologies that have been adapted to the specific demands concerning materials and processing [106;107].

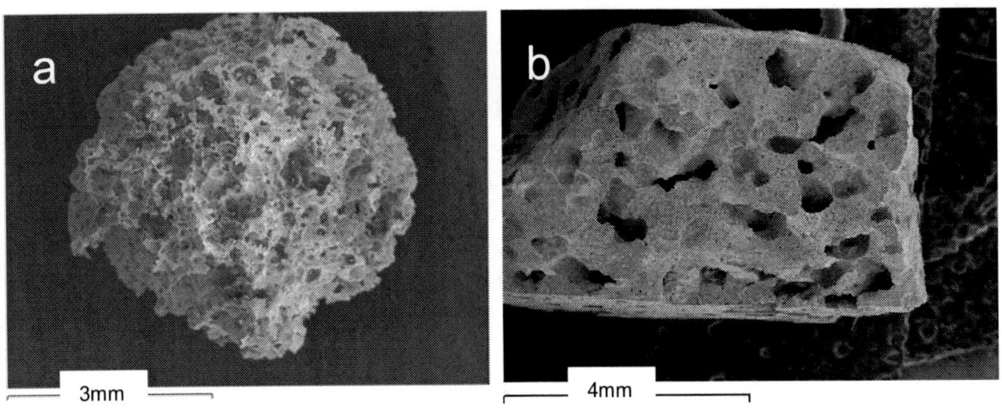

Figure 10. Calcium phosphate glass-ceramic scaffolds elaborated from CaO-P_2O_5-Na_2O-TiO_2 glasses using two different foaming agents, a) albumin and b) H_2O_2.

Metallic Scaffolds

Although metallic materials are not biodegradable, they have also been used in the development of porous structures for bone tissue engineering. The development of porous metallic foams for bone tissue engineering and drug delivery applications has been mostly focused on the used of titanium and titanium alloys. Nevertheless, elaboration of metallic foams using Ta and Mg has also been reported [108].

Titanium and its alloys are highly resistant to corrosion due to the formation of an adhesive TiO_2 oxide layer at their surface; possess a relative low elastic modulus in comparison with other metals used for biomedical applications, and is able to become tightly integrated into bone.

Although numerous studies have been performed on the fabrication processes and design of metallic foams [109;110], their biological response has not been characterized properly and information dealing with the *in vitro* or *in vivo* behaviour of these metallic porous scaffolds is still scarce. However, there are some studies that report good biological behaviour, for instance titanium foams tested *in vitro* with human osteoblasts have shown to efficiently support osteoblast colonization and differentiation into mature bone cells [111]. Other studies have reported titanium fibre meshes (86% porosity and a 250 μm average pore size) that have been used for the *ex vivo* culture of rat bone marrow stromal cells, under static conditions and dynamic conditions (flow perfusion bioreactor), and subsequent implantation in cranial defects in rats [112]. Furthermore, these scaffolds have also found application as delivery systems for transforming growth factor β1 (TGF-β1) and have been used in rabbit cranial defects [113].

The main hurdles associated to these scaffolds is that they are not resorbable and hence their permanent implantation in the body can trigger risks of toxicity caused by the accumulation of metal ions due to corrosion [114], premature failure due to poor wear properties, and higher elastic modulus compared with bone, leading to heterogeneous stress distributions.

Research on shape memory metallic foams is currently carried out. In particular NiTi alloys are under studied, as to reduce stress shielding (removal of bone normal stresses due to the presence of high elastic modulus implants) and increase wear resistance of conventional porous titanium scaffolds. At the moment, only some *in vitro* preliminary studies have been carried out with osteoblastic and endothelial cells to evaluate these materials' cytocompatibility[115].

Surface Functionalization to Improve Cell-Material Interactions

Since most synthetic materials lack osteoinductivity properties, those must be supplied by other sources. It is known that cell response is highly affected by the physicochemical features of the material surface. Thus, tailoring surface characteristics such as surface charges, energy, topography and chemical composition among others may balance this shortage of osteoinductive signals. This chapter will only focus on the chemical approach, specifically on functionalization of surfaces with biomolecules since it is one of the most promising alternatives to provide chemical signals that activate and stimulate cellular events at the molecular level.

Cell adhesion is mediated by protein-cell interactions based on the biological recognition of protein sequences by transmembrane cell receptors known as integrins. In the case of cell adhesion, the most important peptide sequence is RGD (Arginine-Glycine-Aspartic acid) that is found in many ECM proteins and enhances cell attachment. Some of the most commonly employed strategies are the functionalization of biomaterial surfaces using proteins and peptides that mimic the ECM chemistry, in particular RGD sequences [116;117], and the addition of growth factors into the scaffolds composition for a controlled delivery to the surrounding cells. For this last approach, various growth factors have already been used:

fibroblast growth factor (FGF), platelet-derived growth factor, insulin growth factor, transforming growth factor (TGF), bone morphogenetic proteins (BMP), interleukins and interferons [118] (described in detail later in this chapter).

Other functionalization techniques include dip-coating and more sophisticated techniques like the immobilization of biomolecules via self assembled monolayers (SAMs) or grafting from techniques among others.

Dip-coating consists in the impregnation of the material with a solution containing the desired biomolecule. This method is based on the physisorbtion of the molecules to the materials surface. The superficially absorbed molecules are released rapidly within the biological environment. In this case, control over release kinetics is difficult to accomplish.

Surface modification with specific molecular motifs by means of their immobilization via SAMs or via grafted polymer brushes has also been explored.

Self assembled monolayers are surfaces consisting of a single layer of molecules on a substrate. They can be prepared simply by adding a solution of the desired molecule onto the substrate and washing off the excess. The most common example of SAM is an alkane thiol on gold. An alkane with a thiol head group will stick to the gold surface due to the sulphur affinity to gold, and form an ordered assembly with the alkyl chains packing together due to van der Waals forces. SAMs are good platforms for mimicking cell membrane and for immobilizing biomolecules involved in the different cell functioning processes.

Polymer brushes were shown to be a versatile setting for the immobilization of peptide sequences that influence cell behaviour [119;120]. Biointerfaces based on the use of binded polymer chains have several advantages over SAMs: polymer brushes provide multiple binding sites for biomolecules, their flexible nature allows reorganization of the functionalities upon the adhesion of cells and, most importantly, and they can be synthesized very easily and on a variety of substrates [121].

For example, in the case of the cell adhesion process, it not only depends on the receptor/ligand or integrin/RGD interactions but also on the formation of receptor's clusters to form cell focal adhesion points. The flexibility and mobility of peptide sequences coupled to polymer brushes promote the formation of such clusters.

More specific and sophisticated methods for surface functionalization and for modulating cell behaviour are presently being developed. It is predictable that the short term trends in biomaterials intended for tissue engineering applications will involve growing efforts from biochemistry and cell biology fields.

The Use of Growth Factors in Bone Tissue Engineering

In vivo, mammalian cells can communicate in several different ways to coordinate their activities. One of these principal ways is the secretion of soluble signals, called growth factors (GF), by a signalling cell. These growth factors are biochemical signals sent to a target cell in order to guide its biological behaviour [119]. These signals, which are indeed small proteins on the order of 15 to 20 kDa in size, are known as cytokines or chemokines whether they affect cell proliferation and differentiation or cell migration respectively. Numerous studies allowed unravelling the crucial role of a variety of growth factors secreted by different cells that are involved in the formation process of different tissues. The knowledge acquired from these studies has been the starting point of the application of growth factors in tissue engineering (TE) systems. Presently, there is compelling evidence available to indicate that the incorporation of growth factors into in vitro tissue engineering systems has a dramatic effect on tissue generation. The formation, growth and functionality of tissues obtained from cells seeded on 3D constructs are enhanced when exposed to a microenvironment where growth factors are present. More importantly, without a delivery carrier, the growth factor is not able to influence bone regeneration significantly. The direct injection of the growth factor is not efficient enough to enhance tissue regeneration [120].

Requirements for Successful Delivery of Growth Factors

The rationale for using GF into TE systems is to provide the cells with the adequate signals that will guide them toward an appropriate biological response which will lead to the formation of a functional tissue. Therefore, the incorporation of growth factors into a 3D porous scaffold should allow their release and delivery to cells in a timely and prolonged manner that matches biological needs. For a successful delivery, parameters such as i) loading capacity of the matrix, ii) drug distribution into it, iii) binding affinity with it, iv) release kinetics and v) drug stability over a long period of time must be considered. An optimized delivery system for bone application should deliver efficient doses of osteogenic agents while reducing as much as possible the amount of introduced growth factor. This is both for questions of economic viability and safety to avoid potential high-doses side effects. In addition, the biological activity of the growth factors must be preserved during their loading into the delivery carrier.

Different Strategies for Growth Factor Incorporation into a Carrier Material

Growth factors can be either attached to the scaffold or entrapped into it. The attachment of growth factors is mainly performed onto preformed scaffolds, whereas the entrapment generally involves the mixing of the scaffold material (most often polymeric material) with the growth factors during the scaffold fabrication step. Growth factors can be attached through covalent attachment via chemical cross-linking between the growth factors and the material scaffold [121;122], or by simple physical adsorption via ionic complexes formation generally achieved by soaking the scaffold into a growth factor-containing solution (with opposite charges to that of the scaffold) [123].

If the incorporation into the carrier material is done during the fabrication process, the challenge is to avoid denaturing the growth factors because of organic solvents use or other harsh conditions during processing. The use of supercritical fluids to fabricate scaffolds enables to overcome these problems preserving the biological activity of the loaded growth factors [124]. This technique generally consists of dissolving the polymer/growth factor complex in supercritical CO_2, and of expanding it rapidly in a low temperature and pressure environment [125]. In addition, growth factors can be mixed directly with polymeric or ceramic materials using the different techniques of scaffold fabrication detailed anteriorly, such as solvent casting/particulate leaching or electrospinning methods.

The fabrication of microspheres or microparticules is a common way for entrapping growth factors for a controlled delivery [123;126-130]. Microspheres as drug carriers have the advantages of sustained or controlled release, passive or active drug targeting to specific tissues, which will notably reduce the side effects of drugs and improve their bioavailability [131]. Therefore, microspheres as drug delivery system have drawn much attention in pharmaceutical field and have been successfully applied in some clinical trials. These microspheres can either be incorporated into a scaffold, by homogeneous mixing, or be fused to form a scaffold.

The encapsulation of growth factors into microspheres, or alternatively plasmid DNA of growth factors [132], can be performed by different techniques. Either they can be incorporated after the fabrication of the microspheres, via soaking, similarly to the attachment onto scaffold, or they can be entrapped into the microspheres during their fabrication process. For this latter method, various strategies have been used. Two popular methods involving solvent extraction are commonly used, namely double-emulsion and single-emulsion methods.

The double-emulsion method is based on the formation of a double water/oil/water emulsion (W/O/W) [129;130;133-135]. In this method, an aqueous solution of the growth factors is dispersed in a polymer-containing organic solution, forming the primary water-in-oil emulsion (W/O). This primary W/O emulsion is then dispersed via continuous mechanical agitation into a stabilizer-containing aqueous solution, forming a secondary oil-in-water emulsion (O/W). After evaporation of the organic solvent contained in the emulsion droplets, the microspehres loaded with growth factors are obtained. In the single-emulsion technique, the growth factors are directly dispersed into the organic solution of the polymer, forming a unique oil-in-water emulsion [136].

Another technique that can be used for encapsulation into microspheres is based on the phase separation of two water-soluble polymers in an all-aqueous system and subsequent emulsified condensation polymerization [126;127;137].

A simplified diagram representing the different strategies available for the incorporation of growth factors into a matrix is shown in figure 11.

Depending on the incorporation method (attachment or entrapment) and on the nature of the carrier matcrial (microspheres, hydrogels or 3D matrices), the release rate of the growth factors will vary, and should be tailored as a function of the requirements of the application.

Principal Growth Factors Used in Bone Tissue Engineering

The principal growth factors used in bone tissue engineering are i) Transforming Growth Factor β (TGF-β1), ii) Bone Morphogenetic Proteins (BMP-2, BMP-6, BMP-7, and BMP-9), iii) Insulin-like Growth Factor (IGF-1), iv) Fibroblast Growth Factor (FGF-2), and v) Vascular endothelial Growth Factor (VEGF).

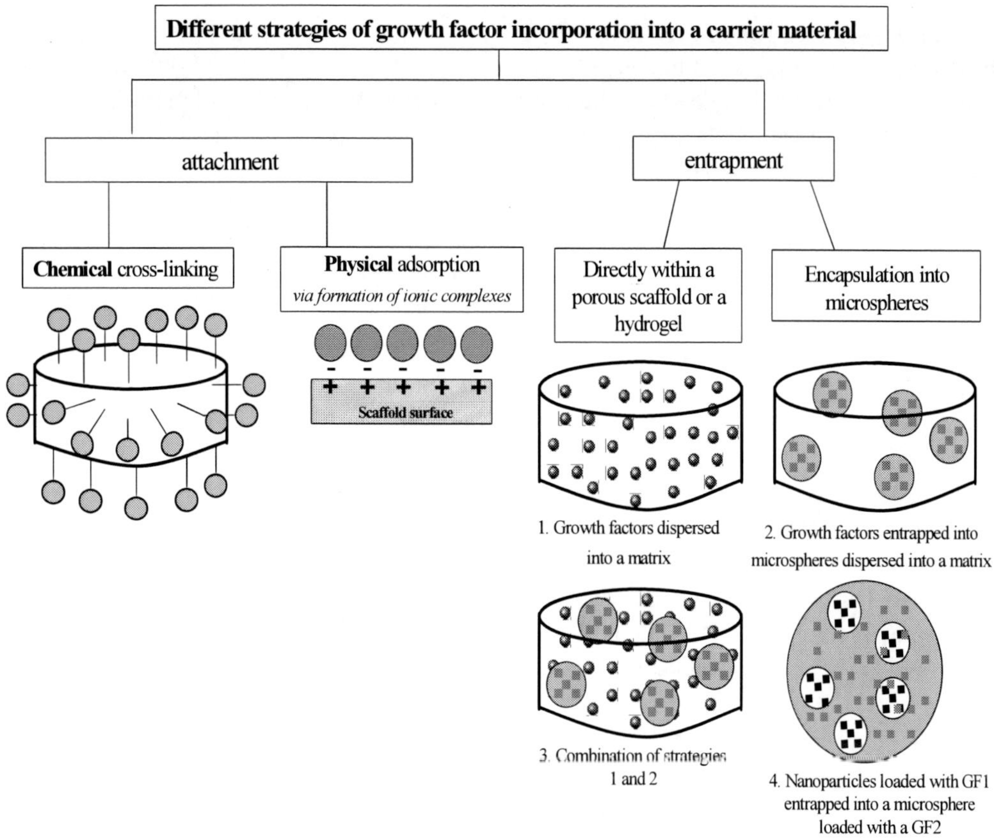

Figure 11. Different strategies for growth factors incorporation (for details, please refer to the text).

Transforming Growth Factor β (TGFβ)

TGFβ is a superfamily of about forty polypeptidic growth factors. They are pleiotropic factors which share similarities in their structure and play a major role in developmental biology. TGFβ are classified into two groups. The first one includes the bone morphogenetic proteins (BMP) and the growth differentiation factors (GDF), while the second one is constituted by TGFβ, activin and nodal [138]. These factors are broadly expressed throughout the body and regulate many cellular pathophysiological processes including cell fate, cell proliferation, cell senescence and tissue repair.

Among the three isoforms of TGFβ that have been described so far, the most interesting one for bone tissue engineering is TGFβ1. It modulates the bone healing process and regulates fracture repair in humans. Moreover, it is locally produced during bone development and regeneration [139-142]. It has been shown to increase the expression of many genes associated with osteoblast activity and to activate osteoblast progenitors [143]. In addition, some other studies showed that TGFβ1 increases alkaline phosphatase activity, collagen, and osteocalcin production by human osteoblasts [144]. TGFβ1-mediated stimulation of *in vivo* and *in vitro* bone formation is linked with an increase in Type I collagen, fibronectin and osteonectin production [142;145;146].

Bone Morphogenetic Proteins (BMPs)

BMPs have been implicated in a variety of functions [147-149]. They not only induce the formation of both cartilage and bone, but also play a role in a number of non-osteogenic developmental processes. BMP may be the most widely used inductive factor in bone tissue engineering applications. They play an important role in embryonic development. Since their discovery in the 60s by Urist [149], at least 20 types of BMPs have been identified and characterized in humans. BMP signal transduction is induced via interaction with the heterodimeric complex of two transmembrane serine/threonine kinase receptors. Although the molecular mechanisms underlying osteoblastic differentiation remain to be identified, BMPs play an important role in regulating osteoblast differentiation and subsequent bone formation. The BMPs with greatest osteogenic capacity are BMP-2, -4, -5, -6, -7, and-9 [147].

BMPs mainly act as a differentiation stimulating factor to direct endochondral ossification and chondrogenesis of mesenchymal stem cells. Recombinant forms (i.e. cloned) of BMPs, particularly BMP-2, BMP-4 and BMP-7, have the ability to heal critical sized bone defects in rodents, dogs, sheep and non-human primates when combined with a carrier of collagen, guanidine-extracted demineralized bone matrix, hydroxyapatite or biodegradable polymers. Of these, recombinant human rhBMP-2 and rhBMP-7 (or osteogenic protein-1, BMP-7) are now commercially available. Other BMP-containing osteoinductive materials are currently being evaluated in animal and clinical studies.

Insulin-Like Growth Factor (IGF)

IGF is a strong stimulant of proliferation and chemotactic migration of many cell populations, and it is involved in bone metabolism [138;150]. The correlation between local IGF expression and new bone formation and mineralization has been reported in many studies. It seems that IGF regulates bone formation through a different mechanism that of BMPs [148].

Fibroblast Growth Factor (FGF)

FGF2, or basic FGF, is part of a family of at least 19 structurally related growth factors. This family has been implicated in a variety of normal and pathologic processes such as repair processes in neural, cartilage, and soft tissue injury among others [151]. FGF2 is also a potent mitogenic factor and, with its receptors, can be found in essentially all tissues of the body [152;153]. In bone, FGF2 is produced intracellularly by osteoblasts, and can be found both in their nucleus and nucleoli, and secreted into the surrounding matrix, hence regulating cell function at the transcriptional level [154-157]. FGF2 has been shown to induce osteoblast expression of TGFβ1 [158] and TGFβ2 [159]. Fibroblast growth factors play a critical role in bone growth [160] and development affecting both chondrogenesis and osteogenesis. During the process of intramembranous ossification, which leads to the formation of the flat bones of the skull, unregulated FGF signaling can produce premature suture closure or craniosynostosis and other craniofacial deformities. Indeed, many human craniosynostosis disorders have been linked to activating mutations in FGF receptors (FGFR) 1 and 2 [161].

As a consequence, FGF-2 has been actively studied as a stimulant of bone healing [148]. Although several lines of evidence indicate that FGFs regulate both proliferation and differentiation in osteoblasts [162], the effects of FGF remain unclear. Apparently, conflicting reports on the exact nature of these effects exist in the literature. This is probably due to differential effects of FGF depending on the stage of osteoblast maturation [159;163]:

whereas there has been general agreement that FGF stimulates the proliferation of osteoblasts in vitro, FGF appears to inhibit expression of differentiation markers such as phosphatase alkaline (ALP) and osteocalcin (OC) in human and rat calvarial cells [162], but increases the levels of ALP, OC, and differentiation of osteogenic precursors in the bone marrow [163]. The mechanism by which FGF signaling increases proliferation in immature osteoblasts, and paradoxically inhibits DNA synthesis and increases apoptosis in differentiating cells is not clear at the moment.

Vascular Endothelial Growth Factor (VEGF)

Vascularisation is a crucial issue in bone tissue engineering [164]. The formation of blood vessels is primordial for the transport of oxygen, nutrients and growth factors into the newly formed tissue, and hence for the viability of the organ. During bone development and fracture healing, vascularisation is observed before bone formation. In intramembranous bone formation, osteoblasts arise from mesenchymal precursors and differentiate into mature osteoblasts. At the transition of preosteoblasts to osteoblasts extensive vascularisation is observed. In endochondral bone formation a cartilage scaffold is formed, which is subsequently replaced by bone. In the latter process bone formation is also accompanied by invasion of blood vessels [165]. Cross-talk between endothelial cells and chondrocytes or osteoblasts has been demonstrated in cocultures [166;167]. In these cultures, endothelial cells stimulated the differentiation of chondrocytes and osteoblasts through production of endothelial-derived growth factors [167;168], whereas osteoblasts stimulated the proliferation of endothelial cells [167]. In addition, it has been demonstrated that BMPs induce the production of VEGF by osteoblasts, which then stimulates angiogenesis. Therefore, BMP-induced VEGF production in osteoblastic cells play a role in the coupling of bone formation and angiogenesis by acting as a chemoattractant for neighboring endothelial cells.

VEGF stimulates proliferation and migration of endothelial cells, which then assemble to form tubular blood vessels. Although the main function of VEGF is related to vasculature, some studies pointed out its implication in the recruitment, survival and activity of bone forming cells [169]. It has been shown that VEGF stimulated the migration and differentiation of primary human osteoblasts [170;171]. In addition, it seems that VEGF is a mediator of different other osteoinductive factors such as TGF-β1, IGF and FGF-2, which in turn regulate the expression pattern of VEGF. Although VEGF is somewhat implicated in the bone formation process, its application in bone tissue engineering is mainly related to its capacity to promote angiogenesis. The combined use of VEGF with other growth factors could synergistically enhance bone formation and fracture healing.

Table 1 presents a summary of some of the recent works that deal with the incorporation of growth factors into a synthetic or natural matrix as a delivery carrier for tissue engineering applications, and particularly for bone tissue engineering.

Table 1. Summary of different types of incorporation technique of different growth factors used in bone tissue engineering applications and of the carrier materials used to incorporate them

Carrier material	Matrix Type	Growth Factor	Incorporation technique	References
PEG	Hydrogel	bFGF	Crosslinking	[121]
Gelatin	Microsphere	TGF□1	Soaking	[172]
Chitosan-based	Hydrogel	rhBMP	Mixing before processing	[173]
PLA	Porous scaffold	VEGF	Supercritical processing	[174]
PLA	Porous scaffold	BMP2	Supercritical processing	[175]
Silk fibroin	Porous scaffold	BMP2	Electrospinning	[176]
PLGA	Microspheres	BMP2	Double-emulsion	[133]
PLGA	Microspheres	TGF□1	Double-emulsion	[135]
PLGA/PEG	Microspheres	VEGF	Single-emulsion	[136]
Collagen	Sponge	BMP2	Soaking	[177;178]
Alginate	Hydrogel	BMP2, TGFb3	Soaking	[179]
Dextran-co-gelatin	Microspheres	IGF1	Soaking	[180]
Dextran-co-gelatin	Microspheres	BMP2	Phase-separation Double-phase emulsion	[126]
Gelatin	Hydrogel	TGF□1	Soaking	[181]
Gelatin	Hydrogel	BMP2	Soaking	[182]
Gelatin	Hydrogel	TGF, IGF	Soaking	[183]

BIOREACTORS: MIMICKING IN VIVO PHYSIOLOGICAL CONDITIONS

Bioreactor Definition

Mammalian tissues have complex three-dimensional and organized structures that give them their specific functionality. Therefore, it is now evident that 2D culturing systems are not appropriate for mimicking tissue regeneration process in vitro, and that 3D systems are a prerequisite for the success of the TE strategy. However, a key question in 3D culture is the maintenance of specific cell phenotype and functions for a long period of time. The development of *in vitro* systems that mimic as close as possible the physiological conditions of the cells cultured, such as nutrients and metabolic environment among others, is therefore crucial. For that purpose, devices known as bioreactors have been developed, and it is presently accepted that they are essential for a controlled fabrication of reproducible engineered constructs.

Bioreactors are *in vitro* culture vessels that provide a controlled environment, which generally include physical, biochemical and mechanical signals (e.g. pH, temperature, pressure, fluidic and mechanical stimuli, nutrient supply and waste removal), to a 3D hybrid cell-scaffold construct in order to achieve the development and/or the physical conditioning and testing of a functional tissue. In addition, new designs of these devices are presently under development as to allow a real-time monitoring of cellular behaviour in response to specific stimuli applied to the hybrid construct. Real-time evaluation will provide essential information on the formation process of 3D tissues, which is a primordial issue if large scale industrial applications are meant.

Bioreactor Principal Functions: Requisites and Limitations

Four of the most common functions of bioreactors are i) the establishment of a spatially uniform cell distribution throughout the 3D construct, ii) the maintenance of desired concentrations of gas (particularly O_2 and CO_2) and nutrients in the culture medium, iii) the supply of efficient mass transfer to the growing tissue and iv) the exposure of the growing tissue to mechanical stimuli.

The first function regards the cell seeding procedure of the 3D scaffold, which is the first crucial step of bioreactor cultivation of engineered tissues [184]. Cell seeding can be done previously to the introduction of the 3D constructs inside the bioreactor, in a different in vitro vessel, or directly inside the same bioreactor where the construct will be kept for tissue generation. This will basically depend on the bioreactor design.

Cell seeding might play a critical role in determining the progression of tissue formation [185]. A successful cell seeding is achieved with:

i) A high percentage of attached cells from the initial solution. High efficiency of seeding is important to maximize cell utilization (particularly interesting when working with stem cells) and to achieve high cell densities into scaffolds after seeding. High cell density has been associated with enhanced tissue formation in 3D constructs, including increased bone mineralization [186].
ii) A high kinetic rate to minimize the time in suspension for anchorage-dependent and shear-sensitive cells, and hence to limit the number of death cells.
iii) A high and spatially uniform distribution of attached cells for rapid and uniform tissue growth. The initial distribution of cells within the scaffold after seeding has been related to the distribution of tissue that is subsequently formed. This suggests that uniform cell-seeding could be the basic parameter for uniform tissue generation [186-189]. Even with a small 3D construct, it can be a significant challenge to distribute a high density of cells efficiently and uniformly throughout the scaffold volume [190].

In spite of several studies that pointed out low seeding efficiencies [186;189;191-193] and non-uniform cell distributions within scaffolds [186;189;192;194-199], static loading remains the most common technique for cell seeding. Significantly higher efficiencies and uniformities can be obtained by dynamic loading. This will be described in detail in the corresponding part of the different bioreactor designs that allows dynamic seeding (please refer to the state-of-the-art paragraph).

The second key point in a properly-designed bioreactor is the controlled delivery of nutrients and gas to the cells, which is typically a mass-transfer issue [200]. Some of the design challenges to be considered here are:

1. the achievement of adequate flux of oxygen at physiological concentration. The oxygen has to be delivered at the same rate it is being consumed, however it has a low solubility in culture media. Low concentration of oxygen leads to cell death whereas high concentration of oxygen can inhibit cell functions and produce cytotoxic free radicals [186]. Moreover, the oxygen flux has to be proportional to the number of cells and therefore should increase with culture time. Oxygen is commonly delivered in two different ways: i) via fluid flow, where the oxygen concentration and flow rate are the variable parameters to achieve adequate amount of oxygen delivery or ii) *in situ* over an oxygenation membrane. In the first case, heterogeneous delivery can derived from significant differences in oxygen concentration between the inlet and outlet stream. On the other hand, the oxygenation membrane is a better way to achieve uniform delivery of oxygen.
2. a uniform flow distribution.
3. the provision of cyto and chemokines and other important molecules (growth factors, proteins, etc.) that must be properly balanced at the same time as oxygen and nutrients are delivered.

As important as delivery, removal of wastes such as carbon dioxide and urea is essential for an optimum process. However, the monitoring of urea concentration in reactors is rarely done, partly due to the unavailability of on-line sensors.

Each one of these key issues has to be tailored as a function of the dimensional and functional requirements of the tissue to be engineered. Mathematical models based on mass transfer theory can be applied for an accurate design as a function of the cellular needs [200;201].

As mentioned earlier, fluid flow is an important aspect of bioreactor's design. With a properly controlled fluid flow, uniform cell seeding (through perfusion systems), nutrients and gas delivery, efficient mass transfer and even mechanical stimuli can be achieved. However, uniform fluid flow can be difficult to achieve with some bioreactor designs, which can lead to an inhomogeneous tissue distribution and growth over the 3D construct. In addition, taking into account the complexity of interconnected porous structures inside a 3D scaffold, it can be difficult to achieve flow uniformity throughout the whole 3D structure. However, this is only relevant when uniform seeding has been previously achieved throughout the 3D scaffold.

Moreover, the flow-induced shear stress applied to the cells inside the scaffold is another crucial issue that has to be rigorously controlled. It is widely accepted that shear stress has a determinant impact on tissue function and viability, and there is a maximal value of sustainable shear stress for every type of cells [201]. An adequate shear stress value can promote cell growth, whereas higher values can provoke cell damage and necrosis of the tissues. In addition, it has been suggested that a homogeneous pore distribution in a 3D scaffold could enable a more precise control over shear stress [202]. However, deeper studies have to be carried out as to gain knowledge on shear stress thresholds to be used in the bioreactor as a function of cell type.

State-of-the-Art of Principal Bioreactor Designs

Presently, there are few models of bioreactors commercially available. Most of these devices are only suitable for experimental studies and not for clinical applications due to their small size and hence to the small size of the engineered tissue. Generally, researchers into TE and bioreactor fields develop their own in-house reactors depending on what kind of experimental studies they want to perform.

In this part, some of the most commonly used working principles applied to bioreactor designs are presented as to give the reader a general idea on the state-of-the-art of bioreactor development.

First Generation Bioreactors

1. Petri dishes that were orbitally mixed were firstly used because they enhanced tissue growth comparing with static dishes [197;203]. However, the thickest bone tissue reported in such reactors were of 0.5 mm [204], which is not big enough to serve as a graft for tissue replacement (it should be at least a few millimetres).
2. Stirred-tank reactor:

In this case, mixing of oxygen and nutrients within the culture medium is achieved by magnetic agitation of the fluid into the flask, in which the scaffolds seeded with the cells are suspended (figure 12).

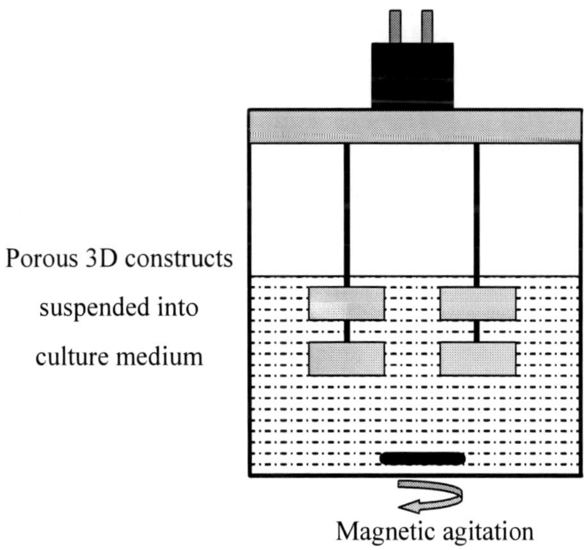

Figure 12. simplified view of a typical stirred-tank bioreactor.

The first advantage of this type of reactors for cell seeding is that they enable significantly higher efficiencies and uniformities of cell seeding compared to static loading, due to the convective transport of the cells into the scaffold [197]. However, some studies have pointed out low seeding efficiencies [189;205] and heterogeneous distributions of cells [194;198], with a higher density of cells on the outer surface of the scaffold [185] when using stirred-tank reactors. This is possibly due to ineffective convection of cells inside the scaffolds.

Rotating-Wall Vessel Reactor

Three common models can be found in the literature:

1. The Slow Turning Lateral Vessel (STLV, figure 13a), that consists in two concentric cylinders. One inner cylinder that is generally static and provides gas exchange through a gas exchange membrane; and one outer cylinder that rotates at a determined speed as to maintain the constructs motionless in a state of continuous free-fall (the gravitational, centrifugal and drag forces applied to the scaffold are balanced.)
2. The Rotating Wall Perfused Vessel (RWPV, figure 13b) is similar to the STLV except that it also includes a recirculation system that allows periodic medium exchange and hence a better control of nutrients and other molecules delivery.
3. The High Aspect Ratio Vessel (HARV, figure 13c), which principle of operation is similar to both previous reactors. However, its particular design allows reducing the rotational speed of the cylinder, and it includes a larger gas exchange membrane to enhance gas exchange.

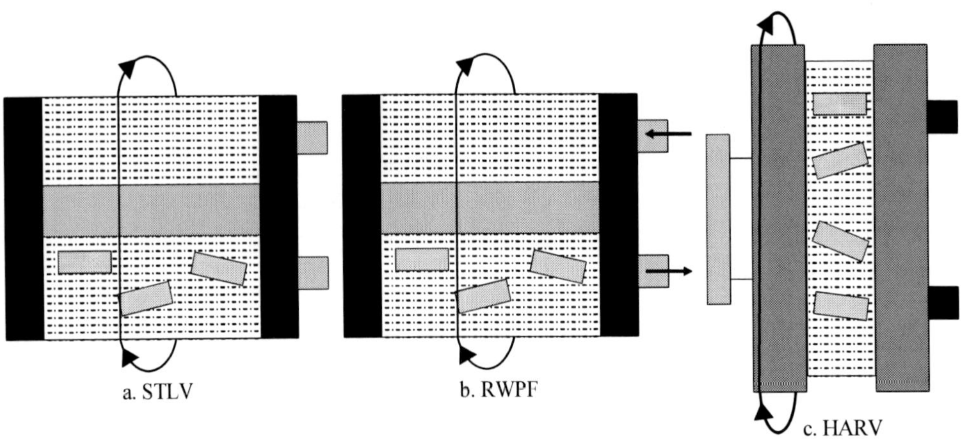

Figure 13. Three types of rotating-wall vessel bioreactors. a/ the Slow Turning Lateral Vessel; b/ the Rotating-Wall Perfused Vessel; c/ The High Aspect Ratio Vessel (see text for details).

The hydrodynamic environment present in these rotating vessel bioreactors seems to be more advantageous for growing engineered tissues than simply stirred-tanks [184].

Perfusion Reactor

Two configurations of perfusion bioreactor can be encountered:

1. The perfusion column (figure 14a), in which the same fluid flow goes through more than one 3D construct.
2. The perfusion chamber (figure 14b), in which the medium flows directly through the pores of a unique scaffold.

In both configurations, there is a continuous medium recirculation. The fluid can be forced to pass through the porous constructs either by a pumping system (figure 14b) or an up-and-down motion of the scaffold through the fluid (figure 14c) [206].

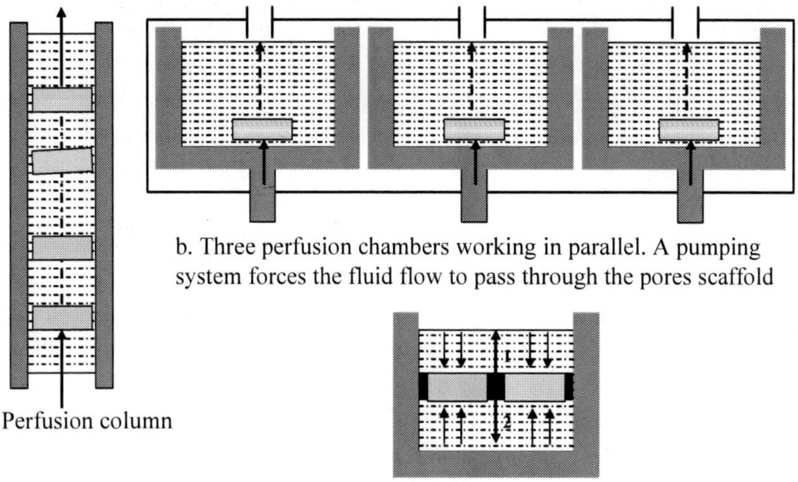

a. Perfusion column

b. Three perfusion chambers working in parallel. A pumping system forces the fluid flow to pass through the pores scaffold

c. Example of perfusion chamber in which an up-and-down motion (1 and 2) is applied to the scaffolds into the fluid as to force the fluid to pass through the pores of the constructs

Figure 14. Example of bioreactors based on a perfusion system. a/ perfusion column reactor; b/ perfusion chamber-based bioreactor; c/ similar to b/ but with different fluid flow system.

In that kind of bioreactors, the dynamic seeding process leads to high efficiency and homogeneous cell distribution. Continuous recirculation of fresh medium enables maintaining physiological concentrations of nutrients and paracrine factors, as well as reducing the accumulation of toxic metabolic products. These factors play undoubtedly a major role in the enhancement of osteogenesis of bone engineered constructs. Several studies have shown that perfusion systems enable to enhance cell viability, alkaline phosphatase activity and osteocalcin production by osteoblastic cells comparing with conventional systems [207;208] as well as mineralized matrix deposition by bone cells [209;210]. However, the flow rate of the medium must be carefully optimized to be the good balance between “too high” and “too low” fluid flows. From one hand, removal of newly synthesized extracellular matrix components off the construct and harmful effects of high shear stresses can derive from high fluid flows and hence have to be avoided. On the other hand, inappropriate mass transfer of nutrients and waste products can be a consequence of low fluid flows.

Rotating-wall vessel reactors and perfusion reactors have efficient designs as to allow excellent transport properties through the scaffold. However, they still need improvement to better control mechanical stimuli; although it is true that the benefits observed for perfusion systems on osteogenesis are partly due to the flow-induced shear stress derived from these systems. For that purpose, a new generation of bioreactors are under development. This is a generation of devices that enable the application of different controllable mechanical loads onto the hybrid construct (cell-seeded scaffold), and that is discussed below.

Bioreactor with Mechanical Stimulation

The increasing interest in bioreactors that include mechanical stimuli control over the 3D constructs derives from several studies that unravelled the direct effect of mechanical forces on biosynthetic cell activity. Other studies provide evidence that differentiation of multi-potent cells along specific lineages could be directed by the application of specific mechanical forces applied. For example, it has been demonstrated that human bone marrow stromal cells (hBMSCs), in the absence of specific ligament growth and regulatory factors, could differentiate into ligament-like cells through the application of physiologically relevant cyclic multi-dimensional mechanical strain [211]. In addition, mechanical stimulation of Bone Marrow Stem Cells induced increased levels of alkaline phosphatase activity, alkaline phosphatase and osteopontin gene expression and mineralized matrix production [212].. However, the nature and intensity of the mechanical forces that are stimulatory for a specific engineered tissue are still unclear and need to be investigated more deeply. Moreover, a modulation of the applied mechanical stimuli is another requirement derived from the variable "needs" of the growing tissue depending on its developmental stage.

Bioreactors including a mechanical system able to apply controllable dynamic compression, and multidimensional strains (simultaneous compression and torsion) to the growing tissue housed within the bioreactor have already been studied [190;213]. In addition to be used for enhancing biosynthetic cell activity for in vitro tissue regeneration, these kinds of bioreactors could be valuable tools for the study of pathophysiological effects of physical forces on developing tissues, and of the effects of physiological forces the engineered tissue will have to bear after implantation in the body.

Future Perspective

Despite bioreactors are widely viewed as essential for the controlled fabrication of reproducible functional engineered tissues, the recent advances made in the field are generally limited to research-scale production. Mass transfer issue still remains one of the major limitations before bioreactors enable the production of engineered tissues with relevant size for clinical applications [214]. This problem could be overcome with the development of innovative designs of circulatory systems that mimic *in vivo* vascularisation of tissues, which is mainly dependant on scaffold production methodologies. One of the solutions proposed to achieve a microfluidic circulatory system is the incorporation of microchannels into the 3D scaffold [215;216], or the promotion of scaffold vascularisation by means of co-culturing vascular cells with the specific cells of the tissue to be generated. This could require the fabrication of different microenvironments inside the 3D constructs as to allow both vascular cells and other specific cells to grow and differentiate. However, the success of the tissue engineering strategy via bioreactor-mediated in vitro culture of hybrid constructs (cell-seeded scaffold) not only depends on innovative bioreactor development but also on multiple interdependent parameters that together form a complex chain which final goal is to achieve mass-production of reproducible functional engineered tissues for clinical routine (figure 15).

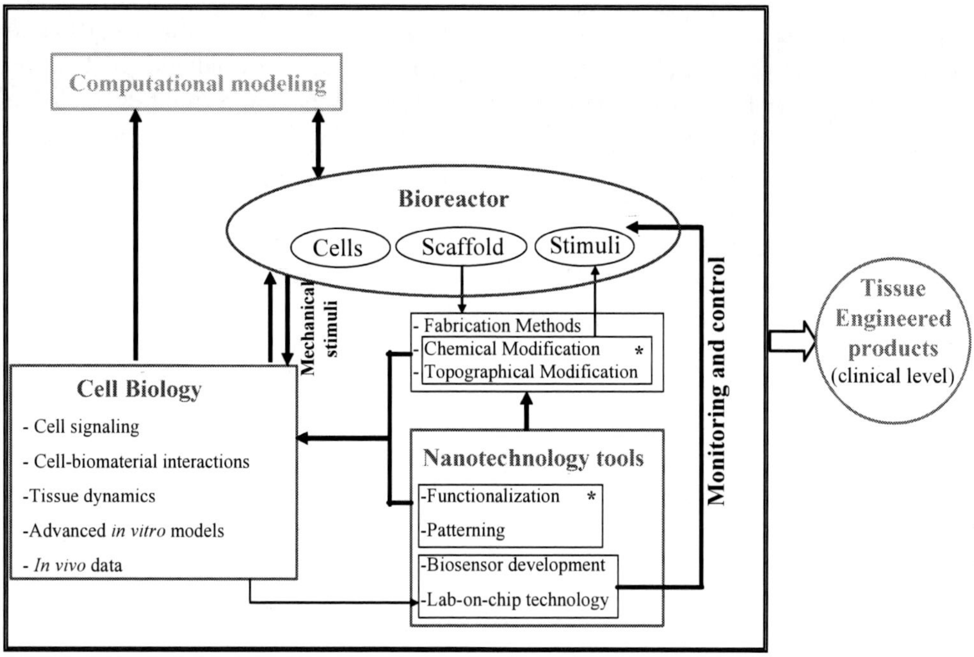

Figure 15. Interdependence of multiple scientific research fields that play a major role in the development of tissue engineered products (see text for details).

CONCLUSION

As mentioned in the introduction, the tissue engineering approach is an interdisciplinary issue that relies on three basic components which are i) human cells, ii) 3D porous scaffold for supporting cell growth and tissue development, and iii) stimuli aimed at creating a favourable environment for in vitro tissue generation, and which include physical, chemical and mechanical cues for cells (figure 2). Acquisition of new data and knowledge in cell biology (including cell-biomaterial interactions, cell signalling and tissue dynamics), scaffold development (including innovative methods of production, surface chemical and topographical modification methods and new materials study) and the effect of chemical, physical and mechanical stimuli on tissue growth and development is primordial for bioreactor technology development. Bioreactor technology advances are in turn a crucial factor to enable progress in these different research fields. The passage of laboratory-scale to industrial-scale engineered tissue products is therefore a complex and interdisciplinary issue that could be described by highlighting four interdependent research fields, as described in Figure 15:

1. Bioreactor design and development:

Bioreactors are considered as tools for studying the effect of cell microenvironment, including mechanical stimulation, on the formation of tissues. For that purpose, a tightly controlled-environment and accurate real-time monitoring is required. This can be achieved through biosensor development and lab-on-chip technology (see Nanotechnology paragraph).

The integration of these devices into bioreactors would enable the monitoring and detection of specific cellular processes and the automatic regulation of culture conditions by feedback loops, which would improve tissue regeneration considerably. However, progress in cell biology is essential to identify specific cellular processes that would be considered excellent indicators of the tissue formation process and hence that would be interesting to monitor.

2. Cell biology research:

Deeper knowledge is still needed regarding cell-biomaterial interactions. How physicochemical parameters, and particularly nanotopography and chemical cues, influence cell-signalling and hence biological response and tissue regeneration remains unclear. For this purpose, advanced in vitro models need to be developed. Microarray technology, nanopatterning and surface functionalization of biomaterials (see Nanotechnology paragraph) appear to be among the most promising tools for elucidating cell biological pathways and hence allowing the design and optimization of biomaterials to serve as a construct for cell culture.

3. Computational modelling:

Future bioreactor designs will undoubtedly integrate considerations from computational modelling, including fluid mechanics for a specific design, as well as mechanics of the scaffold material and tissue dynamics. This last issue is more difficult to address because it requires the development of multi-level and multi-scale computational modelling to capture spatio-temporal dynamics of tissue [214]. These models would enable predicting the development of the engineered tissue [190]. Therefore, it would be possible to define accurately the physical, chemical and mechanical parameters that are to be imposed on the bioreactor system to create a favourable microenvironement at each stage of tissue development. In addition, it would also allow the planning of timely surgery. However, the implementation of these computational models requires the well-characterized and validated experimental data to produce the underlying rules of the model. These experimental data could be obtained by means of in vitro and *in vivo* cell experiments.

4. Nanotechnology:

The application of nanotechnology tools to the development of structures at the molecular level enables the study and the improvement of the interactions between material surfaces and biological entities. Nanotechnologies provide the possibility to produce surfaces, structures and materials with nanoscale features that can mimic the natural environment of cells, to promote certain functions, such as cell adhesion, cell mobility and cell differentiation. Biomaterial surface modifications by means of nanostructuration techniques and chemical functionalisation, particularly self-assembly, are some examples of the application of nanotechnology tools to regenerative medicine [217]. The use of nanotechnology in tissue engineering also seems to be promising for the development of nanodevices, such as nanobiosensors, that could be incorporated on laboratory-on-a-chip (Lab-on-Chip) in order to detect biological, chemical, mechanical or electrical reactions as indicative of specific biological response taking place in the bioreactor. Examples of nanosensors that could be included in bioreactors would be pH sensors or oxygen sensors that use nanotechnology for locally discrete measurements or molecule-release sensors (such as a calcium or potassium detector).

In summary, the prospect of regenerating damaged or non-functional tissues by using an off-the-shelf synthetic product is far from being achieved and economically-viable in the near future. However, this perspective would definitely and radically impact and change medical science. Within this context, fundamental interdisciplinary research needs to be carried out to provide a comprehensive control over specific microenvironmental factors in 3D cultures as to achieve an economically viable approach to obtain reproducible and functional tissue grafts at industrial-scale.

REFERENCES

[1] Banwart JC, Asher MA, Hassanein RS. Iliac Crest Bone-Graft Harvest Donor Site Morbidity - A Statistical Evaluation. *Spine*. 1995; 20(9):1055-1060.

[2] Fernyhough JC, Schimandle JJ, Weigel MC, Edwards CC, Levine AM. Chronic Donor Site Pain Complicating Bone-Graft Harvesting from the Posterior Iliac Crest for Spinal-Fusion. *Spine*. 1992; 17(12):1474-1480.

[3] Goulet JA, Senunas LE, DeSilva GL, Greenfield MLVH. Autogenous iliac crest bone graft - Complications and functional assessment. *Clinical Orthopaedics and Related Research.* 1997;(339):76-81.

[4] Hardouin P, Anselme K, Flautre B, Bianchi F, Bascoulenguet G, Bouxin B. Tissue engineering and skeletal diseases. *Joint Bone Spine*. 2000; 67(5):419-424.

[5] Kuo CK, Tuan RS. Tissue Engineering with Mesenchymal Stem Cells. *IEEE Engineering in Medicine and Biology Magazine*. 2003; September/October:51.

[6] Logeart-Avramoglou D, Anagnostou F, Bizios R, Petite H. Engineering Bone:Challenges and Obstacles. *Tissue Engineering Review Series*. 2005; 9(1):72-84.

[7] Sipe JD. Tissue engineering and reparative medicine. Reparative Medicine: *Growing Tissues and Organs*. 2002; 961:1-9.

[8] Verfaillie CM. Adult stem cells: assessing the case for pluripotency. *Trends in Cell Biology*. 2002; 12(11):502-508.

[9] Stoltz JF, Bensoussan D, Decot V, Netter P, Ciree A, Gillet P. Cell and tissue engineering and clinical applications: An overview. *Bio-Medical Materials and Engineering*. 2006; 16(4):S3-S18.

[10] Pittenger MF, Mackay AM, Beck SC, Jaiswal RK, Douglas R, Mosca JD et al. Multilineage potential of adult human mesenchymal stem cells. *Science.* 1999; 284(5411):143-147.

[11] Chim H, Schantz JT. Human circulating peripheral blood mononuclear cells for calvarial bone tissue engineering. *Plastic and Reconstructive Surgery*. 2006; 117(2):468-478.

[12] Fuchs JR, Hannouche D, Terada S, Zand S, Vacanti JP, Fauza DO. Cartilage engineering from ovine umbilical cord blood mesenchymal progenitor cells. *Stem Cells.* 2005; 23(7):958-964.

[13] Kern S, Eichler H, Stoeve J, Kluter H, Bieback K. Comparative analysis of mesenchymal stem cells from bone marrow, umbilical cord blood, or adipose tissue. *Stem Cells*. 2006; 24(5):1294-1301.

[14] Friedenstein A, Petrakova KV. Osteogenesis in transplants of bone marrow cells. The Journal of Embryological Exp*erimental Morphology*. 1966; 16:381-390.

[15] Goshima J, Goldberg VM, Caplan AI. Osteogenic Potential of Culture-Expanded Rat Marrow-Cells As Assayed Invivo with Porous Calcium-Phosphate Ceramic. *Biomaterials*. 1991; 12(2):253-258.

[16] Haynesworth SE, Goshima J, Goldberg VM, Caplan AI. Characterization of Cells with Osteogenic Potential from Human Marrow. *Bone*. 1992; 13(1):81-88.

[17] Bruder SP, Jaiswal N, Haynesworth SE. Growth kinetics, self-renewal, and the osteogenic potential of purified human mesenchymal stem cells during extensive subcultivation and following cryopreservation. *Journal of Cellular Biochemistry*. 1997; 64(2):278-294.

[18] Jaiswal N, Haynesworth SE, Caplan AI, Bruder SP. Osteogenic differentiation of purified, culture-expanded human mesenchymal stem cells *in vitro*. *Journal of Cellular Biochemistry*. 1997; 64(2):295-312.

[19] Krebsbach PH, Kuznetsov SA, Satomura K, Emmons RVB, Rowe DW, Robey PG. Bone formation *in vivo*: Comparison of osteogenesis by transplanted mouse and human marrow stromal fibroblasts. *Transplantation*. 1997; 63(8):1059-1069.

[20] Fukumoto T, Sperling JW, Sanyal A, Fitzsimmons JS, Reinholz GG, Conover CA et al. Combined effects of insulin-like growth factor-1 and transforming growth factor-beta 1 on periosteal mesenchymal cells during chondrogenesis *in vitro*. *Osteoarthritis and Cartilage*. 2003; 11(1):55-64.

[21] Nakahara H, Bruder SP, Haynesworth SE, Holecek JJ, Baber MA, Goldberg VM et al. Bone and Cartilage Formation in Diffusion-Chambers by Subcultured Cells Derived from the Periosteum. *Bone*. 1990; 11(3):181-188.

[22] Zarnett R, Salter RB. Periosteal Neochondrogenesis for Biologically Resurfacing Joints - Its Cellular-Origin. *Canadian Journal of Surgery*. 1989; 32(3):171-174.

[23] Noth U, Osyczka AM, Tuli R, Hickok NJ, Danielson KG, Tuan RS. Multilineage mesenchymal differentiation potential of human trabecular bone-derived cells. *Journal of Orthopaedic Research*. 2002; 20(5):1060-1069.

[24] Sottile V, Halleux C, Bassilana F, Keller H, Seuwen K. Stem cell characteristics of human trabecular bone-derived cells. *Bone*. 2002; 30(5):699-704.

[25] Tuli R, Seghatoleslami MR, Tuli S, Wang ML, Hozack WJ, Manner PA et al. A simple, high-yield method for obtaining multipotential mesenchymal progenitor cells from trabecular bone. *Molecular Biotechnology*. 2003; 23(1):37-49.

[26] Dragoo JL, Choi JY, Lieberman JR, Huang J, Zuk PA, Zhang J et al. Bone induction by BMP-2 transduced stem cells derived from human fat. *Journal of Orthopaedic Research*. 2003; 21(4):622-629.

[27] Gronthos S, Franklin DM, Leddy HA, Robey PG, Storms RW, Gimble JM. Surface protein characterization of human adipose tissue-derived stromal cells. Journal of Cellular Physiology. 2001; 189(1):54-63.

[28] Wickham MQ, Erickson GR, Gimble JM, Vail TP, Guilak F. Multipotent stromal cells derived from the infrapatellar fat pad of the knee. *Clinical Orthopaedics and Related Research*. 2003;(412):196-212.

[29] De Bari C, Dell'Accio F, Tylzanowski P, Luyten FP. Multipotent mesenchymal stem cells from adult human synovial membrane. *Arthritis and Rheumatism*. 2001; 44(8):1928-1942.

[30] Jankowski RJ, Deasy BM, Huard J. Muscle-derived stem cells. *Gene Therapy.* 2002; 9(10):642-647.

[31] Noort WA, Kruisselbrink AB, in't Anker PS, Kruger M, van Bezooijen RL, de Paus RA et al. Mesenchymal stem cells promote engraftment of human umbilical cord blood-derived CD34(+) cells in NOD/SCID mice. *Experimental Hematology.* 2002; 30(8):870-878.

[32] Miura M, Gronthos S, Zhao MR, Lu B, Fisher LW, Robey PG et al. SHED: *Stem cells from human exfoliated deciduous teeth.* Proceedings of the National Academy of Sciences of the United States of America 2003; 100(10):5807-5812.

[33] Lee JY, Musgrave D, Pelinkovic D, Fukushima K, Cummins J, Usas A et al. Effect of bone morphogenetic protein-2-expressing muscle-derived cells on healing of critical-sized bone defects in mice. *Journal of Bone and Joint Surgery-American Volume.* 2001; 83A(7):1032-1039.

[34] Peng HR, Wright V, Usas A, Gearhart B, Shen HC, Cummins J et al. Synergistic enhancement of bone formation and healing by stem cell-expressed VEGF and bone morphogenetic protein-4. *Journal of Clinical Investigation.* 2002; 110(6):751-759.

[35] Wright VJ, Peng HR, Usas A, Young B, Gearhart B, Cummins J et al. BMP4-expressing muscle-derived stem cells differentiate into osteogenic lineage and improve bone healing in immunocompetent mice. *Molecular Therapy.* 2002; 6(2):169-178.

[36] Gazit D, Turgeman G, Kelley P, Wang E, Jalenak M, Zilberman Y et al. Engineered pluripotent mesenchymal cells integrate and differentiate in regenerating bone: A novel cell-mediated gene therapy. *Journal of Gene Medicine.* 1999; 1(2):121-133.

[37] Lieberman JR, Daluiski A, Stevenson S, Wu L, McAllister P, Lee YP et al. The effect of regional gene therapy with bone morphogenetic protein-2-producing bone-marrow cells on the repair of segmental femoral defects in rats. *Journal of Bone and Joint Surgery-American Volume.* 1999; 81A(7):905-917.

[38] Aslan H, Zilberman Y, Arbeli V, Sheyn D, Matan Y, Liebergall M et al. Nucleofection-based *ex vivo* nonviral gene delivery to human stem cells as a platform for tissue regeneration. *Tissue Engineering.* 2006; 12(4):877-889.

[39] Aslan H, Zilberman Y, Kandel L, Liebergall M, Oskouian RJ, Gazit D et al. Osteogenic differentiation of noncultured immunoisolated bone marrow-derived CD105(+) cells. *Stem Cells.* 2006; 24(7):1728-1737.

[40] Moutsatsos IK, Turgeman G, Zhou SH, Kurkalli BG, Pelled G, Tzur L et al. Exogenously regulated stem cell-mediated gene therapy for bone regeneration. *Molecular Therapy.* 2001; 3(4):449-461.

[41] Peterson B, Zhang J, Iglesias R, Kabo M, Hedrick M, Benhaim P et al. Healing of critically sized femoral defects, using genetically modified mesenchymal stem cells from human adipose tissue. *Tissue Engineering.* 2005; 11(1-2):120-129.

[42] Gafni Y, Pelled G, Zilberman Y, Turgeman G, Apparailly F, Yotvat H et al. Gene therapy platform for bone regeneration using an exogenously regulated, AAV-2-based gene expression system. *Molecular Therapy.* 2004; 9(4):587-595.

[43] Aluigi M, Fogli M, Curti A, Isidori A, Gruppioni E, Chiodoni C et al. Nucleofection is an efficient nonviral transfection technique for human bone marrow-derived mesenchymal stem cells. *Stem Cells.* 2006; 24(2):454-461.

[44] Yaszemski MJ, Payne RG, Hayes WC, Langer R, Mikos AG. Evolution of bone transplantation: Molecular, cellular and tissue strategies to engineer human bone. *Biomaterials*. 1996; 17(2):175-185.

[45] Meinel L, Hofmann S, Karageorgiou V, Zichner L, Langer R, Kaplan D et al. Engineering cartilage-like tissue using human mesenchymal stem cells and silk protein scaffolds. *Biotechnology and Bioengineering*. 2004; 88(3):379-391.

[46] Sharma B, Elisseeff JH. Engineering structurally organized cartilage and bone tissues. *Annals of Biomedical Engineering*. 2004; 32(1):148-159.

[47] Hench LL, Polak JM. Third-generation biomedical materials. *Science*. 2002; 295(5557):1014-+.

[48] Roether JA, Gough JE, Boccaccini AR, Hench LL, Maquet V, Jerome R. Novel bioresorbable and bioactive composites based on bioactive glass and polylactide foams for bone tissue engineering. *Journal of Materials Science-Materials in Medicine*. 2002; 13(12):1207-1214.

[49] Spaans CJ, Belgraver VW, Rienstra O, de Groot JH, Veth RPH, Pennings AJ. Solvent-free fabrication of micro-porous polyurethane amide and polyurethane-urea scaffolds for repair and replacement of the knee-joint meniscus. *Biomaterials*. 2000; 21(23):2453-2460.

[50] Bruder SP, Fox BS. Tissue engineering of bone - Cell based strategies. *Clinical Orthopaedics and Related Research*. 1999;(367):S68-S83.

[51] Temenoff JS, Mikos AG. Review: tissue engineering for regeneration of articular cartilage. *Biomaterials*. 2000; 21(5):431-440.

[52] Salgado AJ, Coutinho OP, Reis RL. Bone tissue engineering: State of the art and future trends. *Macromolecular Bioscience*. 2004; 4(8):743-765.

[53] Olivier V, Faucheux N, Hardouin P. Biomaterial challenges and approaches to stem cell use in bone reconstructive surgery. *Drug Discovery Today*. 2004; 9(18):803-811.

[54] Pachence JM, Kohn J. Biodegradable Polymers. In: Langer R, Lanza RP, Vacanti JP, editors. *Principles of Tissue Engineering*. San Diego: Academic Press, 2000: 263-274.

[55] Vonrecum HA, Cleek RL, Eskin SG, Mikos AG. Degradation of Polydispersed Poly(L-Lactic Acid) to Modulate Lactic-Acid Release. *Biomaterials*. 1995; 16(6):441-447.

[56] Grizzi I, Garreau H, Li S, Vert M. Hydrolytic Degradation of Devices Based on Poly(Dl-Lactic Acid) Size-Dependence. *Biomaterials*. 1995; 16(4):305-311.

[57] Kohn J, Langer R. Bioresorbable and bioerodible materials. In: Ratner BD, Schoen FJ, Lemons JE, editors. *Biomaterials Science*. New York: Academic Press, 1996: 64-72.

[58] Shalaby SW, Johnson RA. Synthetic absorbable polyesters. In: Shalaby SW, editor. Biomedical biopolymers. *Designed to degrade systems*. New York: Hanser, 1994: 1-34.

[59] Chu CC. The Invitro Degradation of Poly(Glycolic Acid) Sutures - Effect of Ph. *Journal of Biomedical Materials Research*. 1981; 15(6):795-804.

[60] Chu CC. An Invitro Study of the Effect of Buffer on the Degradation of Poly(Glycolic Acid) Sutures. *Journal of Biomedical Materials Research*. 1981; 15(1):19-27.

[61] Vert M, Li SM, Garreau H. Attempts to Map the Structure and Degradation Characteristics of Aliphatic Polyesters Derived from Lactic and Glycolic Acids. *Journal of Biomaterials Science-Polymer Edition*. 1994; 6(7):639-649.

[62] Achindler A, Jeffcoat R, Kimmel GL, Pitt CG, Wall ME, Zwiedinger R. Biodegradable polymers for sustained druf delivery. *Contemporary Topics in Polymer Science*. 1977; 11:711-719.

[63] Vert M, Chabot F, Leray J, Christel P. Stereoregular Bioresorbable Polyesters for Orthopedic-Surgery. *Makromolekulare Chemie-Macromolecular Chemistry and Physics.* 1981;30-41.

[64] Nakamura T, Hitomi S, Watanabe S, Shimizu Y, Jamshidi K, Hyon SH et al. Bioabsorption of Polylactides with Different Molecular-Properties. *Journal of Biomedical Materials Research.* 1989; 23(10):1115-1130.

[65] Navarro M, Ginebra MP, Planell JA, Zeppetelli S, Ambrosio L. Development and cell response of a new biodegradable composite scaffold for guided bone regeneration. *Journal of Materials Science-Materials in Medicine.* 2004; 15(4):419-422.

[66] Navarro M, Ginebra MP, Planell JA, Barrias CC, Barbosa MA. *In vitro* degradation behavior of a novel bioresorbable composite material based on PLA and a soluble CaP glass. *Acta Biomaterialia.* 2005; 1(4):411-419.

[67] Yang S, Leong K, Du Z, Chua C. The design of scaffolds for tissue engineering. Part I. Traditional factors. *Tissue Engineering.* 2001; 7:679-689.

[68] Mikos AG, Bao Y, Cima LG, Ingber DE, Vacanti JP, Langer R. Preparation of Poly(Glycolic Acid) Bonded Fiber Structures for Cell Attachment and Transplantation. *Journal of Biomedical Materials Research.* 1993; 27(2):183-189.

[69] Kasuga T, Maeda H, Kato K, Nogami M, Hata K, Ueda M. Preparation of poly(lactic acid) composites containing calcium carbonate (vaterite). *Biomaterials.* 2003; 24(19):3247-3253.

[70] Marra KG, Szem JW, Kumta PN, DiMilla PA, Weiss LE. *In vitro* analysis of biodegradable polymer blend/hydroxyapatite composites for bone tissue engineering. *Journal of Biomedical Materials Research.* 1999; 47(3):324-335.

[71] Liu Q, De Wijn JR, van Blitterswijk CA. Composite biomaterials with chemical bonding between hydroxyapatite filler particles and PEG/PBT copolymer matrix. *Journal of Biomedical Materials Research.* 1998; 40(3):490-497.

[72] Nam YS, Park TG. Porous biodegradable polymeric scaffolds prepared by thermally induced phase separation. *Journal of Biomedical Materials Research.* 1999; 47(1):8-17.

[73] Schugens C, Maquet V, Grandfils C, Jerome R, Teyssie P. Polylactide macroporous biodegradable implants for cell transplantation .2. Preparation of polylactide foams by liquid-liquid phase separation. *Journal of Biomedical Materials Research.* 1996; 30(4):449-461.

[74] Ma PX, Zhang RY. Microtubular architecture of biodegradable polymer scaffolds. *Journal of Biomedical Materials Research.* 2001; 56(4):469-477.

[75] Zhang RY, Ma PX. Poly(alpha-hydroxyl acids) hydroxyapatite porous composites for bone-tissue engineering. I. Preparation and morphology. *Journal of Biomedical Materials Research.* 1999; 44(4):446-455.

[76] Boccaccini AR, Notingher I, Maquet V, Jerome R. Bioresorbable and bioactive composite materials based on polylactide foams filled with and coated by Bioglass (R) particles for tissue engineering applications. *Journal of Materials Science-Materials in Medicine.* 2003; 14(5):443-450.

[77] Cima LG, Langer R, Vacanti JP. Polymers for Tissue and Organ-Culture. *Journal of Bioactive and Compatible Polymers.* 1991; 6(3):232-240.

[78] Hutmacher DW, Sittinger M, Risbud MV. Scaffold-based tissue engineering: rationale for computer-aided design and solid free-form fabrication systems. *Trends in Biotechnology* 2004; 22(7):354-362.

[79] Artzi Z, Weinreb M, Givol N, Rohrer MD, Nemcovsky CE, Prasad HS et al. Biomaterial resorption rate and healing site morphology of inorganic bovine bone and beta-tricalcium phosphate in the canine: A 24-month longitudinal histologic study and morphometric analysis. *International Journal of Oral & Maxillofacial Implants.* 2004; 19(3):357-368.

[80] Legeros RZ, Lin S, Rohanizadeh R, Mijares D, Legeros JP. Biphasic calcium phosphate bioceramics: preparation, properties and applications. *Journal of Materials Science-Materials in Medicine.* 2003; 14(3):201-209.

[81] Bauer TW, Muschler GF. Bone graft materials - An overview of the basic science. *Clinical Orthopaedics and Related Research.* 2000;(371):10-27.

[82] Yoshikawa H, Myoui A. Bone tissue engineering with porous hydroxyapatite ceramics. *Journal of Artificial Organs.* 2005; 8:131-136.

[83] Tampieri A, Celotti G, Sprio S, Delcogliano A, Franzese S. Porosity-graded hydroxyapatite ceramics to replace natural bone. *Biomaterials.* 2001; 22(11):1365-1370.

[84] Furuichi K, Oaki Y, Ichimiya H, Komotori J, Imai H. Preparation of hierarchically organized calcium phosphate-organic polymer composites by calcification of hydrogel. *Science and Technology of Advanced Materials.* 2006; 7(2):219-225.

[85] Almirall A, Larrecq G, Delgado JA, Martinez S, Planell JA, Ginebra MP. Fabrication of low temperature macroporous hydroxyapatite scaffolds by foaming and hydrolysis of an alpha-TCP paste. *Biomaterials.* 2004; 25(17):3671-3680.

[86] Gong WL, Abdelouas A, Lutze W. Porous bioactive glass and glass-ceramics made by reaction sintering under pressure. *Journal of Biomedical Materials Research.* 2001; 54(3):320-327.

[87] Hench.L.L, Splinter RJ, Allen WC. Bonding mechanisms at the interface of ceramic prosthetic materials. *Journal of Biomedical Materials Research.* 1971; 2:117-141.

[88] Hench LL. *Bioceramics. Journal of the American Ceramic Society.* 1998; 81(7):1705-1728.

[89] Lu HH, Tang A, Oh SC, Spalazzi JP, Dionisio K. Compositional effects on the formation of a calcium phosphate layer and the response of osteoblast-like cells on polymer-bioactive glass composites. *Biomaterials.* 2005; 26(32):6323-6334.

[90] Gatti AM, Valdre G, Andersson OH. Analysis of the In-Vivo Reactions of A Bioactive Glass in Soft and Hard-Tissue. *Biomaterials.* 1994; 15(3):208-212.

[91] Xynos ID, Edgar AJ, Buttery LDK, Hench LL, Polak JM. Ionic products of bioactive glass dissolution increase proliferation of human osteoblasts and induce insulin-like growth factor II mRNA expression and protein synthesis. *Biochemical and Biophysical Research Communications.* 2000; 276(2):461-465.

[92] Navarro M, Ginebra MP, Clement J, Martinez S, Avila G, Planell JA. Physicochemical degradation of titania-stabilized soluble phosphate glasses for medical applications. *Journal of the American Ceramic Society.* 2003; 86(8):1345-1352.

[93] Navarro M, Ginebra MP, Planell JA. Cellular response to calcium phosphate glasses with controlled solubility. *Journal of Biomedical Materials Research Part A 2003;* 67A(3):1009-1015.

[94] Dean-Mo L. Fabrication of hydroxyapatite ceramic with controlled porosity. *Journal of Materials Science-Materials in Medicine.* 1997;(8):-227.

[95] Sepulveda P. Gelcasting foams for porous ceramics. *American Ceramic Society Bulletin.* 1997; 76(10):61-65.

[96] Tuck C, Evans JRG. Porous ceramics prepared from aqueous foams. *Journal of Materials Science Letters.* 1999; 18(13):1003-1005.

[97] Tian JT, Tian JM. Preparation of porous hydroxyapatite. *Journal of Materials Science.* 2001; 36(12):3061-3066.

[98] Jones JR, Hench LL. Factors affecting the structure and properties of bioactive foam scaffolds for tissue engineering. *Journal of Biomedical Materials Research Part B-Applied Biomaterials 2004;* 68B(1):36-44.

[99] Gibson LJ, Ashby MF. *Cellular solids: structure and properties.* 2nd ed. Oxford: Pergamon, 1999.

[100] Callcut S, Knowles JC. Correlation between structure and compressive strength in a reticulated glass-reinforced hydroxyapatite foam. *Journal of Materials Science-Materials in Medicine.* 2002; 13(5):485-489.

[101] Kim HW, Knowles JC, Kim HE. Hydroxyapatite porous scaffold engineered with biological polymer hybrid coating for antibiotic Vancomycin release. *Journal of Materials Science-Materials in Medicine.* 2005; 16(3):189-195.

[102] Sepulveda P, Jones JR, Hench LL. Bioactive sol-gel foams for tissue repair. *Journal of Biomedical Materials Research.* 2002; 59:340-348.

[103] Seitz H, Rieder W, Irsen S, Leukers B, Tille C. Three-dimensional printing of porous ceramic scaffolds for bone tissue engineering. *Journal of Biomedical Materials Research Part B-Applied Biomaterials.* 2005; 74B(2):782-788.

[104] Pfister A, Landers SR, Laib A, Hubner U, Schmelzeisen R. Biofunctional Rapid Prototyping for Tissue-Engineering Applications: 3D Bioplotting versus 3D Printing. *Journal of Polymer Science: Part A:Polymer Chemistry.* 2004; 42:624-638.

[105] Levine BR, Sporer S, Poggie RA, Della Valle CJ, Jacobs JJ. Experimental and clinical performance of porous tantalum in orthopedic surgery. *Biomaterials.* 2006; 27(27):4671-4681.

[106] Banhart J. Manufacture, characterisation and application of cellular metals and metal foams. *Progress in Materials Science*. 2001; 46(6):559-5U3.

[107] Korner C, Singer RF. Processing of metal foams - Challenges and opportunities. *Advanced Engineering Materials.* 2000; 2(4):159-165.

[108] St Pierre JP, Gauthier M, Lefebvre LP, Tabrizian M. Three-dimensional growth of differentiating MC3T3-E1 pre-osteoblasts on porous titanium scaffolds. *Biomaterials.* 2005; 26(35):7319-7328.

[109] Sikavitsas VI, van den Dolder J, Bancroft GN, Jansen JA, Mikos AG. Influence of the *in vitro* culture period on the *in vivo* performance of cell/titanium bone tissue-engineered constructs using a rat cranial critical size defect model. *Journal of Biomedical Materials Research Part A.* 2003; 67A(3):944-951.

[110] Vehof JWM, Haus MTU, de Ruijter AE, Spauwen PHM, Jansen JA. Bone formation in Transforming Growth Factor beta-I-loaded titanium fiber mesh implants. *Clinical Oral Implants Research.* 2002; 13(1):94-102.

[111] Rubin JP, Yaremchuk MJ. Complications and toxicities of implantable biomaterials used in facial reconstructive and aesthetic surgery: A comprehensive review of the literature. *Plastic and Reconstructive Surgery.* 1997; 100(5):1336-1353.

[112] Gu YW, Li H, Tay BY, Lim CS, Yong MS, Khor KA. *In vitro* bioactivity and osteoblast response of porous NiTi synthesized by SHS using nanocrystalline Ni-Ti reaction agent. *Journal of Biomedical Materials Research Part A.* 2006; 78A(2):316-323.

[113] Garcia AJ, Ducheyne P, Boettiger D. Cell adhesion strength increases linearly with adsorbed fibronectin surface density. *Tissue Engineering.* 1997; 3(2):197-206.

[114] Hersel U, Dahmen C, Kessler H. RGD modified polymers: biomaterials for stimulated cell adhesion and beyond. Biomaterials 2003; 24(24):4385-4415.

[115] Ripamonti U, Tasker JR. Advances in biotechnology for tissue engineering of bone. *Curr. Pharm. Biotech.* 2000; 1:47-55.

[116] Tugulu S, Silacci P, Stergiopulos N, Klok HA. RGD - Functionalized polymer brushes as substrates for the integrin specific adhesion of human umbilical vein endothelial cells. *Biomaterials.* 2007; 28(16):2536-2546.

[117] Singh N, Cui XF, Boland T, Husson SM. The role of independently variable grafting density and layer thickness of polymer nanolayers on peptide adsorption and cell adhesion. *Biomaterials.* 2007; 28(5):763-771.

[118] Zhao B, Brittain WJ. Polymer brushes: surface-immobilized macromolecules. Progress in Polymer Science 2000; 25(5):677-710.

[119] Palsson BO, Sangeeta NB. Coordination of cellular-fate processes. *Tissue Engineering.* USA: Pearson Prentice Hall, 2004: 105-130.

[120] Hong L, Tabata Y, Miyamoto S, Yamamoto M, Yamada K, Hashimoto N et al. Bone regeneration at rabbit skull defects treated with transforming growth factor-beta 1 incorporated into hydrogels with different levels of biodegradability. *Journal of Neurosurgery.* 2000; 92(2):315-325.

[121] Delong SA, Moon JJ, West JL. Covalently immobilized gradients of bFGF on hydrogel scaffolds for directed cell migration. *Biomaterials.* 2005; 26(16):3227-3234.

[122] Mann BK, Schmedlen RH, West JL. Tethered-TGF-beta increases extracellular matrix production of vascular smooth muscle cells. *Biomaterials.* 2001; 22(5):439-444.

[123] Holland TA, Tabata Y, Mikos AG. *In vitro* release of transforming growth factor-beta 1 from gelatin microparticles encapsulated in biodegradable, injectable oligo(poly(ethylene glycol) fumarate) hydrogels. *Journal of controlled release.* 2003; 91(3):299-313.

[124] Sokolsky-Papkov M, Agashi K, Olaye A, Shakesheff K, Domb AJ. Polymer carriers for drug delivery in tissue engineering. *Advanced Drug Delivery Reviews.* 2007; 59(4-5):187-206.

[125] Vasita R, Katti DS. Growth factor-delivery systems for tissue engineering: a materials perspective. *Expert Review of Medical Devices.* 2006; 3(1):29-47.

[126] Chen FM, Wu ZF, Wang QT, Wu H, Zhang YJ, Nie X et al. Preparation of recombinant human bone morphogenetic protein-2 loaded dextran-based microspheres and their characteristics. *Acta Pharmacologica Sinica.* 2005; 26(9):1093-1103.

[127] Chen FM, Zhao YM, Sun HH, Jin T, Wang QT, Zhou W et al. Novel glycidyl methacrylated dextran (Dex-GMA)/gelatin hydrogel scaffolds containing microspheres loaded with bone morphogenetic proteins: Formulation and characteristics. *Journal of controlled release.* 2007; 118(1):65-77.

[128] Green DW, Leveque I, Walsh D, Howard D, Yang XB, Partridge K et al. Biomineralized polysaccharide capsules for encapsulation, organization, and delivery of

human cell types and growth factors. *Advanced Functional Materials*. 2005; 15(6):917-923.

[129] Meinel L, Illi OE, Zapf J, Malfanti M, Merkle HP, Gander B. Stabilizing insulin-like growth factor-I in poly(D,L-lactide-co-glycolide) microspheres. *Journal of controlled release*. 2001; 70(1-2):193-202.

[130] Wei GB, Pettway GJ, McCauley LK, Ma PX. The release profiles and bioactivity of parathyroid hormone from poly(lactic-co-glycolic acid) microspheres. *Biomaterials*. 2004; 25(2):345-352.

[131] Chen FM, Wu ZF, Sun HH, Wu H, Xin SN, Wang QT et al. Release of bioactive BMP from dextran-derived microspheres: A novel delivery concept. *International Journal of Pharmaceutics*. 2006; 307(1):23-32.

[132] Storrie H, Mooney DJ. Sustained delivery of plasmid DNA from polymeric scaffolds for tissue engineering. *Advanced Drug Delivery Reviews*. 2006; 58(4):500-514.

[133] Hiraoka Y, Yamashiro H, Yasuda K, Kimura Y, Inamoto T, Tabata Y. In situ regeneration of adipose tissue in rat fat pad by combining a collagen scaffold with gelatin microspheres containing basic fibroblast growth factor. *Tissue Engineering*. 2006; 12(6):1475-1487.

[134] Nakahara T, Nakamura T, Kobayashi E, Inoue M, Shigeno K, Tabata Y et al. Novel approach to regeneration of periodontal tissues based on in situ tissue engineering: Effects of controlled release of basic fibroblast growth factor from a sandwich membrane. *Tissue Engineering*. 2003; 9(1):153-162.

[135] Tabata Y, Ikada Y, Morimoto K, Katsumata H, Yabuta T, Iwanaga K et al. Surfactant-free preparation of biodegradable hydrogel microspheres for protein release. *Journal of Bioactive and Compatible Polymers*. 1999; 14(5):371-384.

[136] King TW, Patrick CW. Development and *in vitro* characterization of vascular endothelial growth factor (VEGF)-loaded poly(DL-lactic-co-glycolic acid)/poly(ethylene glycol) microspheres using a solid encapsulation/single emulsion/solvent extraction technique. *Journal of Biomedical Materials Research*. 2000; 51(3):383-390.

[137] Stenekes RJH, Franssen O, van Bommel EMG, Crommelin DJA, Hennink We. The preparation of dextran microspheres in an all-aqueous system: Effect of the formulation parameters on particle characteristics. *Pharmaceutical Research*. 1998; 15(4):557-561.

[138] Puceat M. TGF beta in the differentiation of embryonic stem cells. *Cardiovascular Research*. 2007; 74(2):256-261.

[139] Andrew JG, Hoyland J, Andrew SM, Freemont AJ, Marsh D. Demonstration of Tgf-Beta-1 Messenger-Rna by Insitu Hybridization in Normal Human Fracture-Healing. *Calcified Tissue International*. 1993; 52(2):74-78.

[140] Duneas N, Crooks J, Ripamonti U. Transforming growth factor-beta 1: Induction of bone morphogenetic protein genes expression during endochondral bone formation in the baboon, and synergistic interaction with osteogenic protein-1 (BMP-7). *Growth Factors*. 1998; 15(4):259-+.

[141] McKinney L, Hollinger JO. A bone regeneration study: Transforming growth factor-beta(1) and its delivery. *Journal of Craniofacial Surgery*. 1996; 7(1):36-45.

[142] Rosier RN, O'Keefe RJ, Hicks DG. The potential role of transforming growth factor beta in fracture healing. *Clinical Orthopaedics and Related Research*. 1998;(355):S294-S300.

[143] Mohan S, Baylink DJ. Bone-Growth Factors. *Clinical Orthopaedics and Related Research.* 1991;(263):30-48.

[144] Lilli C, Marinucci L, Stabellini G, Belcastro S, Becchetti E, Balducci C et al. Biomembranes enriched with TGF beta(1) favor bone matrix protein expression by human osteoblasts *in vitro. Journal of Biomedical Materials Research.* 2002; 63(5):577-582.

[145] Centrella M, Horowitz MC, Wozney JM, Mccarthy TL. Transforming Growth-Factor-Beta Gene Family Members and Bone. *Endocrine Reviews.* 1994; 15(1):27-39.

[146] Erlebacher A, Filvaroff EH, Ye JQ, Derynck R. Osteoblastic responses to TGF-beta during bone remodeling. *Molecular Biology of the Cell.* 1998; 9(7):1903-1918.

[147] Gautschi OP, Frey SP, Zellweger R. Bone morphogenetic proteins in clinical applications. *Anz. Journal of Surgery*. 2007; 77(8):626-631.

[148] Lee SH, Shin H. Matrices and scaffolds for delivery of bioactive molecules in bone and cartilage tissue engineering. Advanced Drug Delivery Reviews 2007; 59(4-5):339-359.

[149] Xiao YT, Xiang LX, Shao JZ. Bone morphogenetic protein. *Biochemical and Biophysical Research Communications.* 2007; 362(3):550-553.

[150] Werner H, Katz J. The emerging role of the insulin-like growth factors in oral biology. *Journal of Dental Research.* 2004; 83(11):832-836.

[151] Xu JS, Liu ZH, Ornitz DM. Temporal and spatial gradients of Fgf8 and Fgf17 regulate proliferation and differentiation of midline cerebellar structures. *Development.* 2000; 127(9):1833-1843.

[152] Baird A, Walicke PA. Fibroblast Growth-Factors. *British Medical Bulletin.* 1989; 45(2):438-452.

[153] Nimni ME. Polypeptide growth factors: targeted delivery systems. *Biomaterials.* 1997; 18(18):1201-1225.

[154] Bodo M, Carinci F, Baroni T, Becchetti E, Bellucci C, Giammarioli M et al. Interleukin pattern of Apert fibroblasts *in vitro. European Journal of Cell Biology.* 1998; 75(4):383-388.

[155] Hurley MM, Abreu C, Gronowicz G, Kawaguchi H, Lorenzo J. Expression and Regulation of Basic Fibroblast Growth-Factor Messenger-Rna Levels in Mouse Osteoblastic Mc3T3-E1 Cells. *Journal of Biological Chemistry.* 1994; 269(12):9392-9396.

[156] Katoh R, Urist MR. Surface-Adhesion and Attachment Factors in Bone Morphogenetic Protein-Induced Chondrogenesis In-Vitro. *Clinical Orthopaedics and Related Research.* 1993;(295):295-304.

[157] Yamaguchi TP, Rossant J. Fibroblast Growth-Factors in Mammalian Development. *Current Opinion in Genetics & Development.* 1995; 5(4):485-491.

[158] Noda M, Vogel R. Fibroblast Growth-Factor Enhances Type-Beta-1 Transforming Growth-Factor Gene-Expression in Osteoblast-Like Cells. *Journal of Cell Biology.* 1989; 109(5):2529-2535.

[159] Debiais F, Hott M, Graulet AM, Marie PJ. The effects of fibroblast growth factor-2 on human neonatal calvaria osteoblastic cells are differentiation stage specific. *Journal of Bone and Mineral Research.* 1998; 13(4):645-654.

[160] Yu K, Xu JS, Liu ZH, Sosic D, Shao JS, Olson EN et al. Conditional inactivation of FGF receptor 2 reveals an essential role for FGF signaling in the regulation of osteoblast function and bone growth. *Development.* 2003; 130(13):3063-3074.

[161] Moursi AM, Winnard PL, Winnard AV, Rubenstrunk JM, Mooney MP. Fibroblast growth factor 2 induces increased calvarial osteoblast proliferation and cranial suture fusion. *Cleft Palate-Craniofacial Journal.* 2002; 39(5):487-496.

[162] Mansukhani A, Bellosta P, Sahni M, Basilico C. Signaling by fibroblast growth factors (FGF) and fibroblast growth factor receptor 2 (FGFR2)-activating mutations blocks mineralization and induces apoptosis in osteoblasts. *Journal of Cell Biology.* 2000; 149(6):1297-1308.

[163] Pitaru S, Kotevemeth S, Noff D, Kaffuler S, Savion N. Effect of Basic Fibroblast Growth-Factor on the Growth and Differentiation of Adult Stromal Bone-Marrow Cells - Enhanced Development of Mineralized Bone-Like Tissue in Culture. *Journal of Bone and Mineral Research.* 1993; 8(8):919-929.

[164] Deckers MML, van Bezooijen RL, van der Horst G, Hoogendam J, van der Bent C, Papapoulos SE et al. Bone morphogenetic proteins stimulate angiogenesis through osteoblast-derived vascular endothelial growth factor A. *Endocrinology.* 2002; 143(4):1545-1553.

[165] Erlebacher A, Filvaroff EH, Gitelman SE, Derynck R. Toward A Molecular Understanding of Skeletal Development. *Cell.* 1995; 80(3):371-378.

[166] Bittner K, Vischer P, Bartholmes P, Bruckner P. Role of the subchondral vascular system in endochondral ossification: Endothelial cells specifically derepress late differentiation in resting chondrocytes *in vitro. Experimental Cell Research.* 1998; 238(2):491-497.

[167] Wang DS, Miura M, Demura H, Sato K. Anabolic effects of 1,25-dihydroxyvitamin D-3 on osteoblasts are enhanced by vascular endothelial growth factor produced by osteoblasts and by growth factors produced by endothelial cells. *Endocrinology.* 1997; 138(7):2953-2962.

[168] Kasperk CH, Borcsok I, Schairer HU, Schneider U, Nawroth PP, Niethard FU et al. Endothelin-1 is a potent regulator of human bone cell metabolism *in vitro. Calcified Tissue International.* 1997; 60(4):368-374.

[169] Hsiong SX, Mooney DJ. Regeneration of vascularized bone. *Periodontology.* 2000 2006; 41:109-122.

[170] Mayr-Wohlfart U, Waltenberger J, Hausser H, Kessler S, Gunther KP, Dehio C et al. Vascular endothelial growth factor stimulates chemotactic migration of primary human osteoblasts. *Bone.* 2002; 30(3):472-477.

[171] Orlandini M, Spreafico A, Bardelli M, Rocchigiani M, Salameh A, Nucciotti S et al. Vascular endothelial growth factor-D activates VEGFR-3 expressed in osteoblasts inducing their differentiation. *Journal of Biological Chemistry.* 2006; 281(26):17961-17967.

[172] Park H, Temenoff JS, Holland TA, Tabata Y, Mikos AG. Delivery of TGF-beta 1 and chondrocytes via injectable, biodegradable hydrogels for cartilage tissue engineering applications. *Biomaterials.* 2005; 26(34):7095-7103.

[173] Chenite A, Chaput C, Wang D, Combes C, Buschmann MD, Hoemann CD et al. Novel injectable neutral solutions of chitosan form biodegradable gels in situ. *Biomaterials.* 2000; 21(21):2155-2161.

[174] Kanczler JM, Barry J, Ginty P, Howdle SM, Shakesheff KM, Oreffo ROC. Supercritical carbon dioxide generated vascular endothelial growth factor encapsulated

poly(DL-lactic acid) scaffolds induce angiogenesis *in vitro*. *Biochemical and Biophysical Research Communications*. 2007; 352(1):135-141.

[175] Yang XBB, Whitaker MJ, Sebald W, Clarke N, Howdle SM, Shakesheff KM et al. Human osteoprogenitor bone formation using encapsulated bone morphogenetic protein 2 in porous polymer scaffolds. *Tissue Engineering*. 2004; 10(7-8):1037-1045.

[176] Li CM, Vepari C, Jin HJ, Kim HJ, Kaplan DL. Electrospun silk-BMP-2 scaffolds for bone tissue engineering. *Biomaterials*. 2006; 27(16):3115-3124.

[177] Welch RD, Jones AL, Bucholz RW, Reinert CM, Tjia JS, Pierce WA et al. Effect of recombinant human bone morphogenetic protein-2 on fracture healing in a goal tibial fracture model. *Journal of Bone and Mineral Research*. 1998; 13(9):1483-1490.

[178] Hsu HP, Zanella JM, Peckham SM, Spector M. Comparing ectopic bone growth induced by rhBMP-2 on an absorbable collagen sponge in rat and rabbit models. *Journal of Orthopaedic Research*. 2006; 24(8):1660-1669.

[179] Simmons CA, Alsberg E, Hsiong S, Kim WJ, Mooney DJ. Dual growth factor delivery and controlled scaffold degradation enhance *in vivo* bone formation by transplanted bone marrow stromal cells. *Bone*. 2004; 35(2):562-569.

[180] Chen FM, Zhao YM, Wu H, Deng ZH, Wang QT, Zhou W et al. Enhancement of periodontal tissue regeneration by locally controlled delivery of insulin-like growth factor-I from dextran-co-gelatin microspheres. *Journal of controlled release*. 2006; 114(2):209-222.

[181] Ehrhart NP, Hong L, Morgan AL, Eurell JA, Jamison RD. Effect of transforming growth factor-beta 1 on bone regeneration in critical-sized bone defects after irradiation of host tissues. *American Journal of Veterinary Research*. 2005; 66(6):1039-1045.

[182] Yamamoto M, Takahashi Y, Tabata Y. Enhanced bone regeneration at a segmental bone defect by controlled release of bone morphogenetic protein-2 from a biodegradable hydrogel. *Tissue Engineering*. 2006; 12(5):1305-1311.

[183] Srouji S, Rachmiel A, Blumenfeld I, Livne E. Mandibular defect repair by TGF-beta and IGF-1 released from a biodegradable osteoconductive hydrogel. *Journal of Cranio-Maxillofacial Surgery*. 2005; 33(2):79-84.

[184] Freed LE, Vunjak-Navakovic G. Tissue Engineering Bioreactors. In: Lanza RH, Langer R, Vacanti J, editors. *Principles of Tissue Engineering*. USA: Academic Press, 2000: 143-156.

[185] Vunjak-Novakovic G, Obradovic B, Martin I, Bursac PM, Langer R, Freed LE. Dynamic cell seeding of polymer scaffolds for cartilage tissue engineering. *Biotechnology Progress*. 1998; 14(2):193-202.

[186] Holy CE, Shoichet MS, Davies JE. Engineering three-dimensional bone tissue *in vitro* using biodegradable scaffolds: Investigating initial cell-seeding density and culture period. *Journal of Biomedical Materials Research*. 2000; 51(3):376-382.

[187] Ishaug-Riley SL, Crane-Kruger GM, Yaszemski MJ, Mikos AG. Three-dimensional culture of rat calvarial osteoblasts in porous biodegradable polymers. *Biomaterials*. 1998; 19(15):1405-1412.

[188] Freed LE, Hollander AP, Martin I, Barry JR, Langer R, Vunjak-Novakovic G. Chondrogenesis in a cell-polymer-bioreactor system. *Experimental Cell Research*. 1998; 240(1):58-65.

[189] Kim BS, Putnam AJ, Kulik TJ, Mooney DJ. Optimizing seeding and culture methods to engineer smooth muscle tissue on biodegradable polymer matrices. *Biotechnology and Bioengineering.* 1998; 57(1):46-54.

[190] Martin I, Wendt D, Heberer M. The role of bioreactors in tissue engineering. *Trends in Biotechnology.* 2004; 22(2):80-86.

[191] Bruinink A, Siragusano D, Ettel G, Brandsberg T, Brandsberg F, Petitmermet M et al. The stiffness of bone marrow cell-knit composites is increased during mechanical load. *Biomaterials.* 2001; 22(23):3169-3178.

[192] Li Y, Ma T, Kniss DA, Lasky LC, Yang ST. Effects of filtration seeding on cell density, spatial distribution, and proliferation in nonwoven fibrous matrices. *Biotechnology Progress.* 2001; 17(5):935-944.

[193] Xiao YL, Riesle J, Van Blitterswijk CA. Static and dynamic fibroblast seeding and cultivation in porous PEO/PBT scaffolds. *Journal of Materials Science-Materials in Medicine.* 1999; 10(12):773-777.

[194] Burg KJL, Delnomdedieu M, Beiler RJ, Culberson CR, Greene KG, Halberstadt CR et al. Application of magnetic resonance microscopy to tissue engineering: A polylactide model. *Journal of Biomedical Materials Research.* 2002; 61(3):380-390.

[195] Glicklis R, Shapiro L, Agbaria R, Merchuk JC, Cohen S. Hepatocyte behavior within three-dimensional porous alginate scaffolds. *Biotechnology and Bioengineering.* 2000; 67(3):344-353.

[196] Nehrer S, Breinan HA, Ramappa A, Young G, Shortkroff S, Louie LK et al. Matrix collagen type and pore size influence behaviour of seeded canine chondrocytes. *Biomaterials.* 1997; 18(11):769-776.

[197] VunjakNovakovic G, Freed LE, Biron RJ, Langer R. Effects of mixing on the composition and morphology of tissue-engineered cartilage. *Aiche Journal.* 1996; 42(3):850-860.

[198] Wendt D, Marsano A, Jakob M, Heberer M, Martin I. Oscillating perfusion of cell suspensions through three dimensional scaffolds enhances cell seeding efficiency and uniformity. *Biotechnology and Bioengineering.* 2003; 84(2):205-214.

[199] Xie YB, Yang ST, Kniss DA. Three-dimensional cell-scaffold constructs promote efficient gene transfection: Implications for cell-based gene therapy. *Tissue Engineering.* 2001; 7(5):585-598.

[200] Palsson BO, Sangeeta NB. Scaling up for *ex vivo* cultivation. *Tissue Engineering.* USA: Pearson Prentice Hall, 2004: 223-243.

[201] Martin Y, Vermette P. Bioreactors for tissue mass culture: Design, characterization, and recent advances. *Biomaterials.* 2005; 26(35):7481-7503.

[202] Raimondi MT, Boschetti F, Falcone L, Fiore GB, Remuzzi A, Marinoni E et al. Mechanobiology of engineered cartilage cultured under a quantified fluid-dynamic environment. *Biomechanics and Modeling in Mechanobiology.* 2002; 1(1):69-82.

[203] Freed LE, Marquis JC, VunjakNovakovic G, Emmanual J, Langer R. Composition of Cell-Polymer Cartilage Implants. *Biotechnology and Bioengineering.* 1994; 43(7):605-614.

[204] Martin I, Padera RF, Vunjak-Novakovic G, Freed LE. *In vitro* differentiation of chick embryo bone marrow stromal cells into cartilaginous and bone-like tissues. *Journal of Orthopaedic Research.* 1998; 16(2):181-189.

[205] Carrier RL, Papadaki M, Rupnick M, Schoen FJ, Bursac N, Langer R et al. Cardiac tissue engineering: Cell seeding, cultivation parameters, and tissue construct characterization. *Biotechnology and Bioengineering*. 1999; 64(5):580-589.

[206] Timmins NE, Scherberich A, Fruh JA, Heberer M, Martin I, Jakob M. Three-dimensional cell culture and tissue engineering in a T-CUP (Tissue Culture Under Perfusion). *Tissue Engineering*. 2007; 13(8):2021-2028.

[207] Cartmell SH, Porter BD, Garcia AJ, Guldberg RE. Effects of medium perfusion rate on cell-seeded three-dimensional bone constructs *in vitro*. *Tissue Engineering*. 2003; 9(6):1197-1203.

[208] Wang YC, Uemura T, Dong R, Kojima H, Tanaka J, Tateishi T. Application of perfusion culture system improves *in vitro* and *in vivo* osteogenesis of bone marrow-derived osteoblastic cells in porous ceramic materials. *Tissue Engineering*. 2003; 9(6):1205-1214.

[209] Bancroft GN, Sikavitsast VI, van den Dolder J, Sheffield TL, Ambrose CG, Jansen JA et al. Fluid flow increases mineralized matrix deposition in 3D perfusion culture of marrow stromal osteloblasts in a dose-dependent manner. *Proceedings of the National Academy of Sciences of the United States of America*. 2002; 99(20):12600-12605.

[210] Goldstein AS, Juarez TM, Helmke CD, Gustin MC, Mikos AG. Effect of convection on osteoblastic cell growth and function in biodegradable polymer foam scaffolds. *Biomaterials*. 2001; 22(11):1279-1288.

[211] Altman G, Horan R, Martin I, Farhadi J, Stark P, Volloch V et al. Cell differentiation by mechanical stress. *FASEB J*. 2002; 16(2):270-272.

[212] Mauney JR, Sjostorm S, Blumberg J, Horan R, O'Leary JP, Vunjak-Novakovic G et al. Mechanical stimulation promotes osteogenic differentiation of human bone marrow stromal cells on 3-D partially demineralized bone scaffolds *in vitro*. *Calcified Tissue International*. 2004; 74(5):458-468.

[213] Altman GH, Lu HH, Horan RL, Calabro T, Ryder D, Kaplan DL et al. Advanced bioreactor with controlled application of multi-dimensional strain for tissue engineering. *Journal of Biomechanical Engineering-Transactions of the Asme*. 2002; 124(6):742-749.

[214] Pancrazio JJ, Wang F, Kelley CA. Enabling tools for tissue engineering. *Biosensors & Bioelectronics*. 2007; 22(12):2803-2811.

[215] Bettinger CJ, Weinberg EJ, Kulig KM, Vacanti JP, Wang YD, Borenstein JT et al. Three-dimensional microfluidic tissue-engineering scaffolds using a flexible biodegradable polymer. *Advanced Materials*. 2006; 18(2):165-+.

[216] Leclerc E, Furukawa KS, Miyata F, Sakai Y, Ushida T, Fujii T. Fabrication of microstructures in photosensitive biodegradable polymers for tissue engineering applications. *Biomaterials*. 2004; 25(19):4683-4690.

[217] Engel E, Michiardi A, Navarro M, Lacroix D, Planell JA. Nanotechnology in regenerative medicine: the materials side. *Trends in Biotechnology*. 2007; In press.

In: Tissue Engineering: Roles, Materials and Applications ISBN: 978-1-60456-293-4
Editors: S. J. Barnes, L. P. Harris, pp. 67-82

Chapter 3

COLLAGEN BIOMATERIALS FOR CARDIAC TISSUE ENGINEERING APPLICATIONS

Pingchuan Zhang[1], Marc Ruel[1, 2] and Erik Suuronen[1, 2]
[1] Division of Cardiac Surgery, University of Ottawa Heart Institute;
[2] Department of Cellular and Molecular Medicine, University of Ottawa

1. ABSTRACT

Cardiac diseases are leading causes of morbidity and mortality in men and women worldwide. Extensive damage to the heart is presently treatable by surgical reconstruction, implantable devices, and ultimately transplantation. These current treatments are "mechanical" and do not cure heart disease. The interdisciplinary field of tissue engineering has evolved to offer the hope of biologic, natural solutions for tissue regeneration and repair. While engineering an entire heart is an ambitious goal, the use of growth factors, cell therapy, and biomaterial scaffolds offers the promise of rebuilding the injured heart from its component parts. Of the many biomaterials that have been tested for such applications, collagen-based scaffolds are considered as one of the most attractive options. Collagens are highly biocompatible, biodegradable and possess weak antigenecity; and the physical properties of their gels can be controlled by regulating cross-linking reactions. In addition to these highly desirable characteristics, the heart's extracellular matrix is comprised mostly of collagens, making collagen a highly suitable choice for cardiac engineering purposes. This review discusses current progress made in the use of collagen biomaterials in the development of vascular grafts and heart valves, and in cell delivery and myocardial tissue engineering for the treatment of cardiac disease.

2. INTRODUCTION

Collagen is the most abundant protein within the human body. With continuous discoveries of chemically distinct collagens, 28 types of collagen have been identified to date [1]. The family of collagens is mainly comprised of fibrillar collagens, collagens with globular domains, and other collagen structures. Three α-chains form the polypeptide structure of each collagen type and the α-chains are unique to each collagen type. Functioning together with other components of the extracellular matrix, collagens not only confer structural support, but also act as adhesion and signaling molecules, interacting with various types of cells during cell migration, adhesion and differentiation [2].

In the healthy human heart, collagen content has been reported to be 41.5-52.5mg/g dry weight by using hydroxyproline measurements [3]. Connective tissue constituents, containing collagen in particular, form a network within the heart [4]. Within the meshwork, cardiomyocytes are connected together by intercalated discs, which permit a coupled contraction. The systolic contraction is transmitted by the extracellular matrix scaffold through the heart. Immediately after contraction, an active relaxation process follows. As Weber et al. [5] have noted, the tensile strength of the scaffold is similar to that of steel, and the extracellular matrix network is a major determinant of myocardial stiffness during the relaxation. Changes in the extracellular components profoundly affect heart function [6].

2.1. Characteristics of Collagen Type I

As the leading component of extracellular matrix, the bulk of collagen is constituted by fibrillar type I collagen, which is the most investigated protein in the collagen family. In vivo, water-soluble procollagen I are synthesized by fibroblasts and secreted into extracellular space. After proteolytic removal of C- and N- propeptides, procollagen I is converted to insoluble collagen. Due to its cell and tissue biocompatibility and its weak antigenicity, collagen I has been widely researched in tissue engineering. Medically, collagen I has been used extensively in humans as a natural protein [7, 8]. For example, collagen-coated Dacron grafts have been employed in vascular surgery since 1980s [9, 10]. In this application, collagen coating is used as a sealant to make grafts impervious to blood.

The degradation of collagen is a process as complicated as the synthesis of collagen, and a collagen's structure in vivo is more complex than those in vitro. Enzymes which are able to degrade collagen in solution in vitro are unable to break down polymeric insoluble collagen in vivo [2]. The degradation of collagen in vivo is due to the effects of metalloproteinases (MMPs). It is well known that MMP-1, MMP-8, MMP-13, and MMP-14 can target and degrade collagen type 1 [11-13]. Therefore, with considerable stability, collagen is still biodegradable, and amenable to remodeling, which is an important characteristic in tissue engineering applications.

2.2. Isolation, Purification and Crosslinking of Collagen I

As early as 1955, Schmitt et al. [14] developed a method to isolate and purify collagen in high yield from bovine skin by using the enzyme pepsin. In 1958, Gross et al. [15] demonstrated that acidic solutions of collagen could form brittle gels after adjusting to physiological ionic strength and pH at less than 10°C, and then warming to 30-34°C. Following this pioneering work, current collagen preparations from skin, tendons, hide or other animal parts, typically involve extraction and purification of collagen by using enzymes and/or diluted organic acids, precipitation with salts, and gelation and/or lyophilization. After treating hydrous collagen with acetone to remove water, a finely divided fibrous collagen is harvested, and the material can then be adjusted to neutral pH and dried [16].

Cross-linking is a process occurring during maturation of collagen molecules, during which intra- or inter-molecular covalent bonds are formed and loss of water-solubility is subsequently achieved. The mechanism of cross-linking in the human body is associated with deamination of ε-amino groups in lysine or hydroxylysine residue sites, and the lysine or hydroxylysine allysine formed contribute to reactions between allysine and lysine, hydroxylysine or another allysine [2]. In an attempt to reconstitute human tissue or organs in vitro, cross-linking is employed to facilitate the research of tissue engineering. Among various cross-linking strategies, the use of glutaraldehyde is very effective and the most well-studied [17]. However, due to potential cytotoxicity and calcification induced by the use of glutaraldehyde [18], alternative chemical reagents, such as 1-ethyl-3-(3-dimethyl aminopropyl) carbodiimide (EDC), have been investigated to cross-link collagen I [17, 19, 20]. Use of EDC can form peptide-like bonds in biomaterials without incorporation of cross-linking reagents. Moreover, recent results demonstrated that biocompatible materials can be manufactured from the use of EDC [21] with increased resistance against enzymatic degradation [19]. As a substitute for chemical cross-linking, alternative physical methods are under investigation, which include heat [22], dehydrothermal treatment [23], ultraviolet- and γ-irradiation [23, 24], and photo-chemical [25] cross-linking. Since physical cross-linking tends to be weak [22-25], further assessment of these methods is needed.

3. Collagen and Cardiac Tissue Engineering

3.1. The Interaction of Collagen and Cells

As the most abundant proteins in human body, collagens are not only the main components contributing to biomechanical properties of tissues, but are also important for growth, differentiation, morphogenesis, and repair of injured tissue. The interactions between collagens and cells, especially during cell adhesion and migration, play an important role in such events.

For example, the migration of endothelial cells is critical to angiogenesis (formation of blood vessels). In addition to chemotactic endothelial cell (EC) migration induced by various cytokines, chemotaxis-independent EC migration toward collagen has been reported [26]. This type of migration is defined as haptotaxis, namely directed cell crawling toward extra cellular matrix proteins [27]. The haptotactic migration of EC depends on the interaction between collagens and its cell surface receptor. Integrins, and in particular integrin $\alpha 1\beta 1$ and $\alpha 2\beta 1$, play an important role in EC-collagen interactions [26]. Integrins initiate migration by enabling endothelial cells to adhere to the extracellular matrix. The activation of integrins initiates the binding of focal adhesions and cytoskeleton actin, which is essential for cell migration [28, 29]. In addition, Herbst et al. [30] demonstrated that endothelial cell adhesion and migration was enhanced by type IV collagen compared with laminin, and that extracellular matrix may be especially important for endothelial cell migration during the repair of large vessels.

Endothelial progenitor cells (EPCs) are also important for the angiogenic process, and the extracellular matrix is reported to play an important role in EPC migration. Unfortunately, the information regarding these processes is still not well-defined. Among the limited information, Bloch et al. [31] reported that without β1 integrin, fewer blood vessels developed in a teratoma assay. The result suggested that integrin was a crucial extracellular matrix receptor for EPCs in vascular development.

Smooth muscle cells (SMC) also contribute to angiogenesis during vascular development and repair of injured blood vessels. Compared to chemotaxis, collagen I and IV elicited the most potent effects on SMC migration [32]. Adiguzel et al. [33] demonstrated that the migration of SMCs toward a haptotactic gradient was greater in type VIII collagen than in type I collagen. Furthermore, Sibinga et al. [34] reported that the expression of type VIII collagen was upregulated after vascular injury and in atherosclerosis. These studies suggested that SMCs can modify the composition of the matrix by producing type VIII collagen to favor their own migration.

Studies of tissue engineered myocardium and cell transplantation have demonstrated that collagen-based matrices can improve the behavior of neonatal cardiomyocytes. In Zimmermann's model, neonatal cardiomyocytes were cultured with collagen-based matrices and other growth factors [35]. After 7 days of culture and another 7 days of pre-conditioning by stretching, structures that are absent in neonatal myocardium were observed on matrix-cultured cardiomyocytes. These included the presence of: adherens junctions, gap junctions, desmosomes, a well-developed T-tubular system, and a basement membrane surrounding cardiac myocytes. Compared to native tissue from newborn, 6-day-old, and adult rats, histological analysis showed that cardiac cells in engineered heart tissue interconnected and oriented like mature native myocardium rather than immature tissue. Furthermore, Zhang et al. [36] and Kutschka et al. [37] reported that collagen-based matrices enhanced survival of transplanted cells in ischemic sites, which led to the improvement of injured left ventricular function.

3.2. Collagen in Tissue Engineering Vascular Grafts

Despite the reported success of synthetic materials in large blood vessel grafts (diameter > 6mm), synthetics have performed poorly in small caliber conduits [38]. Although autografts are still the gold standard for small vessel revascularization procedures, they are not always available under clinical circumstances, and additional surgeries are usually required. Regarding these issues, there is a clear need for small-caliber blood vessel grafts, which account for the majority of the demand. To be successful, these grafts will require patency that is equivalent to that of native blood vessels. Hopefully, the progress of tissue engineering offers a new strategy to construct a small caliber blood vessel in vitro (figure 1), which has been expected since the widespread use of native vein segments in coronary bypass surgery in late 1960s [39, 40].

Since Weinberg and Bell's pioneering work [41], two decades worth of efforts have been made by various research groups. During the last 20 years, different types of materials have been tested in the fabrication of artificial conduits. These include decellularized tissues, biodegradable polymer scaffolds, cell sheets, biopolymer scaffolds (including collagen) and novel biomimetic materials [42]. In order to realize adequate functional grafts, three requirements have to be satisfied: (1) mechanical support provided by a biocompatible component with high tensile strength; (2) suitable compliance provided by a biocompatible elastic component; and (3) an anti-coagulant intra-luminal surface covered by non-activated and confluent endothelium [43].

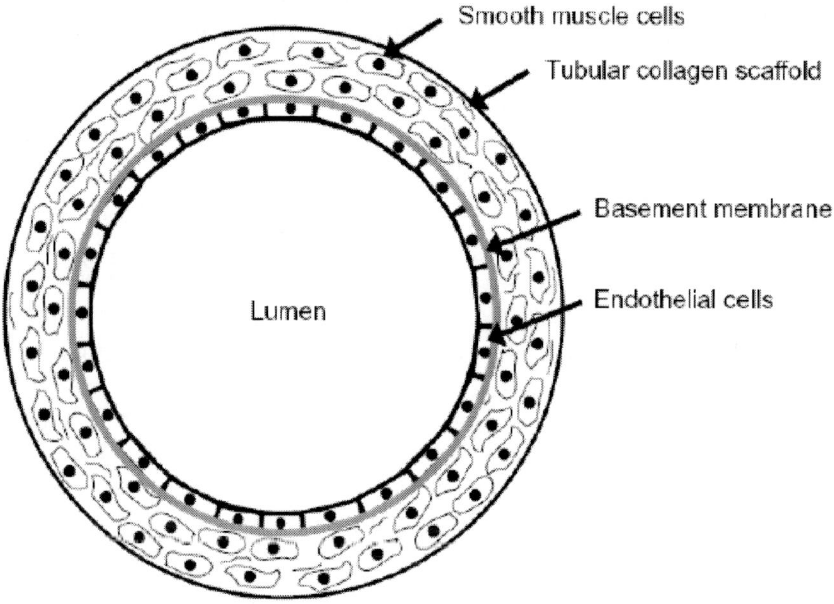

Figure 1. Schematic of the required elements in the basic design of a tissue-engineered vascular graft. A tubular collagen scaffold is seeded with SMCs, and the luminal surface is coated with a membrane for the formation of a complete endothelium.

With a pivotal role in the development of tissue engineering small caliber grafts, collagen has been investigated thoroughly as a primary scaffold. Conceptually, it is extremely attractive to construct such a collagen-based graft for three main reasons. Firstly, collagen is a major component of arterial walls [2]. The U.S. Food and Drug Administration has already approved the use of bovine type I collagen in tissue engineering skin products [44]. Secondly, contrasting sharply to synthetic polymer systems, collagen allows direct cellularization by mixing cells with liquid hydrogels during gelation. And finally, the desired circumferential alignment can be achieved by providing anisotropic strain using a non-adhesive mandrel. This technique has been used on an initial isotropic collagen network [45], to provide a directional scaffold to guide the remodeling process and similar alignment of extracellular matrix with native blood vessels [42].

Encouraged by the promising results of Weinberg and Bell's early work [41], a number of researchers followed the strategy to engineer complete artificial vascular grafts, containing vascular smooth muscle cells, fibroblasts and type I collagen gel. Typically the gel is cast into an annular mold. Endothelial cells are then seeded on the luminal surface of the reconstituted collagen tube [45-48]. However, due to gel weakness, these early graft attempts were inadequate for implantation in humans. In the case of Weinberg and Bell's grafts, they were forced to use a Dacron® (Invista, Wilmington, DE) mesh to support the artificial vessel, which inevitably reintroduced foreign materials and compliance mismatch [41]. With regard to this problem, intense efforts have been made to improve the strength of collagen scaffolds over the last decade. In 1993, L'Heureux reported that the artificial artery graft can be strengthened by cell-generated contraction force around the non-adhesive mandrel [46]. Although the graft modulus was increased significantly, the construct still did not satisfy clinical application requirements. Subsequently, Tranquilo et al. [49] introduced the method of using magnetic prealignment of collagen fibrils to increase circumferential tensile strength. Girton et al. [50, 51] used ribose and lysyl oxidase to stiffen and enhance collagen gel constructs. Moreover, Seliktar et al. [48] and Isenberg et al. [52] achieved dynamic conditioning by a pressurized medium method. However, despite a 265% increase in tensile strength for these conditioned grafts, the ultimate wall stress was only around 40-250 kPa, and the equivalent rupture strength was below 500 mm Hg. This is far below the rupture strength of native arterials (which is above 2000 mmHg [53]), and, therefore, these types of blood vessels were also not adequate for implantation in humans.

More recently, Wu et al. [60] developed a bi-layer type I collagen membrane, with which SMCs and endothelial cells were seeded on the lower and upper side of the porous membrane, respectively. As subcutaneous implantation tests demonstrated, the tissue engineered vascular grafts had adequate tensile strength, good biocompatibility, and regenerative capabilities. In another study, Lee et al. [61] reported that electrospun scaffolds were fabricated by using a mixture of 45% collagen, 15% elastin, and 40% synthetic polymer to mimic the ratio of collagen and elastin in native blood vessels. These non-toxic and stable grafts possessed controlled mechanical properties which were suitable for reconstituting an ultrastructure similar to that of native blood vessels.

In an attempt to address the limitation of collagen-based artificial vessels, fibrin has been used as an alternative [54, 55]. While there is a decrease in the synthesis of extracellular matrix (ECM) by collagen-entrapped fibroblasts [56, 57], these cells increase ECM synthesis and deposition in fibrin-based gels. Furthermore, Swartz et al. [58] developed ovine fibrin based artificial vessel which was implanted into the external jugular vein of sheep and remained patent for up to 15 weeks. However, the most impressive completely biological tissue engineered grafts have been fabricated by using a cell sheet approach. Without using an ECM, L'Heureux's group [59] developed a new edition tissue construct with burst strength above 3000 mmHg, which is much higher than the human saphenous vein, the standard clinical bypass graft. An early-stage clinical trial is under way that is focused on arteriovenous fistula in hemodialysis patients.

3.3. Collagen in Tissue Engineering Heart Valves

Every year, around 95,000 valve procedures are performed in the United States [62], and heart valve replacement procedures account for more than 60% of the cases because the tissue is not amenable to surgical repair. There are some concerns associated with the use of prosthetic valves for valve replacement. Patients transplanted with tissue valves may require re-operation due to the deterioration of the bioprosthetic valves. Therefore, durability is reported as a concern related to the use of tissue valves. On the other hand, tissue valves remain a favourable option compared to the use of mechanical valves, which require mandatory anticoagulant treatment to prevent thrombosis and embolism in patients. Tissue engineering aims at developing a new living valve to address these issues. However, because of unremitting success made by bioprosthetic valves, surgeons may disagree as to whether there is an immediate need for constructing tissue engineered heart valves in vitro. As clinical reports have shown, tissue valves have had a good long-term performance [63, 64]. Bioprosthetic valves currently being used have a lifespan of 15 to 20 years [65]. Moreso, due to a remarkable decrease of reoperative mortality in recent decades, surgeons are leaning to the belief that the reoperative risk of bioprosthetic valves is lower than the cumulative risk arising from the use of mechanical valves [66]. Given the excellent performance of bioprosthetic valves, there seems no reason to implant a novel, untested valve in adults. In contrast to adults, the results of surgical procedures related to valve defects in children were diverse and dependant on the age [67, 68], and therefore an opportunity for tissue engineering of valves exists in this cohort.

In the design of tissue engineered heart valves, the following 2 principal approaches have been applied: (1) decellularization of xenogenic tissues with or without cell seeding; and (2) bioresorbable scaffolds [66]. Among the latter, collagen-based constructs are the least addressed area. The strategy of collagen-based tissue engineering blood vessels was logically extended to the construction of tissue engineering artificial valves, in which collagen and cells were cast to a mold with inner and outer components [69]. Similar to the collagen-based vessels, the valve gels contracted and the fibrils aligned in the direction of strain [70]. More recently, Tranquillo's group used this technique and achieved both commissure-to-commissure alignment of collagen fibers in the leaflets and circumferential alignment in the root. However, due in part to the lack of extracellular matrix production by fibroblasts trapped in the collagen, the tests showed the mechanical strength of their artificial valves was much lower than native valves [71]. Rather than following the technique previously mentioned, Vesely's group focused on constructing a simple 1D string which could be used for surgical repair of mitral chordae. By using neonatal rat aortic smooth muscle cells and type I rat-tail collagen, single [72, 73] and branched [74] collagen-cell constructs were fabricated. Although these constructs were not ready for human use, it was claimed that they were not far away from animal model use [66].

3.4. Collagen in Biomaterials for Cell Delivery

Due to limited regenerative capacity of adult cardiomyocytes and the need for improved vascularization of ischemic tissue, cell-based therapy, including local and systemic transplantation of stem and progenitor cells, has been attempted for more than a decade [75, 76]. However, homing and retention of the transplanted cells within the target area depend on adhesion characteristics of the donor and host cells, and the survival rate of cells in the myocardium is less than 15% [77]. In addition, the non-specific delivery of cells to recipient body tissues and organs may lead to unwanted side effects [78]. With regard to improving the administration, retention and long-term survival of cellular therapy, tissue engineered matrices have been investigated to support transplanted cells (figure 2) [79]. Various types of liquid fibrin [80-82] and Matrigel matrices [36, 83] have been reported to support transplanted cells and restore injured myocardium. Noticeably, collagen is the major component of Matrigel. These matrices facilitated targeted tissue restoration after myocardial injury including increased cell transplant survival, increased blood flow and improved cardiac function. An extra benefit of matrix-assisted cell delivery is that the use of extracellular matrix components themselves may help to improve matrix remodeling, stabilize the tissue and restore function [84, 85].

Recently, collagen-based matrices were used to facilitate delivery of endothelial progenitor cells in a rat ischemic hindlimb model [86, 87]. This injectable matrix was vascularized by the host, and compared to cells injected alone, delivery of cells within the matrix enhanced their implantation and vasculogenic potential (figure 3). Another study reported the use of a type I collagen glycosaminoglycan (GAG) scaffold [88] for transplantation of bone marrow-derived mesenchymal stem cells into the infarcted region of rat hearts. Three weeks after cell transplantation, the majority of the collagen-GAG scaffolds degraded, and substantial new blood vessel growth was observed.

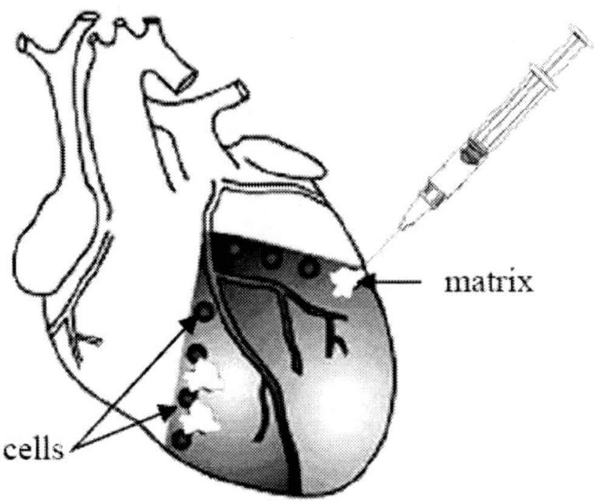

Figure 2. Schematic of delivery of stem cells within injectable matrix that gels in situ for improved cell survival.

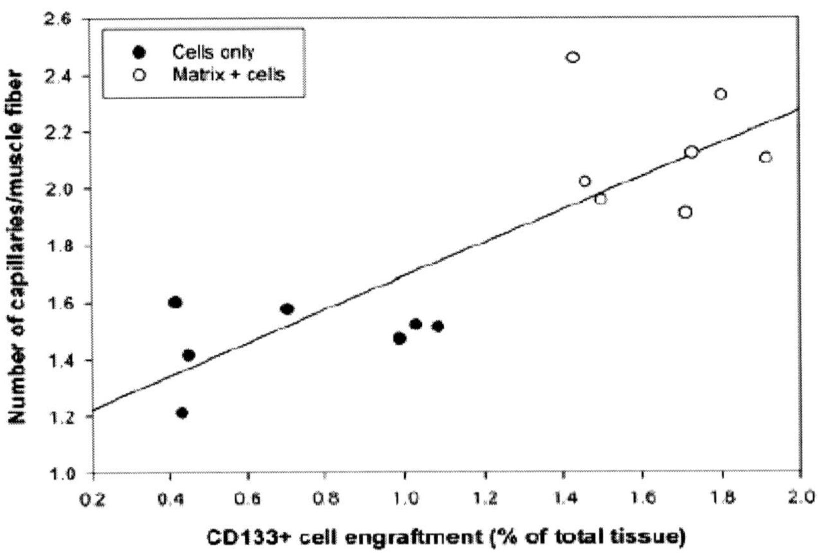

Figure 3. Cell engraftment correlated with the number of arterioles within the ischemic muscle, and the use of a matrix increased cell engraftment.

Despite these encouraging results, a potential limitation to this strategy may be that without further surgery, only a single application of matrix aided cell delivery is feasible. Regarding this concern, the use of matrix alone that can recruit endogenous progenitor cells from the circulation in situ may constitute an alternative or concomitant strategy to cell transplantation therapies. In this approach, host progenitor cells could home to the site of injury for improved longer-term regeneration. The addition of biomolecules within matrices has been demonstrated to assist in directing tissue regeneration [89-93]. Based on these achievements, a current study aims at developing matrices with specific ligands for improved recruitment and homing of endogenous circulating progenitor cells to sites of regeneration (figure 4), in essence creating a novel acellular "cellular" therapy. To this end, an acellular matrix was developed containing the oligosaccharide sialyl LewisX (sLe^X), which binds L-selectin, an adhesion molecule expressed on circulating progenitor cells and involved in their homing and adhesion [94]. Injection of the sLe^X-collagen matrix into rat ischemic hindlimbs increased the numbers of recruited progenitor cells, enhanced neovascularization, and improved perfusion of the hindlimb muscle [95]. This suggests that progenitor cell-based therapy may be possible without actual cell transplantation.

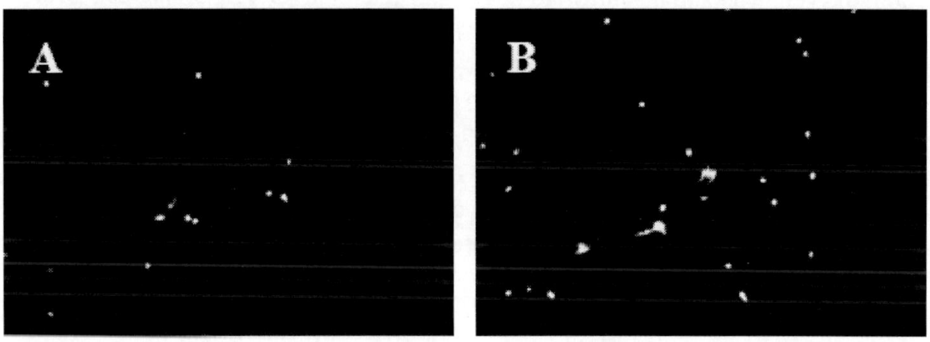

Figure 4. The addition of sLe^X to a collagen-based matrix (A) improved the adhesion of circulating progenitor cells compared to collagen matrix alone (B).

3.5. Collagen in Tissue Engineered Myocardium

As an alternative to cell transplantation, tissue engineered myocardium aims at restoring, maintaining, or enhancing impaired heart function. As previously mentioned, the strategy of cell transplantation currently suffers from some limitations. First, the low survival rate of cells (<15%) after transplantation [77] leads to new muscle formation with inadequate thickness needed for restoration of injured heart function. Second, the number, density, orientation, and electromechanical connections of cells cannot be controlled. Finally, the lack of graft vascularity limits cell survival and graft size [96]. Regarding these critical obstacles, new advances in tissue engineering may provide innovative solutions for myocardial regeneration procedures.

At the present time, three different strategies have been developed to construct artificial myocardium equivalents in vitro [97]. These consist of: (1) seeding cardiomyocytes on solid biomaterials [98-103], (2) entrapping cardiomyocytes in type I liquid collagen [35, 104-108], and (3) stacking cardiac myocyte monolayers to form multi-layered heart muscle constructs [109-111]. During the last decade, the second strategy, namely engineered heart tissue (EHT), has been investigated extensively by Eschenhagen and Zimmermann's group.

In 1997, Eschenhagen et al. [104] mixed liquid extracellular matrix, mainly composed of collagen type I and Matrigel, with freshly isolated heart cells to constitute EHT. The reconstituted mixture was cast to molds of different size and shape, and the artificial tissue with lattice sheet [104, 105, 107] and circular [35, 105, 106, 108] geometries were fabricated. After culturing for 5-7 days, the mixture consolidated and contracted, and EHT was capable of contracting spontaneously and synchronously [104]. Under conditions that physically stretched the tissue, successive culturing of EHT improved morphological, functional, and mechanical properties [107]. By using syngeneic rats, EHTs were transplanted either into the peritoneal cavity or onto the surface of healthy beating heart [108]. Under immunosuppressant treatment, EHTs remained viable and contractile for at least 8 weeks in vivo. In the absence of immediate blood vessel access, it was observed that EHTs were vascularized quickly after implantation [97]. More recently, Eschenhagen and Zimmermann have reported the development of stacked EHTs and multiloop EHTs [106]. During fabrication, oxygen, load and insulin were optimized quantitatively. After harvest, EHTs were transplanted into the ischemic area of an infarct in a rat model. Upon evaluation 28 days later, there was no delayed electrical coupling between grafts and native myocardium, and no arrhythmia was recorded by ambulatory ECG telemetry. Compared to controls, EHT prevented further dilation, induced systolic wall thickening of infarcted myocardial segments and improved heart function. In terms of potential clinical application, immunogenicity of EHTs and a suitable cell source are two major concerns. Investigation is under way in terms of searching for cardiogeneic autologous stem cells and establishing tissue culturing technique without serum and Matrigel [106].

4. Conclusion

In summary, understanding the interaction of extracellular matrix and various inoculating cells are key points in developing cardiac tissue engineering. As the most abundant protein in the ECM, collagen I has been investigated for more than half a century. With extensive investigation in tissue engineered blood vessel, heart valves, and myocardium, collagen I has proven to be a promising matrix candidate. Furthermore, advances in stem cell research indicate that control of the regenerating environment is needed to direct tissue repair, and matrix scaffolds, such as collagens, can help achieve this goal. However, inconsistency between the native and engineered tissue remain, and further investigation is needed before collagen-based engineered materials are ready for use in first line therapy in the clinic.

Acknowledgements

The authors thank Drew Kuraitis for his editorial assistance. Dr. Pingchuan Zhang is a Lawrence Soloway endowed research fellow in cardiac surgery, and his work is supported by

University of Ottawa Heart Institute Foundation. The authors research is supported by grants from the Canadian Institutes of Health Research, the Heart and Stroke Foundation of Canada.

References

[1] Plumb, D.A., et al., Collagen XXVII is developmentally regulated and forms thin fibrillar structures distinct from those of classical vertebrate fibrillar collagens. *J. Biol. Chem,* 2007. 282(17): p. 12791-5.

[2] Kucharz, E.J., The Collagens: Biochemistry and Pathophysiology. 1992, New York: *Springer-Verlag.* 1-3, 187-193.

[3] Caspari, P.G., et al., Collagen in the normal and hypertrophied human ventricle. *Cardiovasc Res,* 1977. 11(6): p. 554-8.

[4] Caulfield, J.B. and T.K. Borg, The collagen network of the heart. *Lab. Invest,* 1979. 40(3): p. 364-72.

[5] Shirwany, A. and K.T. Weber, Extracellular matrix remodeling in hypertensive heart disease. *J. Am. Coll. Cardiol,* 2006. 48(1): p. 97-8.

[6] Berk, B.C., K. Fujiwara, and S. Lehoux, ECM remodeling in hypertensive heart disease. *J. Clin. Invest,* 2007. 117(3): p. 568-75.

[7] Stegman, S.J., et al., A light and electron microscopic evaluation of Zyderm collagen and Zyplast implants in aging human facial skin. A pilot study. *Arch. Dermatol,* 1987. 123(12): p. 1644-9.

[8] Stegman, S.J., [Zyderm collagen. Clinical efficacy and tolerance]. *Rev. Laryngol. Otol. Rhinol.* (Bord), 1987. 108(1): p. 23-6.

[9] Guidoin, R., et al., Collagen coatings as biological sealants for textile arterial prostheses. *Biomaterials,* 1989. 10(3): p. 156-65.

[10] Tolan, M., et al., Clinical experience with a collagen impregnated woven Dacron graft. *J. Cardiovasc. Surg.* (Torino), 1995. 36(4): p. 323-7.

[11] Shapiro, S.D., Matrix metalloproteinase degradation of extracellular matrix: biological consequences. *Curr. Opin. Cell. Biol,* 1998. 10(5): p. 602-8.

[12] Wong, T.T., et al., Matrix metalloproteinases in disease and repair processes in the anterior segment. *Surv. Ophthalmol,* 2002. 47(3): p. 239-56.

[13] Chung, L., et al., Collagenase unwinds triple-helical collagen prior to peptide bond hydrolysis. *Embo J,* 2004. 23(15): p. 3020-30.

[14] Gross, J., J.H. Highberger, and F.O. Schmitt, Extraction of Collagen from Connective Tissue by Neutral Salt Solutions. *Proc. Natl. Acad. Sci. U S A,* 1955. 41(1): p. 1-7.

[15] Gross, J. and D. Kirk, The heat precipitation of collagen from neutral salt solutions: some rate-regulating factors. *J. Biol. Chem,* 1958. 233(2): p. 355-60.

[16] Gunasekaran, S., Preparation of purified and biocompatible collagen using two proteolytic enzyme treatments and a reducing agent, U.S.P.a.T. Office, Editor. 2000: the United States of America.

[17] Charulatha, V. and A. Rajaram, Influence of different crosslinking treatments on the physical properties of collagen membranes. *Biomaterials,* 2003. 24(5): p. 759-67.

[18] Simmons, D.M. and J.N. Kearney, Evaluation of collagen cross-linking techniques for the stabilization of tissue matrices. *Biotechnol. Appl. Biochem,* 1993. 17 (Pt 1): p. 23-9.

[19] Olde Damink, L.H., et al., In vitro degradation of dermal sheep collagen cross-linked using a water-soluble carbodiimide. *Biomaterials,* 1996. 17(7): p. 679-84.

[20] Liu, Y., et al., A simple, cross-linked collagen tissue substitute for corneal implantation. *Invest. Ophthalmol. Vis. Sci,* 2006. 47(5): p. 1869-75.

[21] Lee, C.R., A.J. Grodzinsky, and M. Spector, The effects of cross-linking of collagen-glycosaminoglycan scaffolds on compressive stiffness, chondrocyte-mediated contraction, proliferation and biosynthesis. *Biomaterials,* 2001. 22(23): p. 3145-54.

[22] Itoh, S., et al., Evaluation of cross-linking procedures of collagen tubes used in peripheral nerve repair. *Biomaterials,* 2002. 23(23): p. 4475-81.

[23] Billiar, K., et al., Effects of carbodiimide crosslinking conditions on the physical properties of laminated intestinal submucosa. *J. Biomed. Mater Res,* 2001. 56(1): p. 101-8.

[24] Pieper, J.S., et al., Preparation and characterization of porous crosslinked collagenous matrices containing bioavailable chondroitin sulphate. *Biomaterials,* 1999. 20(9): p. 847-58.

[25] Ibusuki, S., et al., Photochemically cross-linked collagen gels as three-dimensional scaffolds for tissue engineering. *Tissue Eng,* 2007. 13(8): p. 1995-2001.

[26] Senger, D.R., et al., The alpha(1)beta(1) and alpha(2)beta(1) integrins provide critical support for vascular endothelial growth factor signaling, endothelial cell migration, and tumor angiogenesis. *Am. J. Pathol,* 2002. 160(1): p. 195-204.

[27] Davis, G.E. and D.R. Senger, Endothelial extracellular matrix: biosynthesis, remodeling, and functions during vascular morphogenesis and neovessel stabilization. *Circ. Res,* 2005. 97(11): p. 1093-107.

[28] Klemke, R.L., et al., Regulation of cell motility by mitogen-activated protein kinase. *J. Cell Biol,* 1997. 137(2): p. 481-92.

[29] Giancotti, F.G., Complexity and specificity of integrin signalling. *Nat. Cell. Biol,* 2000. 2(1): p. E13-4.

[30] Herbst, T.J., et al., Differential effects of laminin, intact type IV collagen, and specific domains of type IV collagen on endothelial cell adhesion and migration. *J. Cell Biol,* 1988. 106(4): p. 1365-73.

[31] Bloch, W., et al., Beta 1 integrin is essential for teratoma growth and angiogenesis. *J. Cell. Biol,* 1997. 139(1): p. 265-78.

[32] Nelson, P.R., S. Yamamura, and K.C. Kent, Extracellular matrix proteins are potent agonists of human smooth muscle cell migration. *J. Vasc. Surg,* 1996. 24(1): p. 25-32; discussion 32-3.

[33] Adiguzel, E., et al., Migration and growth are attenuated in vascular smooth muscle cells with type VIII collagen-null alleles. *Arterioscler. Thromb. Vasc. Biol,* 2006. 26(1): p. 56-61.

[34] Sibinga, N.E., et al., Collagen VIII is expressed by vascular smooth muscle cells in response to vascular injury. *Circ. Res,* 1997. 80(4): p. 532-41.

[35] Zimmermann, W.H., et al., Tissue engineering of a differentiated cardiac muscle construct. *Circ. Res,* 2002. 90(2): p. 223-30.

[36] Zhang, P., et al., Artificial matrix helps neonatal cardiomyocytes restore injured myocardium in rats. *Artif. Organs,* 2006. 30(2): p. 86-93.

[37] Kutschka, I., et al., Collagen matrices enhance survival of transplanted cardiomyoblasts and contribute to functional improvement of ischemic rat hearts. *Circulation,* 2006.

114(1 Suppl): p. I167-73.

[38] Klinkert, P., et al., Saphenous vein versus PTFE for above-knee femoropopliteal bypass. A review of the literature. *Eur. J. Vasc. Endovasc. Surg,* 2004. 27(4): p. 357-62.

[39] Conte, M.S., The ideal small arterial substitute: a search for the Holy Grail? *Faseb J,* 1998. 12(1): p. 43-5.

[40] Kakisis, J.D., et al., Artificial blood vessel: the Holy Grail of peripheral vascular surgery. *J. Vasc. Surg,* 2005. 41(2): p. 349-54.

[41] Weinberg, C.B. and E. Bell, A blood vessel model constructed from collagen and cultured vascular cells. *Science,* 1986. 231(4736): p. 397-400.

[42] Isenberg, B.C., C. Williams, and R.T. Tranquillo, Small-diameter artificial arteries engineered in vitro. *Circ. Res,* 2006. 98(1): p. 25-35.

[43] Mitchell, S.L. and L.E. Niklason, Requirements for growing tissue-engineered vascular grafts. *Cardiovasc. Pathol,* 2003. 12(2): p. 59-64.

[44] Clark, D.P., C.W. Hanke, and N.A. Swanson, Dermal implants: safety of products injected for soft tissue augmentation. *J. Am. Acad. Dermatol,* 1989. 21(5 Pt 1): p. 992-8.

[45] Barocas, V.H., T.S. Girton, and R.T. Tranquillo, Engineered alignment in media equivalents: magnetic prealignment and mandrel compaction. *J. Biomech. Eng,* 1998. 120(5): p. 660-6.

[46] L'Heureux, N., et al., In vitro construction of a human blood vessel from cultured vascular cells: a morphologic study. *J. Vasc. Surg,* 1993. 17(3): p. 499-509.

[47] Hirai, J. and T. Matsuda, Venous reconstruction using hybrid vascular tissue composed of vascular cells and collagen: tissue regeneration process. *Cell Transplant,* 1996. 5(1): p. 93-105.

[48] Seliktar, D., et al., Dynamic mechanical conditioning of collagen-gel blood vessel constructs induces remodeling in vitro. *Ann. Biomed. Eng,* 2000. 28(4): p. 351-62.

[49] Tranquillo, R.T., et al., Magnetically orientated tissue-equivalent tubes: application to a circumferentially orientated media-equivalent. *Biomaterials,* 1996. 17(3): p. 349-57.

[50] Girton, T.S., T.R. Oegema, and R.T. Tranquillo, Exploiting glycation to stiffen and strengthen tissue equivalents for tissue engineering. *J. Biomed. Mater Res,* 1999. 46(1): p. 87-92.

[51] Girton, T.S., et al., Mechanisms of stiffening and strengthening in media-equivalents fabricated using glycation. *J. Biomech. Eng,* 2000. 122(3): p. 216-23.

[52] Isenberg, B.C. and R.T. Tranquillo, Long-term cyclic distention enhances the mechanical properties of collagen-based media-equivalents. *Ann. Biomed. Eng,* 2003. 31(8): p. 937-49.

[53] Dahl, S.L., et al., Decellularized native and engineered arterial scaffolds for transplantation. *Cell Transplant,* 2003. 12(6): p. 659-66.

[54] Grassl, E.D., T.R. Oegema, and R.T. Tranquillo, Fibrin as an alternative biopolymer to type-I collagen for the fabrication of a media equivalent. *J. Biomed. Mater Res*, 2002. 60(4): p. 607-12.

[55] Neidert, M.R., et al., Enhanced fibrin remodeling in vitro with TGF-beta1, insulin and plasmin for improved tissue-equivalents. *Biomaterials,* 2002. 23(17): p. 3717-31.

[56] Thie, M., et al., Aortic smooth muscle cells in collagen lattice culture: effects on ultrastructure, proliferation and collagen synthesis. *Eur. J. Cell Biol,* 1991. 55(2): p. 295-304.

[57] Clark, R.A., et al., Collagen matrices attenuate the collagen-synthetic response of cultured fibroblasts to TGF-beta. *J. Cell Sci,* 1995. 108 (Pt 3): p. 1251-61.

[58] Swartz, D.D., J.A. Russell, and S.T. Andreadis, Engineering of fibrin-based functional and implantable small-diameter blood vessels. *Am. J. Physiol. Heart Circ. Physiol,* 2005. 288(3): p. H1451-60.

[59] L'Heureux, N., et al., Technology insight: the evolution of tissue-engineered vascular grafts--from research to clinical practice. *Nat. Clin. Pract. Cardiovasc. Med,* 2007. 4(7): p. 389-95.

[60] Wu, H.C., et al., Coculture of endothelial and smooth muscle cells on a collagen membrane in the development of a small-diameter vascular graft. *Biomaterials,* 2007. 28(7): p. 1385-92.

[61] Lee, S.J., et al., In vitro evaluation of electrospun nanofiber scaffolds for vascular graft application. *J. Biomed. Mater. Res. A,* 2007.

[62] Thom, T., et al., Heart disease and stroke statistics--2006 update: a report from the American Heart Association Statistics Committee and Stroke Statistics Subcommittee. *Circulation,* 2006. 113(6): p. e85-151.

[63] Marchand, M.A., et al., Fifteen-year experience with the mitral Carpentier-Edwards PERIMOUNT pericardial bioprosthesis. *Ann. Thorac. Surg,* 2001. 71(5 Suppl): p. S236-9.

[64] Biglioli, P., et al., Long-term outcomes of the Carpentier-Edwards pericardial valve prosthesis in the aortic position: effect of patient age. J. *Heart Valve Dis,* 2004. 13 Suppl 1: p. S49-51.

[65] Rahimtoola, S.H., Choice of prosthetic heart valve for adult patients. *J. Am. Coll. Cardiol,* 2003. 41(6): p. 893-904.

[66] Vesely, I., Heart valve tissue engineering. *Circ. Res,* 2005. 97(8): p. 743-55.

[67] Vrandecic, M., et al., The use of bovine pericardium for pulmonary valve reconstruction or conduit replacement: long-term clinical follow up. *J. Heart Valve Dis,* 1998. 7(1): p. 54-61.

[68] Kanter, K.R., et al., One hundred pulmonary valve replacements in children after relief of right ventricular outflow tract obstruction. *Ann. Thorac. Surg,* 2002. 73(6): p. 1801-6; discussion 1806-7.

[69] Tranquillo, R.T., Girton, Timothy S. , Neidert, Michael Tissue equivalent approach to a tissue-engineered cardiovascular valve, U.S.P.a.T. Office, Editor. 2000, Regents of the University of Minnesota (Minneapolis, MN): the United States of America.

[70] Tower, T.T., M.R. Neidert, and R.T. Tranquillo, Fiber alignment imaging during mechanical testing of soft tissues. *Ann. Biomed. Eng,* 2002. 30(10): p. 1221-33.

[71] Neidert, M.R. and R.T. Tranquillo, Tissue-engineered valves with commissural alignment. *Tissue Eng,* 2006. 12(4): p. 891-903.

[72] Shi, Y. and I. Vesely, Fabrication of mitral valve chordae by directed collagen gel shrinkage. *Tissue Eng,* 2003. 9(6): p. 1233-42.

[73] Shi, Y. and I. Vesely, Characterization of statically loaded tissue-engineered mitral valve chordae tendineae. *J. Biomed. Mater Res. A,* 2004. 69(1): p. 26-39.

[74] Shi, Y., L. Rittman, and I. Vesely, Novel geometries for tissue-engineered tendonous collagen constructs. *Tissue Eng,* 2006. 12(9): p. 2601-9.

[75] Koh, G.Y., et al., Long-term survival of AT-1 cardiomyocyte grafts in syngeneic myocardium. *Am. J. Physiol,* 1993. 264(5 Pt 2): p. H1727-33.

[76] Li, R.K., et al., Cardiomyocyte transplantation improves heart function. *Ann. Thorac Surg,* 1996. 62(3): p. 654-60; discussion. 660-1.

[77] Muller-Ehmsen, J., et al., Survival and development of neonatal rat cardiomyocytes transplanted into adult myocardium. *J. Mol. Cell Cardiol,* 2002. 34(2): p. 107-16.

[78] Zwaginga, J.J. and P. Doevendans, Stem cell-derived angiogenic/vasculogenic cells: possible therapies for tissue repair and tissue engineering. *Clin. Exp. Pharmacol. Physiol,* 2003. 30(11): p. 900-8.

[79] Suuronen, E.J., et al., Building in vitro models of organs. *Int. Rev. Cytol,* 2005. 244: p. 137-73.

[80] Christman, K.L., et al., Fibrin glue alone and skeletal myoblasts in a fibrin scaffold preserve cardiac function after myocardial infarction. *Tissue Eng,* 2004. 10(3-4): p. 403-9.

[81] Christman, K.L., et al., Injectable fibrin scaffold improves cell transplant survival, reduces infarct expansion, and induces neovasculature formation in ischemic myocardium. *J. Am. Coll. Cardiol,* 2004. 44(3): p. 654-60.

[82] Liu, J., et al., Autologous stem cell transplantation for myocardial repair. *Am. J. Physiol. Heart Circ. Physiol,* 2004. 287(2): p. H501-11.

[83] Kofidis, T., et al., Injectable bioartificial myocardial tissue for large-scale intramural cell transfer and functional recovery of injured heart muscle. *J. Thorac. Cardiovasc. Surg,* 2004. 128(4): p. 571-8.

[84] Kofidis, T., et al., Novel injectable bioartificial tissue facilitates targeted, less invasive, large-scale tissue restoration on the beating heart after myocardial injury. *Circulation,* 2005. 112(9 Suppl): p. I173-7.

[85] Mizuno, T., et al., Elastin stabilizes an infarct and preserves ventricular function. *Circulation,* 2005. 112(9 Suppl): p. I81-8.

[86] Suuronen, E.J., et al., Tissue-engineered injectable collagen-based matrices for improved cell delivery and vascularization of ischemic tissue using CD133+ progenitors expanded from the peripheral blood. *Circulation,* 2006. 114(1 Suppl): p. I138-44.

[87] Suuronen, E.J., et al., Promotion of angiogenesis in tissue engineering: developing multicellular matrices with multiple capacities. *Int. J. Artif. Organs,* 2006. 29(12): p. 1148-57.

[88] Xiang, Z., et al., Collagen-GAG scaffolds grafted onto myocardial infarcts in a rat model: a delivery vehicle for mesenchymal stem cells. *Tissue Eng,* 2006. 12(9): p. 2467-78.

[89] Suuronen, E.J., et al., Innervated human corneal equivalents as in vitro models for nerve-target cell interactions. *FASEB J,* 2004. 18(1): p. 170-2.

[90] Christman, K.L., et al., Enhanced neovasculature formation in ischemic myocardium following delivery of pleiotrophin plasmid in a biopolymer. *Biomaterials,* 2005. 26(10): p. 1139-44.

[91] Cao, X. and M.S. Shoichet, Photoimmobilization of biomolecules within a 3-dimensional hydrogel matrix. *J. Biomater. Sci. Polym. Ed,* 2002. 13(6): p. 623-36.

[92] Bauer, S.M., et al., Vascular endothelial growth factor-C promotes vasculogenesis, angiogenesis, and collagen constriction in three-dimensional collagen gels. *J. Vasc. Surg,* 2005. 41(4): p. 699-707.

[93] Mi, L., et al., Self-assembling protein hydrogels with modular integrin binding

domains. *Biomacromolecules,* 2006. 7(1): p. 38-47.

[94] Biancone, L., et al., Role of L-selectin in the vascular homing of peripheral blood-derived endothelial progenitor cells. *J. Immunol,* 2004. 173(8): p. 5268-74.

[95] Suuronen, E.J., et al., Cell-based therapy without cell transplantation: tissue engineering of an acellular matrix for the recruitment of endogenous circulating progenitor cells. *Circulation,* 2007. 116 (Suppl): II-399.

[96] Gonen-Wadmany, M., L. Gepstein, and D. Seliktar, Controlling the cellular organization of tissue-engineered cardiac constructs. *Ann. N. Y. Acad. Sci,* 2004. 1015: p. 299-311.

[97] Zimmermann, W.H., et al., Heart muscle engineering: an update on cardiac muscle replacement therapy. *Cardiovasc. Res*, 2006. 71(3): p. 419-29.

[98] Bursac, N., et al., Cardiac muscle tissue engineering: toward an in vitro model for electrophysiological studies. *Am. J. Physiol,* 1999. 277(2 Pt 2): p. H433-44.

[99] Kofidis, T., et al., In vitro engineering of heart muscle: artificial myocardial tissue. *J. Thorac. Cardiovasc. Surg,* 2002. 124(1): p. 63-9.

[100] van Luyn, M.J., et al., Cardiac tissue engineering: characteristics of in unison contracting two- and three-dimensional neonatal rat ventricle cell (co)-cultures. *Biomaterials,* 2002. 23(24): p. 4793-801.

[101] Carrier, R.L., et al., Cardiac tissue engineering: cell seeding, cultivation parameters, and tissue construct characterization. *Biotechnol. Bioeng,* 1999. 64(5): p. 580-9.

[102] Li, R.K., et al., Survival and function of bioengineered cardiac grafts. *Circulation,* 1999. 100(19 Suppl): p. II63-9.

[103] Leor, J., et al., Bioengineered cardiac grafts: A new approach to repair the infarcted myocardium? *Circulation,* 2000. 102(19 Suppl 3): p. III56-61.

[104] Eschenhagen, T., et al., Three-dimensional reconstitution of embryonic cardiomyocytes in a collagen matrix: a new heart muscle model system. Faseb J, 1997. 11(8): p. 683-94.

[105] Zimmermann, W.H., et al., Three-dimensional engineered heart tissue from neonatal rat cardiac myocytes. *Biotechnol. Bioeng,* 2000. 68(1): p. 106-14.

[106] Zimmermann, W.H., et al., Engineered heart tissue grafts improve systolic and diastolic function in infarcted rat hearts. *Nat. Med,* 2006. 12(4): p. 452-8.

[107] Fink, C., et al., Chronic stretch of engineered heart tissue induces hypertrophy and functional improvement. *Faseb. J,* 2000. 14(5): p. 669-79.

[108] Zimmermann, W.H., et al., Cardiac grafting of engineered heart tissue in syngenic rats. *Circulation,* 2002. 106(12 Suppl 1): p. I151-7.

[109] Shimizu, T., et al., Electrically communicating three-dimensional cardiac tissue mimic fabricated by layered cultured cardiomyocyte sheets. *J. Biomed. Mater. Res,* 2002. 60(1): p. 110-7.

[110] Shimizu, T., et al., Fabrication of pulsatile cardiac tissue grafts using a novel 3-dimensional cell sheet manipulation technique and temperature-responsive cell culture surfaces. *Circ. Res,* 2002. 90(3): p. e40.

[111] Furuta, A., et al., Pulsatile cardiac tissue grafts using a novel three-dimensional cell sheet manipulation technique functionally integrates with the host heart, in vivo. *Circ. Res,* 2006. 98(5): p. 705-12.

In: Tissue Engineering: Roles, Materials and Applications ISBN: 978-1-60456-293-4
Editors: S. J. Barnes, L. P. Harris, pp. 83-97

Chapter 4

THE CONTRIBUTION OF ELECTROSPINNING TO TISSUE ENGINEERING AND RELATED FIELDS

Shobana Shanmugasundaram, Yuki Imura, Michael Jaffe and Treena Livingston Arinzeh
Department of Biomedical Engineering, New Jersey Institute of Technology, Newark, New Jersey, USA

1. ELECTROSPINNING PROCESS AND THEORY

Electrospinning is a specialized fiber fabrication technique that uses the potential difference of a static electric field as the driving force to propel a fine, charged stream of polymer solution or melt from a capillary orifice to a grounded collection device. The fibers produced can range from 5 microns to 20 nanometers. Since polymer solution electrospinning is more widely used, the focus of this chapter will be entirely on this method. Solution electrospinning was first patented by A. Formhals in 1934 [1]. A high voltage electrode is placed in contact with the polymer solution contained in a pipette or syringe-like vessel with a capillary tip. This electrode provides a source of charge, which is injected into the solution. The ground electrode of the high voltage source is attached to collector plate, which serves as a target for the electrostatically propelled polymer fluid stream. The potential difference between the capillary tip and the ground is typically on the order of 10 to 30kV. At the capillary tip, the field puts a stress on the charged solution and distorts it into a cone-like shape called the Taylor cone. At a sufficiently high potential difference, the electrical forces overcome the surface tension of the Taylor cone. Then, a stream of polymer fluid escapes and is propelled towards the grounded target. As the ejected stream forms a filament that traverses the distance from the Taylor cone at the capillary tip to the grounded target, the solvent is lost by evaporation and the polymer vitrifies into a coherent filament. The motion of the filament is straight for a relatively short distance then becomes erratic due to an electric field-induced bending instability. The result of this dynamic process is a non-woven filament mat that collects on the grounded target (figure 1).

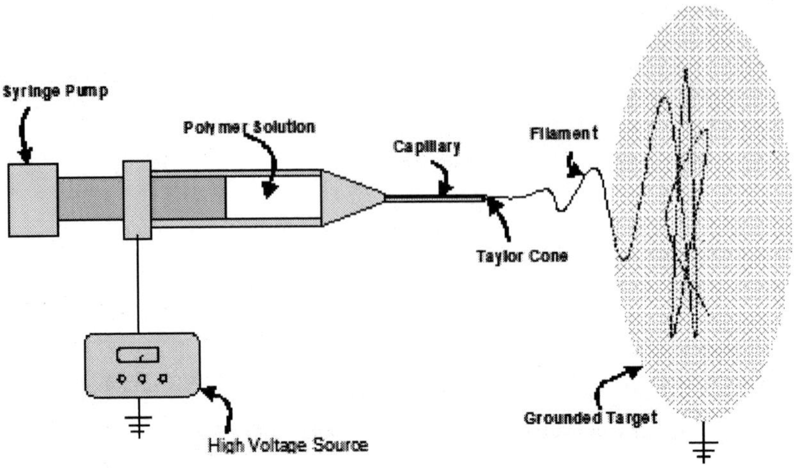

Figure 1. Typical solution electrospinning configuration.

The electrospinning process can be classified into three stages: 1. jet initiation; 2. bending instability; 3. solidification of fiber [2]. Early investigations were done by Taylor where the effect of an electric field on a water drop was studied [3]. It was shown that a liquid forms a cone at the tip of a capillary with an angle of 49.3° at a critical point when equilibrium is reached with the applied potential. Further increase in potential resulted in the initiation of an axial jet of fluid. Taylor did further experiments where he studied the effects of an electric field on liquid jets using various viscous and non-viscous fluids in an electric field and determined the minimum voltage needed to make the fluid jet appear [4]. When a potential difference is applied, charge is injected into the liquid until the resulting electrical force exceeds the surface tension forces holding the bead of liquid together. A hemispherical shape of the bead is transformed into that of a cone, at which point a fluid jet emerges from the vertex in order to eject the excess charge within. Detailed studies focusing on the jet initiation have been done relating the electrohydrodynamic process to the formation of the cone jet. The electric field lines have to be penetrated in the liquid in order for the fluid jet to form. Therefore the liquid must not be a perfect conductor. This penetration will allow for the formation of an electric field component tangent to the surface, which creates the driving force to propel the fluid jet beyond the cone [5].

Once the jet has initiated, the fluid travels along a straight path for a certain distance. The charged fluid travels towards the collector of the opposite polarity. As the fluid traverses toward the collector, the fluid undergoes an electrically driven bending instability, also referred to as "whipping". Different ideas regarding the nature of this instability have been given. The bending may allow for a large elongation to occur in a small region of space. This continuous elongation is strongly influenced by the repulsion between the charges on the fluid. Three steps in the initiation of the bending cycle were observed: 1. the straight jet forms a linear array of bends; 2.as the jet elongates, the linear array of bends spirals out with increasing size; 3. as the spiral grows, the cross-section diameter of the fiber decreases [2]. Although the spiraling fibers appear to be splitting with the naked eye, Baumgarten showed that with the aid of a high speed motion camera, the fibers where actually continuous throughout the process [6]. However, Reneker et al. observed cases where splitting and splaying of the jet due to radial charge repulsion have occurred which provides an alternative mechanism for the reduction in fiber diameter [2]. The process of diameter reduction was characterized by the "area reduction ratio", which is the ratio of the cross-sectional area of the upper end of a jet segment to that of the lower end of the same segment. This ratio is equal to the draw ratio if the volume of the material in the segment is conserved [2]. This study also showed that with bending, a very high reduction of the fiber diameter (as much as 10^5 times) was attained and the resulting increase in jet length occurred in a relatively short region of a few centimeters. This magnitude of the elongation and its high strain rate implies that the molecular chains are oriented along the axis of the jet.

Several researchers also believe that the bending instability is the most important factor in the formation of small diameter fibers in electrospinning. Shin et al. states that the rapid growth of the non-axisymmetric instability causes bending and stretching of the jet [7]. The instability is thought to occur through either: 1. small lateral fluctuations in the centerline of the jet resulting in the redistribution of the dipolar charges; or 2. mutual repulsion of the surface charges on the fiber. Three modes of instability were described: 1. the classical Rayleigh mode (axisymmetric) instability; 2. electric field induced axisymmetric conducting mode; 3. the whipping non-axisymmetric instability. The key factor in understanding the bending process is the competition between these modes of instabilities for a given set of conditions. Depending upon the electrospinning conditions, the resulting bending instability can be caused by one or a combination of these modes.

Due to the nature of this process, there are several process and polymer material parameters that affect the electrospinning process and the quality of the resulting fibers. The process parameters are: 1. magnitude and shape of the driving electric field; 2. flow rate of polymer fluid; 3. distance from the capillary tip to the grounded collector target; and 4. environmental conditions in the space the filament traverses. The polymer material parameters are: 1. molecular weight of the polymer and its solution concentration; 2. solution viscosity; 3. surface tension; and 4. conductivity and dielectric properties of the polymer fluid. The process parameters, along with the polymer solution parameters, influence the quality of the fiber and allow for a range of fibers to be produced. The kind of fibers made will depend on the parameters chosen. Therefore it is important to understand how the final fibers are affected by these parameters.

The applied voltage is a critical element in the electrospinning process due to the fact that it provides the primary physical driving force for the fiber formation. When the polymer solution is subjected to a potential difference, charge is injected into it. Initiation of a fluid jet takes place and is propelled by columbic attraction to the collector. Generally, higher applied voltages create higher electric field strengths, resulting in greater spinline tension to form fibers with smaller diameters. The increase in spinline tension also results in increased fiber velocity as it travels toward the collector. Since sufficient solvent evaporation is another crucial factor in nanoscale fiber formation and is dependent on the time of flight of the fiber as it moves toward the collector, care must given in deciding the amount of space between the capillary tip and the collector to ensure the desired diameter reduction is achieved. Although diameter reduction occurs with initial increases in voltage, the diameter begins to increase beyond a critical voltage level [6]. The corresponding increase in the electrostatically induced spinline stresses serves to draw out more fluid out of the capillary. Increasing the applied voltage also increases the probability of getting beaded fibers. The primary mechanism of charge transport is the flow of the polymer solution from the tip to the collector. Therefore, an increase in current is an indicator of an increase in mass flow rate, with all other parameters held constant. When this increase in mass flow rate out of the tip exceeds the supply rate of the solution to the tip, the Taylor cone becomes increasingly unstable and the bead density in the resulting fiber increases. Beaded fibers are usually not desired for most applications and its occurrence may be reduced by careful control of the mass flow rate to and from the tip [8].

The level of balance between the mass flow rate and the applied voltage is one of the most important factors in determining the fiber diameter. When the mass flow rate is increased for a given voltage, the fiber is increased in diameter because more fluid volume exits the capillary tip. Steps can be taken to prevent this from occurring when increasing the flow rate is necessary by correspondingly increasing the applied voltage to induce further diameter reduction and/or increasing the tip-to-collector distance, which will allow more solvent to evaporate from the fiber, further reducing its diameter. Varying the mass flow rate will also affect the volume charge density in the fluid. As the flow rate is increased, the volume charge density decreases [9]. When this occurs, the different sections of the fiber may merge together in mid-flight in some cases [9]. Furthermore, since the amount of this charge density generally reflects the amount of electrostatic stresses present in the spinline and its subsequent diameter reduction capability, control of the fiber diameter and porosity of the collected non-woven mat may be achieved through careful control of the mass flow rate.

The distance between the tip and collector is also an important process parameter because the electric field strength and flight time of the fiber are directly affected [10]. The electric field strength is determined by dividing the applied voltage by the tip-to-collector distance. Therefore, decreasing this distance for a given voltage will increase the electrostatic stresses exerted on the fiber and will thus increase the rate of deformation of the fiber. Decreasing the distance will also shorten the time it takes for the fiber to reach the collector. This combination of accelerated fiber deformation and reduced flight time may not allow for sufficient solvent loss to take place, thus resulting in merged [10] or beaded fibers.

Solution electrospinning is also highly affected by factors that control the amount and rate of solvent evaporation. Therefore, the environment through which the fiber travels toward the collector plays an important role in determining the final fiber diameter and morphology. Humidity in the electrospinning environment and its temperature will affect the solvent evaporation rate. Electrospinning in excessively humid conditions have been known to yield beaded fibers, where electrospinning with identical parameters in low humidity yielded unbeaded fibers. It is theorized that the lowered rate of solvent evaporation due to the humid environment forces more of the solvent to remain in the fiber longer, enabling solvent surface tension effects to create the beaded morphology. Temperature also affects the viscosity of the solution and therefore a major factor in determining the extent to which the fiber can deform.

The polymer material parameters also play a critical role in determining the final fiber diameter and morphology. The molecular weight of the polymer will affect the viscosity of the solution because of the number of polymer chain entanglements associated with it. It is these entanglements that prevent the solution jet from breaking up during the process and ensure the jet remains continuous [10]. The solution viscosity also depends on the concentration of the polymer with respect to the solvent. The proper ratio is crucial for producing fibers of desired size, morphology, and uniformity. If the concentration is too low, beaded fibers may develop and in extreme cases, fiber formation will not occur and droplets of solution will be produced instead. If the concentration is too high, the corresponding increase in viscosity will prevent sufficient deformation of the fiber to occur, resulting in fiber diameters larger than desired. The presence of surface tension in the solvent component of the solution is also the mechanism for bead formation. However, increasing the viscosity through the increase in polymer concentration increases the interaction of the polymer and solvent. Surface tension is lowered and thus the tendency for beads to form is reduced [10].

The conductivity of the solution also plays a major role in such an electrically driven process as electrospinning. Increasing the conductivity will enable increased charge capacity of the fluid, which will directly affect the extent of fiber diameter reduction. The conductivity can be increased through the addition of ions, such as salt, into the solution. The resulting increase in charge density increases the stretching of the fiber, producing fibers of smaller diameter. Another positive effect on this behavior is the reduction in the formation of beads [11]. Likewise, the dielectric constant of the solvent used plays an important role in determining the way in which fiber formation takes place. In general, a solution with greater dielectric properties yields fibers with smaller diameters and reduced tendencies for bead formation [10].

Although creating electrospun fibers is a fairly simple procedure, one must appreciate that the process of electrospinning is very complex and still not completely understood. However, careful attention to the parameters that affect electrospun fiber formation will increase the probability of attaining fibers of desired size, morphology, and uniformity.

2. Advantages of Electrospun Scaffolds in Tissue Engineering

For Tissue Engineering applications, an ideal scaffold should be able to mimic the structure and biological functions of the natural extracellular matrix (ECM). Therefore, the ECM is used as a model for fabricating scaffolds in Tissue Engineering. The ECM is a complex network of fibrous and adhesive proteins, elastic fibers, glycosaminoglycans to which growth factors, chemokines and cytokines are bound. The function of ECM is to provide scaffolding for cells and tissues, cues for cell proliferation and differentiation, and maintenance of differentiated tissues and modulation of the repair and regeneration process in response to injury [12]. The composition and spatial organization of the components of the ECM is specific to the tissue, and it regulates cell behavior either by cell –ECM interactions that directly regulate cell functions through receptor-mediated signaling or by controlling the mobilization of growth factors, thus modulating cell proliferation and controlling cell phenotype [13].

One of the major applications of electrospun mats is tissue engineering scaffolds. It is believed that the non-woven nanofibrous network mimics the structure of the natural ECM, as large diameter fibers do not mimic the morphological characteristics of the native fibrils. Human cells can attach and organize well around fibers with diameters smaller than those of the cells [14]. The individual fibers provide a large surface to volume ratio due to the smallness in diameter and surface topography. On a larger scale, these properties of the individual fibers are further enhanced by the porous structure and mechanical strength of the non-woven fabric. By varying the parameters in electrospinning it is possible to fabricate scaffolds of synthetic and natural polymers with varied diameter, shape, orientation and porosity. It is also possible to incorporate biological factors such as growth factors in the fibrous mat to provide chemical cues to the cells similar to the ECM.

3. Electrospun Synthetic and Natural Polymers

Scaffolds can be classified into three categories: biologically derived, synthetic, and ceramics or mineral-based matrices. The synthetic and natural polymers that have been electrospun for potential use as scaffolds in tissue engineering are summarized in table 1. Polylactic and polyglycolic acids and their copolymers have been widely investigated. Natural polymers, such as collagens, are also routinely sought.

Table 1. Electrospun polymers used in Tissue Engineering

Polymer	Reference
Poly(Glycolic Acid) Poly(L-Lactic Acid) Poly(D,L-Lactic Acid) Poly(D,L-Lactic-*Co*-Glycolic Acid 50:50) Poly(D,L-Lactic-*Co*-Glycolic Acid 85:15) 75 Wt% Poly(Glycolide-Co-Lactide) (PLA10GA90) + 25 Wt% Poly(L-Lactide) 85wt% Poly(Lactide-Co-Glycolide) (PLA75GA25) + 15wt% Poly(Ethylene Glycol)-B-Poly(D,L-Lactide) Poly(ε-Caprolactone) Poly-L-Lactic Acid/Hydroxyapatite	[15-19]
Poly(3-Hydroxybutyrate-Co-3-Hydroxyvalerate) /Type-I Collagen	[20]
Type I Collagen	[21]
Type II Collagen	[22]
Collagen Gelatin Solubilized Alpha-Elastin Recombinant Human Tropoelastin	[23]
Chitosan / Polyethylene Oxide	[24]
Fibrinogen	[25]
Silk Fibroin	[26]

4. Morphology of Electrospun Polymers

Electrospinning is a versatile process to fabricate scaffolds comprised of oriented fibers, tubular constructs, nonwoven mats with pores on individual fibers, composite mats and coaxial electrospinning of living cells as viable biological threads and scaffolds. The natural ECM can be composed of random and aligned fibrous structures. For example, the collagen fibers in the tendon are uniaxially aligned fiber bundles whereas they are random arrays in the skin. Therefore, current research efforts use the electrospinning technique to fabricate scaffolds consisting of aligned fibers for applications such as tendon/ligament, nerve and vascular tissue engineering (figure 2.b.). In order to obtain aligned fibers, the collector configuration in the electrospinning setup is modified, typically using rotating drums, discs, tubes or parallel electrodes as collectors. Though it is possible to fabricate aligned scaffolds by manipulating the collector setup, the degree of orientation is limited due to the fact that the high voltage charges the polymer jet causing instability in the electrospinning process [27].

Modifying the collector setup also allows for spinning more than one polymer to form a multi-layered mat and tube-like structures. A nanofibrous mat of polyethylene oxide, type I collagen and styrenated gelatin was fabricated by a multilayered electrospinning technique. Each polymer was electrospun sequentially to the same collector to fabricate a trilayered mesh of three polymers. They also fabricated a bilayered construct of type I collagen (inner layer) and segmented polyurethane (outer layer) using a rotating and traversing mandrel-type collector [28]. Recently, the potential of bilayered nonwoven micro-nano tubes of chitosan in nerve regeneration was examined. The results showed that they facilitated humoral permeation that would result in enhanced nerve regeneration [29].

In order to increase surface area, the electrospinning process has been modified to create pores within the fibers (figure 2.a.). Bognitzki et al. fabricated nanofibrous meshes of poly-L-lactide with a regular pore structure, and the pore size was in the order of 100 x 250 nm [30]. The presence of pores in the fibers would be of interest in studying the influence of scaffold topography on cell – substrate interactions.

Coaxial electrospinning has been used to fabricate polymer fibers incorporated with biologically active molecules like growth factors, enzymes etc.. The polymer solution and the biologically active material are in different reservoirs during the spinning process. The high voltage applied during electrospinning charges the polymer solution but not the biologically active molecules which are affected only by the viscous stresses. Smooth muscle cells at a concentration of 10^7cells/ml were electrospun with a coaxial setup with polydimethylsiloxane as the outer core. Flow cytometry results showed that the cells were viable over long periods of time [31]. In another study, hybrid blood vessel construct was fabricated using smooth muscle cells that were electrosprayed concurrently with poly (ester urethane) urea electrospinning. The construct was cytocompatible, mechanically strong, and had compliance values similar to native vessels [32].

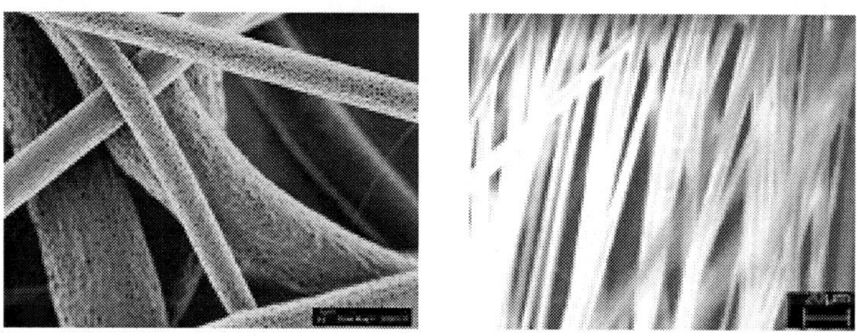

[a] [b]

Figure 2. SEM image of Poly (L-lactide) (a) with pores. Magnification: 3000X (b) aligned fiber. Magnification: 500X.

5. *In Vitro* Studies Using Electrospun Scaffolds

The following section reviews *in vitro* studies investigating electrospun scaffolds according to the specific tissue or organ of interest. Common for all applications are studies comparing the effects of fiber size/dimension and orientation on cell attachment, infiltration, and function.

A. Skin

H. M. Powell et al. investigated the effect of varying fiber diameter, interfiber distance, and porosity of electrospun gelatin on dermal and epidermal tissue regeneration. Results suggested that the interfiber distances between 5 and 10 μm possessed high cell viability, optimal cell organization, and excellent barrier formation [33]. To increase the porosity of the nanofiber mats, Poly (e-carprolactone) fibers were electrospun using an auxiliary electrode and a chemical blowing agent (BA) [34]. The auxiliary electrode stabilized the unstable initial jet and was influenced by the blowing agent. The mats fabricated with this method had high porosity compared to an electrospun nanofiber mat. Cell proliferation studies were carried out using human dermal fibroblasts, and the results showed that scaffolds fabricated by this method had enhanced the cell attachment in comparison to normal electrospun scaffold.

The potential of chitin, a structural polysaccharide, as a scaffold for wound dressing was investigated [35]. The electrospun nanofibers of chitin were compared to commercially available non-woven fibers that were approximately two orders of magnitude larger in diameter. Though the chitin nanofibers degraded faster than the larger diameter mats, the cell attachment and spreading of human keratinocytes and fibroblasts was higher in chitin nanofibers than in the larger diameter mats due to the large surface area of the nanofibers.

B. Musculoskeletal System: Cartilage, Tendon, Ligament, Bone

For cartilage repair, six of the most commonly used poly (α-hydroxy esters) were electrospun to evaluate their potential as tissue engineering scaffolds using chondrocytes and mesenchymal stem cells. The results demonstrated that poly (L-lactic acid) (PLLA) exhibited the optimal structural integrity and cells on PLLA had the highest proliferation as compared to the other five polymers [15]. The ECM of cartilage is composed primarily of Type II collagen. Shields et al. electrospun mats of Type II collagen and assessed the behavior of chondrocytes on these scaffolds. Their results demonstrated that the chondrocytes adhered, proliferated and infiltrated in to the type II collagen scaffolds [36]. For meniscus repair, Baker et al. cultured fibrochondrocytes and mesenchymal stem cells on aligned and random electrospun mats of poly (ε-caprolactone) to investigate the effect of fiber alignment. The results showed that the cell seeded aligned mats had improved mechanical properties than the random scaffolds by 10 weeks [37].

In tendon and ligament tissue engineering, knitted scaffolds have shown enhanced collagen matrix deposition than braided scaffolds. Sahoo et al. fabricated nano-microfibrous scaffolds of Poly(D,L-Lactic-*Co*-Glycolic Acid 10:90) (PLGA 10:90) by electrospinning

PLGA 10:90 on the surface of knitted PLGA 10:90. The fibers were deposited in between the loops of the knitted mats and hence enhanced the surface area of the scaffold. Bone marrow stromal cells secreted a tendon/ligament specific matrix on the hybrid scaffolds [38].

For bone repair and regeneration, electrospinning technique has been used to fabricate composite scaffolds. By adding a bioactive ceramic to the polymer fiber, the expectation is to improve the osteobiologic response, i.e. enhance the osteogenic differentiation of cells. The studies are summarized in table 2. The results thus far have been mixed in *in vitro* conditions.

Table 2. Electrospun composites for bone tissue engineering

Material	Cell Type	Results of the study	Reference
Bioactive Glass Nanofiber/ Poly(D,L-Lactic Acid) (PLA)	MC3T3-E1	Improved osteoblastic responses compared to pure PLA nanofibers	[39]
Poly (Lactide-*Co*-Glycolide) (PLGA)/Amorphous Tricalcium Phosphate	Human mesenchymal stem cells	Osteogenesis and cell proliferation were comparable to pure PLGA scaffolds	[40]
Poly-L-lactic acid (PLLA)/ Hydroxyapatite (HA)	MG-63	Cell adhesion and proliferation were comparable to PLLA mats	[41]
Silk fibroin containing bone morphogenetic protein 2 (BMP-2) and/or nanoparticles of hydroxyapatite (nHAP)	Human mesenchymal stem cells	Calcium deposition and upregulation of BMP-2 transcript levels were high in silk fibroin fibers with BMP-2 and nHAP	[42]
Poly(ε-caprolactone)/ Calcium carbonate	Human osteoblasts	Good cell attachment and proliferation	[43]

C. Circulatory and Digestive System

The potential of electrospun mats as scaffolds has been investigated for aortic valve, cardiac muscle, liver, and blood vessel applications. Lieshout et al compared electrospun scaffolds to knitted scaffolds for tissue engineering the aortic valve. Human myofibroblasts seeded on both types of scaffolds exhibited good cell proliferation and collagen production, but the electrospun scaffolds were not structurally stable in the perfusion conditions. The authors conclude that a combination of electrospun and knitted scaffold would be optimum for aortic valve repair [44]. In another study, the effect of surface properties of electrospun scaffolds on the behavior of cardiac myocytes was assessed. They found that the cells organized themselves with respect to the scaffold orientation, thus the electrospun scaffolds were flexible as well guided the growth of cardiac myocytes [45]. The use of electrospun scaffolds *in vitro* for potential application in the digestive and circulatory system are listed in table 3.

Table 3. Electrospun scaffolds used in the circulatory and digestive system

Material	Cell Type	Results of the study	Reference
poly(L-lactic acid)-*co*-poly(ε-caprolactone)(70:30) [P(LLA-CL)]	Human coronary artery endothelial cells	Supported endothelialization of cells, and improved bioactivity relative to pure P(LLA-CL) nanofibers	[46]
Poly (propylene carbonate) (PPC)	Rat Bone marrow mesenchymal stem cells	Cells integrated into the microstructure of PPC scaffold	[47]
Collagen type I, Elastin, Poly(D,L-lactide-coglycolide) (50:50)	Bovine endothelial and smooth cells	Tissue composition and mechanical properties of the scaffold matched the native vessels	[48]
poly(caprolactone)/gelatin	Endothelial cells	Cells oriented in the direction of aligned fibers	[49]
Polycaprolactone/polyurethane	Human umbilical vein endothelial cells	Supported cell attachment	[50]
poly[(ethyl phenylalanato)1.4 (ethyl glycinato)0.6 phosphazene]	rat neuromicrovascular endothelial cells	Cell proliferation was higher compared to polystyrene tissue-culture plates	[51]
Poly(L-lactide-co-caprolactone) (PLLC)	Porcine esophageal epithelial cells	Fibronectin grafting on the PLLC scaffolds promoted epithelium regeneration	[52]
Galactosylated poly(e-caprolactone-co-ethyl ethylene phosphate)	Rat hepatocytes	Promoted the formation of an integrated spheroid–nanofiber construct	[53]
Ultra-Web™ Synthetic ECM	NIH 3T3 fibroblasts, normal rat kidney (NRK) cells, T47D breast epithelial cells	Cells formed in vivo like morphologies	[54]
Nylon 6/6	C2C12 myoblasts	Maturating myotubes aligned in parallel to the fiber direction	[55]
DegraPol®	C2C12 myoblast, L6 myoblast	Enhanced skeletal myogenesis on aligned fiber scaffolds	[56]

D. Nervous System

Electrospun scaffolds are sought after for nerve regeneration because of the nanoscale dimension and alignment of the fibers to promote neurite outgrowth. Patel et al. fabricated aligned electrospun mats of PLLA then incorporated the ECM protein laminin and basic fibroblast growth factor on the surface of the mats. To test the influence of orientation and bioactivity of the electrospun mats, they cultured rat dorsal root ganglion on the nanofiber scaffolds *in vitro*. The results demonstrated that nanotopographic pattern had an effect on neurite outgrowth and cell migration. Aligned nanofiber scaffolds promoted neurite outgrowth and extension compared to random mats, and it was enhanced by the presence of immobilized protein and growth factor. In this study, the average fiber diameter of the mats was 500 nm as their pilot studies showed that fiber diameter between 100 - 1000 nm had no significant effect on cell alignment and neurite outgrowth [57]. Schnell et al. investigated the use of aligned poly-ε-caprolactone (PCL) and collagen/PCL as scaffolds for nerve implants. Though both scaffolds were favorable for cell attachment, collagen/PCL scaffolds enhanced schwann cell migration and process formation [58].

6. USE OF ELECTROSPUN SCAFFOLDS IN VIVO

Limited studies exist examining electrospun scaffolds *in vivo*. In part, their use may be limited due to the difficulty in preparing a contiguous three-dimensional structure with thicknesses necessary to treat tissue defects of varying shapes and sizes. To overcome this, Srouji et al. maintained human mesenchymal stem cell (MSC) seeded electrospun PCL/Collagen (1:1) scaffolds in dynamic culture for 6 weeks, and then stacked them to form a multilayered construct. These were implanted subcutaneously on nude mice. The MSC – polymer construct integrated well with the surrounding tissues, neovascularized and had good MSC proliferation over time [59]. The electrospun scaffolds stacked to form a multilayered construct could be beneficial in fabricating scaffolds with specific shape and thicknesses for *in vivo* application.

Nanosized electrospun mats may also be limited in use *in vivo* due to their low porosity and small pore size ranges. Conventional electrospinning methods that can achieve fiber diameters down to the nanoscale, produce a corresponding mat that has reduced spacing between fibers. This, in turn, could prevent cell infiltration, vascularization, and three-dimensional tissue formation *in vivo*. Currently, applications that do not require an open structure, such as tubes or sheet-like structures are being investigated. Bilayered electrospun tubes of chitosan with two different deacetylation rates, 78% and 93%, were examined in rat models with injured sciatic nerve. Among the two scaffolds, the 93% deacetylated chitosan tubes were better candidates for nerve regeneration because of their material properties such as degradation rate and mechanical stability [60].

REFERENCES

[1] Formhals, A., U.S. Patent, *1,975,504 (1934)*

[2] *Reneker, DH;* Yarin, AL; Fong, H; Koombhongse, S. Bending instability of electrically charged liquid jets of polymer solutions in electrospinning. *Journal of Applied Physics,* 2000, 87, 4531-4547.

[3] Taylor, G. Disintegration of water drops in an electric field. *Proceedings of the Royal Society of London. A, 1*964, 280, 383-397.

[4] Taylor, G. Electrically driven jets. *Proceedings of the Royal Society of London. A,* 1969, 313, 453-475.

[5] Cloupeau, M; Prunet-Foch, B. Electrohydrodynamic spraying functioning modes: a critical review. *Journal of Aerosol Science,* 1994, 25, 1021-1036.

[6] Baumgarten, P. Electrostatic spinning of acrylic microfibers. *Journal of Colloid and Interface Science,* 1971, 36, 71-79.

[7] Shin, YM; Hohman, MM; Brenner, MP; Rutledge, GC. Electrospinning: A whipping

fluid jet generates submicron polymer fibers. *Applied Physics Letters,* 2001, 78, 1149-1151.

[8] Deitzel, JM; Kleinmeyer, J; Harris, D; Beck Tan, NC. The effect of processing variables on the morphology of electrospun nanofibers and textiles. *Polymer,* 2001, 42, 261-272.

[9] Theron, SA; Zussman, E; Yarin, AL. Experimental investigation of the governing parameters in the electrospinning of polymer solutions. *Polymer,* 2004, 45, 2017-2030.

[10] Ramakrishna, S; Fujihara, K; Teo, W; Lim, TC; Ma, Z. *An Introduction to Electrospinning of Nanofibers*. First Edition. Singapore: World Scientific; 2005.

[11] Zong, X; Kim, K; Fang, D; Ran, S; Hsiao, B; Chu, B. Structure and process relationship of electrospun bioabsorbable nanofiber membranes. *Polymer,* 2002, 43, 4403-4412.

[12] Hynda K Kleinman DP, Matthew P Hoffman. Role of the extracellular matrix in morphogenesis. *Current Opinion in Biotechnology.* 2003;14:526-532.

[13] Francesco Rosso AG, Manlio Barbarisi, Alfonso Barbarisi. From Cell-ECM Interactions to Tissue Engineering. *Journal Of Cellular Physiology.* 2004;199:174-180.

[14] C.T. Laurencin AMAA, M.D. Borden, J.A. Cooper, Jr., . Tissue engineering: orthopedic applications. *Annu. Rev. Biomed. Eng.* 1999;1:19-46.

[15] Li WJ CJJ, Mauck RL, Tuan RS. Fabrication and characterization of six electrospun poly(alpha-hydroxy ester)-based fibrous scaffolds for tissue engineering applications. *Acta Biomater.* 2006;2(4):377-385.

[16] Sui G YX, Mei F, Hu X, Chen G, Deng X, Ryu S. Poly-L-lactic acid/hydroxyapatite hybrid membrane for bone tissue regeneration. *J. Biomed. Mater. Res. A.* 2007;82(2):445-454.

[17] Zong X BH, Chung CY, Yin L, Fang D, Hsiao BS, Chu B, Entcheva E. Electrospun fine-textured scaffolds for heart tissue constructs. *Biomaterials.* 2005;26(26):5330-5338.

[18] Yang F XC, Kotaki M, Wang S, Ramakrishna S. Characterization of neural stem cells on electrospun poly(L-lactic acid) nanofibrous scaffold. *J. Biomater. Sci. Polym. Ed.* 2004;15(12):1483-1497.

[19] Srouji S KT, Suss-Tobi E, Livne E, Zussman E. 3-D Nanofibrous electrospun multilayered construct is an alternative ECM mimicking scaffold. *J. Mater. Sci. Mater.* Med 2007.

[20] Meng W KS, Yuan J, Kim JC, Kwon OH, Kawazoe N, Chen G, Ito Y, Kang IK. Electrospun PHBV/collagen composite nanofibrous scaffolds for tissue engineering. *J. Biomater. Sci. Polym. Ed.* 2007;18(1):81-94. .

[21] S. Zhong WET, X. Zhu, R.W. Beuerman, S. Ramakrishna, L.Y.L.Yung. An aligned nanofibrous collagen scaffold by electrospinning and its effects on in vitro fibroblast culture. J*ournal of Biomedical Materials Research Part A.* 2006;79A:456-463.

[22] Shields KJ BM, Bowlin GL, Wayne JS. Mechanical properties and cellular proliferation of electrospun collagen type II. *Tissue Eng*. 2004;10:1510-1517.

[23] M. Li MJM, M.R. Gandhi, F.K. Ko, A.S. Weiss, P.I. Lelkes. Electrospun protein fibers as matrices for tissue engineering. *Biomaterials.* 2005;26:5999-6008.

[24] Bhattarai N ED, Veiseh O, Matsen FA, Zhang M. Electrospun chitosan-based nanofibers and their cellular compatibility. *Biomaterials.* 2005;26(31):6176-6184.

[25] McManus MC BE, Simpson DG, Barnes CP, Bowlin GL. Electrospun fibrinogen: feasibility as a tissue engineering scaffold in a rat cell culture model. *J. Biomed. Mater.*

Res. A. 2007;81(2):299-309.

[26] B.M. Min GL, S.H. Kim, Y.S. Nam, T.S. Lee, W.H. Park. Electrospinning of silk fibroin nanofibers and its effect on the adhesion and spreading of normal human keratinocytes and fibroblasts in vitro. *Biomaterials.* 2004;25:1289-1297.

[27] R Murugan SR. Design Strategies of Tissue Engineering Scaffolds with Controlled Fiber Orientation. *Tissue Engineering.* 2007;13(8):1845-1866.

[28] Satoru Kidoaki IKK, Takehisa Matsuda. Mesoscopic spatial designs of nano- and microfiber meshes for tissue-engineering matrix and scaffold based on newly devised multilayering and mixing electrospinning techniques. *Biomaterials.* 2005;26:37-46.

[29] Wei Wang SI, Atsushi Matsuda, Shizuko Ichinose, Kenichi Shinomiya, Yuiro Hata, Junzo Tanaka. Influences of mechanical properties and permeability on chitosan nano/microfiber mesh tubes as a scaffold for nerve regeneration. *Journal of Biomedical Materials Research Part A.* 2007.

[30] Michael Bognitzki WC, Thomas Frese, Andreas Schaper, Michael Hellwig, Martin Steinhart, Andreas Greiner, and Joachim H. Wendorff. Nanostructured Fibers via Electrospinning. *Advanced Materials.* 2001;13:70-72.

[31] Jayasinghe SN IS, McEwan JR. Cell electrospinning highly concentrated cellular suspensions containing primary living organisms into cell-bearing threads and scaffolds. *Nanomed.* 2007;2(4):555-567.

[32] John J. Stankusa LS, Kazuro Fujimotoc, Yi Hongc, David A. Vorpb, William R. Wagner. Fabrication of cell microintegrated blood vessel constructs through electrohydrodynamic atomization. *Biomaterials.* 2007;28:2738-2746.

[33] H. M. Powell STB. Fiber density of electrospun gelatin scaffolds regulates morphogenesis of dermal-epidermal skin substitutes. *Journal of Biomedical Materials Research Part A.*

[34] GeunHyung Kim WK. Highly Porous 3D Nanofiber Scaffold Using an Electrospinning Technique. *Journal of Biomedical Materials Research Part B: Applied Biomaterials.* 2007;81(1):104-110.

[35] Noh HK LS, Kim JM, Oh JE, Kim KH, Chung CP, Choi SC, Park WH, Min BM. Electrospinning of chitin nanofibers: degradation behavior and cellular response to normal human keratinocytes and fibroblasts. *Biomaterials.* 2006;27(21):3934-3944.

[36] Shields KJ BM, Bowlin GL, Wayne JS. Mechanical properties and cellular proliferation of electrospun collagen type II. *Tissue Eng.* 2004;10(9-10):1510-1517.

[37] Baker BM MR. The effect of nanofiber alignment on the maturation of engineered meniscus constructs. *Biomaterials.* 2007;28(11):1967-1977.

[38] Sahoo S OH, Goh JC, Tay TE, Toh SL. Characterization of a novel polymeric scaffold for potential application in tendon/ligament tissue engineering. *Tissue Eng.* 2006;12(1):91-99.

[39] Hae-Won Kim H-HL, Gae-Sig Chun. Bioactivity and osteoblast responses of novel biomedical nanocomposites of bioactive glass nanofiber filled poly(lactic acid). *Journal of Biomedical Materials Research Part A.* 2007;Epub 23 Jul 2007.

[40] Ju-Ha Song B-HY, Hyoun-Ee Kim, Hae-Won Kim. Bioactive and degradable hybridized nanofibers of gelatin-siloxane for bone regeneration. *Journal of Biomedical Materials Research Part A.* 2007;Epub 23 Jul 2007.

[41] Gang Sui XY, Fang Mei,Xiaoyang Hu,Guoqiang Chen,Xuliang Deng, Seungkon Ryu. Poly-L-lactic acid/hydroxyapatite hybrid membrane for bone tissue regeneration.

Journal of Biomedical Materials Research Part A. 2007;82(2):445-454.

[42] Li C VC, Jin HJ, Kim HJ, Kaplan DL. Electrospun silk-BMP-2 scaffolds for bone tissue engineering. *Biomaterials.* 2006;27(16):3115-3124.

[43] Fujihara K KM, Ramakrishna S. . Guided bone regeneration membrane made of polycaprolactone/calcium carbonate composite nano-fibers. *Biomaterials.* 2005;26(19):4139-4147. .

[44] M. I. Van Lieshout CMV, M. C. M. Rutten, G. W. M. Peters, F. P. T. Baaijens. Electrospinning versus knitting: two scaffolds for tissue engineering of the aortic valve. *J. Biomater. Sci. Polymer. Edn.* 2006;17(1-2):77-89.

[45] Xinhua Zong HB, Chiung-Yin Chung, Lihong Yin, Dufei Fang,Benjamin S. Hsiao, Benjamin Chu, Emilia Entcheva. Electrospun fine-textured scaffolds for heart tissue constructs. *Biomaterials.* 2005;26:5330-5338.

[46] He W YT, Teo WE, Ma Z, Ramakrishna S. Fabrication and endothelialization of collagen-blended biodegradable polymer nanofibers: potential vascular graft for blood vessel tissue engineering. *Tissue Eng.* 2005;11(9-10):1574-1588.

[47] Zhang J QH, Wang H, Hu P, Ou L, Guo S, Li J, Che Y, Yu Y, Kong D. Engineering of vascular grafts with genetically modified bone marrow mesenchymal stem cells on poly (propylene carbonate) graft. *Artif Organs.* 2006;30(12):898-905.

[48] Stitzel J LJ, Lee SJ, Komura M, Berry J, Soker S, Lim G, Van Dyke M, Czerw R, Yoo JJ, Atala A. Controlled fabrication of a biological vascular substitute. *Biomaterials.* 2006;27(7):1088-1094.

[49] Ma Z HW, Yong T, Ramakrishna S. Grafting of gelatin on electrospun poly(caprolactone) nanofibers to improve endothelial cell spreading and proliferation and to control cell Orientation. *Tissue Eng.* 2005 11(7-8):1149-1158.

[50] Williamson MR BR, Kielty C. PCL-PU composite vascular scaffold production for vascular tissue engineering: attachment, proliferation and bioactivity of human vascular endothelial cells. *Biomaterials.* 2006 27(19):3608-3616.

[51] Carampin P CM, Lora S, Menti AM, Baiguera S, Bellini S, Grandi C, Parnigotto PP. Electrospun polyphosphazene nanofibers for in vitro rat endothelial cells proliferation. *J. Biomed. Mater Res. A.* 2007;80(3):661-668.

[52] Zhu Y LM, Ong WF, Chan-Park MB, Chian KS. Esophageal epithelium regeneration on fibronectin grafted poly(L-lactide-co-caprolactone) (PLLC) nanofiber scaffold. *Biomaterials.* 2007;28(5):861-868.

[53] Chua KN LW, Zhang P, Lu H, Wen J, Ramakrishna S, Leong KW, Mao HQ. Stable immobilization of rat hepatocyte spheroids on galactosylated nanofiber scaffold. *Biomaterials.* 2005;26(15):2537-2547.

[54] Schindler M AI, Kamal J, Nur-E-Kamal A, Grafe TH, Young Chung H, Meiners S. A synthetic nanofibrillar matrix promotes in vivo-like organization and morphogenesis for cells in culture. *Biomaterials.* 2005;26(28):5624-5631.

[55] Huber A PA, Shakesheff KM. Reconstruction of spatially orientated myotubes in vitro using electrospun, parallel microfibre arrays. *Eur. Cell Mater.* 2007;14:56-63.

[56] Riboldi SA SN, Pigini L, Neuenschwander P, Simonet M, Mognol P, Sampaolesi M, Cossu G, Mantero S. Skeletal myogenesis on highly orientated microfibrous polyesterurethane scaffolds. *J. Biomed. Mater. Res. A.* 2007;Epub Aug 8.

[57] Patel S KK, Quigley R, Gao H, Hsiao BS, Poo MM, Li S. Bioactive nanofibers: synergistic effects of nanotopography and chemical signaling on cell guidance. *Nano*

Letters. 2007;7(7):2122-2128.

[58] Schnell EKK, Balzer S, Brook G, Klee D, Dalton P, Mey J. Guidance of glial cell migration and axonal growth on electrospun nanofibers of poly-e-caprolactone and a collagen/poly-e-caprolactone blend. *Biomaterials.* 2007;28:3012-3025.

[59] Srouji S KT, Suss-Tobi E, Livne E, Zussman E. 3-D Nanofibrous electrospun multilayered construct is an alternative ECM mimicking scaffold. *J. Mater. Sci: Mater Med.* 2007;Epub Aug 15, 2007

[60] Wang W SI, Matsuda A, Ichinose S, Shinomiya K, Hata Y, Tanaka J. Influences of mechanical properties and permeability on chitosan nano/microfiber mesh tubes as a scaffold for nerve regeneration. *Journal of Biomedical Materials Research Part A.* 2007;Epub Oct 2007.

In: Tissue Engineering: Roles, Materials and Applications ISBN: 978-1-60456-293-4
Editors: S. J. Barnes, L. P. Harris, pp. 99-118

Chapter 5

DESIGN OF BIOMIMETIC GLYCOPOLYMERS FOR LIVER TISSUE ENGINEERING

***T. Akaike*[1]*, T. Hoshiba*[2]*, S. H. Kim*[3]*, K. Kobayashi*[4]*, M. Goto*[5]*, H. L. Jiang*[6] *and C. S. Cho*[*6]**

[1] Department of Biomolecular Engineering,
Tokyo Institute of Technology, Yokohama 226-8501, Japan
[2] Biomaterials Center, National Institute for Materials Science,
Tsukuba 305-0044, Japan
[3] Biomaterials Research Center, Korea Institute of Science
and Technology, Seoul 136-791, Korea
[4] Graduate School of Engineering, Nagoya University,
Nagoya 464-8603, Japan
[5] Celagix, 109 Tokyo Tech Yokohama Venture Plaza,
Yokohama 226-8501, Japan
[6] Department of Agricultural Biotechnology,
Seoul National University, Seoul 151-921, Korea

ABSTRACT

Recently, the development of biomaterials for liver tissue engineering has focused on the design of biomimetic glycopolymers that can perform specific hepatocyte behaviors and direct liver regeneration through molecular recognition. Molecular recognition of glycopolymers by hepatocytes has been achieved by chemical modification of biomaterials with galactose moiety as a specific ligand that can incur specific interactions with asialoglycoprotein receptors (ASGPR) on the hepatocytes. The biomimetic glycopolymers potentially mimic many roles of extracellular matrix (ECM) in liver tissues, such as attachment, proliferation, migration, differentiation and survival of hepatocytes. This review discusses mechanism of specific interaction between galactose moieties of the galactose-carrying polymers and ASGPR of hepatocytes. Liver tissue engineering typically employs synthetic three-dimensional ECMs to induce long-term and stable differentiated hepatocyte functions. Therefore, the synthetic ECMs can be

* Correspondence: Tel: 82-2-880-4636; Fax: 82-2-875-2494; E-mail: chocs@plaza.snu.ac.kr

fabricated from two classes of biomaterials: natural polymers and synthetic ones. The criteria to design biomimetic glycopolymers such as galactose density, topology of ECM, deposited natural ECM, coculture system and cell source are addressed. The potential mechanisms on hepatocellular signaling by the ECM are also summarized.

Keywords: Biomimetic glycopolymer; Liver; Tissue engineering; Extracellular matrix; Asialoglycoprotein receptors.

1. INTRODUCTION

There are two kinds of cell-biomaterial interactions: one is specific and the other is unspecific. Unspecific interactions are usually difficult to control because the cell membrane has characteristic surface properties such as negative charge, lipophilic membrane proteins and lipophilic proteins of the extracellular matrix (ECM) that mediate unspecific adhesion to biomaterial surface [1].

On the other hand, specific interactions are much more controllable because they are mainly related to the interactions of defined chemical structures, such as specific ligands that interact with their related cell surface receptors [2]. The term 'biomimetic' has been coined to describe materials that are capable of such defined interactions [3]. Especially, Tessmar et al. mentioned that biomimetic materials are materials that mimic a biological environment to elicit a desired cellular response, facilitating the fulfillment of their task [4]. It is clear that specific ligands integrated into the biomaterials play an important role with respect to cell-biomaterial specific interactions. The main task of a biomimetic material is not necessarily the specific interaction with a cell or tissue, but rather the fulfillment of the intended goal such as targeting of a certain cell type or providing a scaffold structure for tissue formation [1]. There are several ways of incorporation of specific ligands to biomimetic materials for enhancing cell adhesion [5] or for impacting cell proliferation or differentiation [6,7].

Recently, tissue engineering, which applies methods from engineering and life sciences to create artificial constructs to direct tissue regeneration [8], has attracted many scientists and medical doctors with a hope to treat patients in a minimally invasive and less painful way [9] and this new techniques of tissue engineering are rapidly emerging and may make inroads into the care of patients in the 21st century [10]. The tissue engineering paradigm is to isolate specific cells through a small biopsy from a patient, to grow them on a three-dimensional biomimetic scaffold under precisely controlled culture conditions, to deliver the construct to the desired site in the patient's body, and to direct new tissue formation into the scaffold that can be degraded over time [11]. Several important parameters should be considered including biomimetic scaffolds that serve as a mechanical support for cell growth, progenitor cells that can be differentiated into specific cell types, and inductive growth factors that can modulate cellular activities [12]. The design of biomimetic glycopolymers is very important for tissue engineering strategies because the biomimetic glycopolymers can serve as a substrate on which cell populations can attach and migrate, be implanted with a combination of specific cell types, and be utilized as a drug carrier to activate specific cellular function through biomolecular recognition [13].

The liver plays an important role in a broad spectrum of physiological functions [14], and represents a significant tissue engineering technique because there are many liver functions such as metabolism, storage, synthesis and release of vitamins, carbohydrates, proteins, lipids and cyclic tetrapyrroles. The liver also detoxifies and inactivates endogeneous and exogeneous substances such as toxins and metals, and activates precursor molecules such as proenzymes and coagulation factors [15]. Furthermore, the liver produces bile, which is involved in intestinal lipid absorptions, and secretes a majority of circulating plasma proteins. Therefore, the design of biomimetic glycopolymers for liver tissue engineering is critically important because hepatocyte growth and function are affected by ECM features. To mimic the biological environment, the incorporation process of old proteins by the liver has been imitated. For this imitation, the interaction between galactose moieties of old proteins and asialoglycoproteins has been focused. This review discusses mechanism of specific interaction between galactose moieties of the galactose-carrying polymers and asialoglycoprotein receptors (ASGPR) on the hepatocytes. Liver tissue engineering typically employs synthetic three-dimensional ECMs to induce long-term and stable differentiated hepatocyte functions. Therefore, the synthetic ECMs can be fabricated from two classes of biomaterials: natural polymers and synthetic ones. The criteria to design biomimetic glycopolymers such as galactose density, topology of ECM, deposited natural ECM, coculture system and cell source are addressed. The potential mechanisms on hepatocellular signaling by the ECM are also summarized.

2. Mechanisms by which Hepatocytes Can Bind with Galactose Residues on De-Sialylated Glycoproteins

The ASGPR was first identified by Pricer et al. [16], who found that circulating asialoglycoproteins (ASGPs) bind to and are degraded by hepatocytes. The hepatocytes possess a cell surface receptor that recognizes and binds molecules having exposed galactose, N-acetylgalactosamine, or glucose residues [17]. After binding receptors, the ASGPs are internalized via coated pits and coated vesicles and subsequently appear in a complex network of tubules and uncoated vesicles before reaching the lysosomes where they are degraded [18]. These receptors are integral membrane glycoproteins localized on the sinusoidal face of hepatocytes. Hepatic plasma membrane receptors mediate the specific binding and uptake ("receptor-mediated endocytosis") of partially deglycosylated glycoproteins [19]. The rat ASGPR contains three types of subunits, rat hepatic lectin 1 (RHL 1) as a major subunit (41.5 kDa) and RHL-2/3 as a minor subunit (49 and 54 kDa) [20]. Each subunit contains one or more Ser and The residues in its cytoplasmic domain that are potential sites of phosphorylation; in addition, RHL 1 also contains one cytoplasmic Tyr [21].

3. ECM

Tissue engineering generally requires an artificial ECM for tissue regeneration, because cell adhesion, proliferation and differentiation, resulting in tissue regeneration, is difficult if such a matrix is not provided that functions as a cell scaffold. Especially, the regeneration of live-tissue substitutes for liver tissue engineering requires a suitable ECM for hepatocyte culture because hepatocytes are anchorage-dependent cells and hepatocytes rapidly lose liver-specific functions and viability when removed from the architectural liver [22]. In this review, natural ECMs are not covered due to space limitation; however, recent reviews on hepatocyte behaviors by natural ECMs are available [23-25].

3.1. Natural Polymer

Natural polymers are composed of polysaccharides, proteins and nucleic acids. Natural polymers have some remarkable merits over synthetic ones, that is, their excellent physiological properties such as cell adhesion, mechanical properties similar to natural tissues, biocompatibility and biodegradability [25]. On the other hand, natural polymers have several disadvantages such as risk of viral infection, antigenicity, instability, deterioration and limited versatility in designing an ECM with specific properties.

Alginate is widely used due to its low toxicity, easy chemical modification and mild gelation formation [26]. Galactose-carrying lactobionic acid was covalently linked with alginate to improve hepatocyte-galactose interactions by Yang et al. [27]. The results indicated that liver functions of hepatocytes entrapped in galactosylated alginate capsules were enhanced compared with those entrapped in alginate ones owing to the biomimetic glycopolymer.

Chitosan is suitable as a substrate for biomimetic glycopolymers because its structure is quite similar to glycosaminoglycans found in native tissue [27] and the free amino groups in the chitosan are easily modified by the covalent attachment of molecules using carbodiimide chemistry. Park et al. covalently bound galactose moieties to chitosan to enhance galactose-specific recognition between galactose in the galactosylated chitosan (GC) and ASGPR of hepatocytes and spheroid formation of hepatocytes in the presence of EGF [28]. Alginate/GC scaffold as the biomimetic material was also used to improve cell adhesion and stability of scaffold, and to retrain better differentiated cell functions [29-31].

Xyloglucan (XG), a polysaccharide derived from tamarind seeds, is composed of glucose units in the main chain and xylose and galactose units in the side chain. When XG is partially degraded by β-galactosidase, the resultant product exhibits thermally-reversible gelation in dilute aqueous solution [32]. Seo et al. studied specific interaction between galactose moieties in the XG and ASGPR on the hepatocytes in the two-dimensional and three-dimensional states because the XG has galactose moieties in the side chain. The results indicated that the hepatocyte adhesion to the XG-coated surface was dependent on the presence of Ca^{2+} ions, whereas that to the XG-coated surface could not be induced by Mg^{+2} ions, indication of specific interaction between XG and hepatocytes [33]. Also, alginate microcapsules prepared with XG enhanced spheroid formation of hepatocytes owing to the specific interaction [34].

Gelatin is a denatured collagen, and is a component of natural ECM. It has been used as the wound dressing due to the biodegradable property [35]. Hong et al. covalently bound galactose moieties to gelatin to increase the specific interaction between hepatocytes and the matrix [36]. The results indicated that hepatic specific metabolic functions were well maintained in the galactosylated gelatin sponge.

Silk fibroin (SF) is a natural fibrous protein created by the *Bombyx mori* silkworm, and mainly consists of the Gly-Ala-Gly-Ala-Gly-Ser sequence having its native molecular weight of 370 kDa [37]. Attachment of hepatocytes onto the galactosylated SF-coated dishes showed about an 8-fold increase as compared with that on uncoated dishes, being comparable to that on collagen-coated dishes [38] due to the hepatoctyte-specific substrate although liver-functions of hepatocytes by the three-dimensional galactosylated SF was not performed.

3.2. Synthetic Polymer

Synthetic polymers can be manufactured reproducibly on a large scale, and can also be processed into synthetic ECM in which the macrostructure and mechanical properties can be readily controlled [25]. Also, degradation time of some synthetic polymers can be manipulated. However, the greatest disadvantage of synthetic polymers is the lack of cell-recognition sites. Therefore, many efforts have been done to incorporate cell-adhesion ligands into the synthetic polymers to design biomimetic polymers.

Firstly, galactose groups were covalently coupled to polyacrylamide as the synthetic polymer by Weigel et al. [39]. The results indicated that rat hepatocytes were bound to beta-galactosylated polyacrylamide in a Ca^{2+}-dependent manner via a patch of ASGPR [40].

Akaike's group reported that galactose-derivatized polystyrene (PS), poly(*N-P*-vinylbenzyl-4-*0-β*-D-galactopyranosyl-D-gluconamide) (PVLA) is an excellent synthetic ECM because it guides hepatocyte adhesion through the unique ASGPR-galactose interaction between ASGPR of hepatocytes and highly concentrated galactose moieties along the polymer chains and the round morphology of hepatocytes on PVLA was found to trigger the spheroid formation in the presence of epidermal growth factor (EGF), which is the first report of spheroid formation of hepatocytes through receptor-mediated mechanism, leading to enhanced cell functions [41-43]. Also, this polymer is stably adsorbed into the tissue culture dish [44] owing to its amphiphilic character as shown in figure 1 [45].

Galactose was covalently coupled to star-shape poly(ethylene oxide) [46], poly (styrene-co-maleic acid) [47] and poly(allyl amine) [48] to target the ASGPR to enhance cell adhesion to biomimetic materials.

Hepatocyte-ECM interaction is a surface phenomenon although bulk properties of the ECM dictate the mechanical properties of the ECM. Galactose ligands were introduced into surface of poly(D, L-lactic-co-glycolic acid) [49], acrylic acid-grafted poly(ethylene terephthalate) (PET) [50], PS [51] and silica [52]. The results indicated that the hepatoctytes cultured on the galactosylated surface exhibited good attachment and promoted spheroid formation of the attached cells.

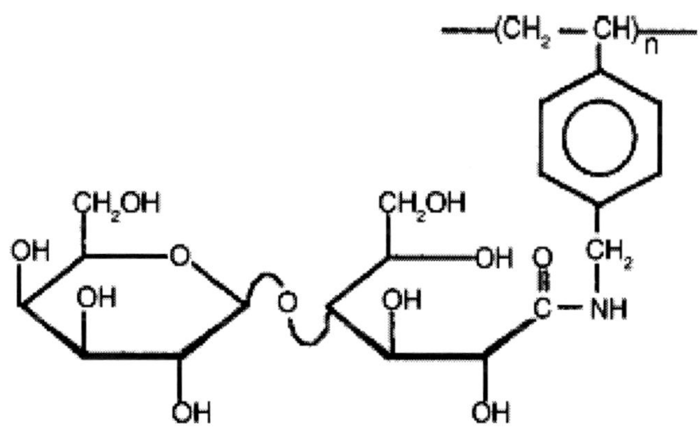

Figure 1. Chemical structure of PVLA. Source: From Ref. [45].

4. Criteria to Design Biomimetic Glycopolymers for Liver Tissue Engineering

The ASGPR binding with the galactose ligand in the biomimetic glycopolymers determines the strength of the cell attachment to the surface, spheroid formation of hepatocytes and differentiation of the hepatocytes. In this part, five important design factors are addressed.

4.1. Galactose Density

The galactose density in the glycopolymers is one of the important parameters to affect attachment, morphology, and functions of the hepatocytes because triantennary molecules with three terminal galactose residues bind to the lectins with the higher affinity than biantennary ones with one or two terminal galactose residues [53]. Akaike's group reported that hepatocytes attached at low concentration (0.5μg/ml) of PVLA were spread with high ^{3}H-thymidine uptake activity, whereas at high concentration (100μg/ml) of PVLA the attached hepatocytes were round with low ^{3}H-thymidine uptake activity [54]. On the other hand, attached hepatocytes were much more spread out on collagen I-coated surface than coated one with PVLA or poly (L-lysine) (PLL) as shown in figure 2 [55]. Hepatocytes attached onto PVLA below 20 ng/ml coating density expressed low levels of ASGPR of hepatocytes and exhibited higher proliferative ability [56]. Mao's group reported that hepatocytes formed spheroid formation with higher albumin synthesis at a high galactose density in the galactosylated PET [57] and galactosylated poly(vinylidene difluoride) [58] owing to easy migration of hepatocytes.

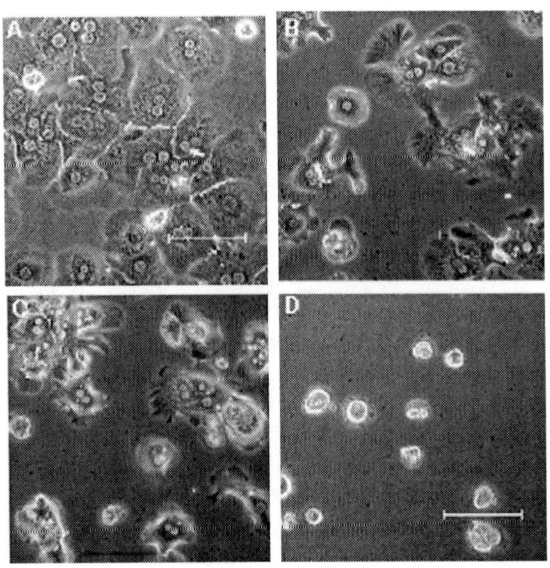

Figure 2. Hepatocyte morphology depending on adhesion matrix. Hepatocytes were cultured for 16 h in WE base on PS dishes coated with (A) 10 μg/ml collagen type I, (B) 5 mg/ml PLL, (C) 0.5 μg/ml PVLA, or (D) 100 μg/ml PVLA; scale bar=100 μm. Source: From Ref. [55].

4.2. Topology of ECM

Topology of ECM is an important parameter to affect cell morphology, function, and physiological responsiveness because three-dimensional ECMs induce more effectively differentiated hepatocyte functions than two-dimensional ones [59]. Also, many researches have been focused on the design of ECM topology for the spheroid formation of hepatocytes because this structure resembles the tight cell-cell contact in native liver and tight junctions, which mimic the morphology and ultrastructure of the native lobule [60]. Therefore, the design of three-dimensional scaffold to form hepatocyte spheroids is a powerful approach for enhancing liver-specific functions in liver tissue engineering. However, the cells in large spheroids experience mass transfer limitations in the absence of vascular network.

4.3. Deposited Natural ECM

It was reported that natural ECMs are deposited onto the surface of synthetic ECMs during the cell culture and results in affecting cellular behaviors [61] although it is not fully known what kinds of ECMs are deposited on the surface of biomaterials. Hoshiba et al. compared deposited natural ECM between PVLA as a hepatocyte-specific ECM and PLL as a non-specific one during the culture of hepatocytes *in vitro* [62]. The results indicated that laminin was dominantly deposited on PVLA-coated surface whereas fibronectin was dominantly deposited on PLL-coated one and the expression level of ECM genes was different between the hepatocytes cultured on PVLA- and PLL-coated surfaces, suggesting that the adsorption behavior of ECM molecules is different between PVLA- and PLL-coated surfaces. However, the exact mechanism of deposition of natural ECM on the surface of biomaterials should be cleared because the cellular behaviors are affected by the deposited ECM.

4.4. Coculture System

It is already reported that coculture systems enhanced liver-specific functions of hepatocytes and regulated cell growth, migration, and/or differentiation although the precise mechanisms remain unclear [63]. Seo et al. compared liver-specific functions of hepatocytes between coculture with hepatocytes and 3T3 fibroblasts in the alginate/galactosylated chitosan scaffold and monoculture with hepatocytes [64]. The results showed that the stabilized liver-specific functions such as albumin synthesis, ammonia elimination and cytochrome P450 activity of hepatocytes cocultured with hepatocytes and 3T3 fibroblasts were improved compared with those monocultured with hepatocytes as shown in figure 3. Also, the liver functions of the hepatocytes were enhanced under the coculture with mouse embryonic fibroblasts [65], liver epithelial cells [66], sinusoidal cells [67], stellate cells [68], bone marrow cells [69], Kupper cells [70], fetal hepatocytes [71], and pancreatic islet cells [72].

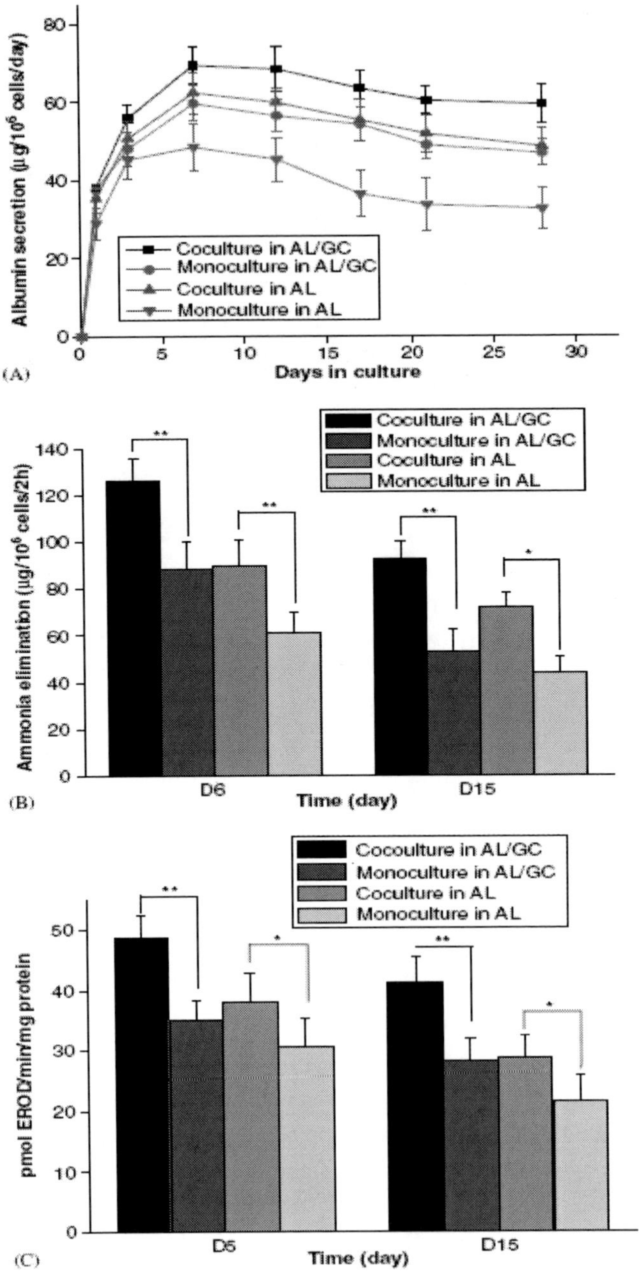

Figure 3. Comparision of liver-specific functions such as albumin secretion (A), ammonia elimination (B), and EROD activities (C) in AL and AL/GC sponges for monoculture and coculture of hepatocytes with NIH3T3. A: Albumin secretion rates for both systems were measured over 28 days in culture. ELISA specific for mouse albumin were used to quantify protein in 24 h conditioned medium. B: Ammonia elimination was compared between both systems on day 5 and 16 of culture. C: EROD deethylase activity by cytochrome P4501A1 was compared between both systems on day 6 and 16 of culture. EROD activity was measured using the highly fluorescent resorufin in both systems. All data are represented as mean$\pm SD_{(n-1)}$ (n=6). Statistics (two-tailed paired Student's t-test) compared monoculture with coculture; *P<0.05 and **P<0.01. Source: From Ref. [64].

4.5. Cell Source

In the liver tissue engineering, choosing cell type and cell source is very important because it is necessary to choose cells that demonstrate the particular phenotype of interest [73]. The various cell types are stem cells, hematopoietic cells, oval progenitor cells, and mature hepatocytes [74]. A summary of the various potential cell sources for liver tissue engineering is shown in table 1 [22] and discussed below.

Table 1. Characteristics of various cell sources available for liver tissue engineering [modified from 22]

Cell Source	Supply	Liver-Specific Function	Host Compatibility	Risks to Patient	Ref.
Primary human Hepatocytes	Scarce	Unstable, but limited experience with cell culture	Excellent	Small	22
Stem-cell derived human hepatocytes	Unlimited	Low efficiency of differentiation to hepatocytes	Excellent	Small	75-79
Immortalized human hepatocytes	Unlimited	Function poorly characterized	Excellent	Transmission of tumorigenic products	80,81
Human hepatoma cell line	Unlimited	Reduced level compared to primary cells	Excellent	Transmission of tumorigenic products	80,81
Primary xenogenic hepatocytes	Unlimited	Can express high level but generally decreases over time	Excellent	Transmission of zoonosis	74,82

Human hepatocytes appear to be the natural choice for liver tissue engineering although they are scare owing to a competing demand of liver transplantation. However, liver-specific functions of the daughter cells after replication of adult human hepatocytes *in vitro* is generally unstable although they are very biocompatible to host.

A new promising avenue is the discovery of liver stem cells. Recent results suggest that there are stem cells present within [75] as well as outside the liver [76], which can differentiate into fully mature hepatocytes. A major extrahepatic source is stem cells of the bone marrow which may take part in normal tissue renewal as well as in liver regeneration after severe induced hepatic injury [77]. Hepatic progenitor cells (oval or Ito cells) found within the liver has already begun to differentiate, but still have several options before becoming destined to a specific cell line [73]. Recently, the hepatic differentiation potential of commercially available human mesenchymal stem cells [78] and human umbilical cord blood-derived cells [79] were reported.

Kobayashi et al. adopted a reversible immortalization system with the SV40 T antigen and Cre-Lox recombination targeting human hepatocytes in their final differentiated state [80]. This system allowed them to establish a safe human hepatocyte line that is capable of differentiation al low cost and on a large scale although immortalized C3A cells have lower levels of P450A1 activity, ammonia removal, and amino acid metabolism [81] and there are concerns with the possibility of transmission of tumorigenic products into the patient [22].

Xenotransplantation utilized hepatocytes not of human origin, but that of animal. Porcine hepatocytes in particular have been the focus of many studies [73]. Porcine hepatocytes exhibit P-450 function, synthesis of urea, albumin, and other proteins, and are activated by growth factors as similar to human cells [82]. However, there are several disadvantages such as the risk of xenozoonosis [74] and immune rejection by the patient although the cells are easily obtained in large quantities and demonstrate the same qualities and therapeutic effects of human hepatocytes [73].

5. Hepatocellular Signaling Pathway

There have been many reports on hepatocellular signaling pathways because liver diseases are closely related with hepatocellular signaling pathways. Signaling pathways in liver health and diseases are not included in this review due to space limitation. Hepatocytes grow and function in the galactose carrying polymers as the biomimetic ECMs through specific interaction. The hepatocyte-ECM interaction regulates adhesion, proliferation, migration, differentiation, and survival of the hepatocytes. This part discusses current knowledge of potential mechanisms by which the ECM affects adhesion, differentiation and survival of hepatocytes.

5.1. Adhesion Signaling

Adhesive interaction between cells and their surrounding ECMs plays important roles in tissue morphogenesis, cell migration, proliferation, and the differentiation of a variety of cell types [83] although how intracellular cytoskeleton and signaling proteins connect and communicate with ECM remains a fundamental question [84]. Kim et al. compared phosphorylation of focal adhesion kinase (FAK) which is a key molecule involved in the formation of focal contact between PVLA as the synthetic galactose-carrying ECM and fibronectin as the natural one in the hepatocytes [85]. Phosphorylated FAK was not detected in hepatocytes cultured on PVLA, even at 12 h after seeding whereas it was detected in hepatocytes cultured on fibronectin as shown in figures 4 B and 4 C although the hepatocyte adhesion rate was almost the same for both matrices at 1 h post seeding as shown in figure 4 A, indicating that the down-regulation of FAK phosphorylation in hepatocytes cultured on PVLA does not come from difference of cell adhesion. Also, they reported that vinculin and phosphorylated FAK in hepatocytes adhered to PVLA were not localized at the points of focal adhesion whereas they were localized in hepatocytes adhered on collagen as shown in figure 5, suggesting that non-integrin ASGPR-mediated cell adhesion do not initiate integrin-mediated signaling.

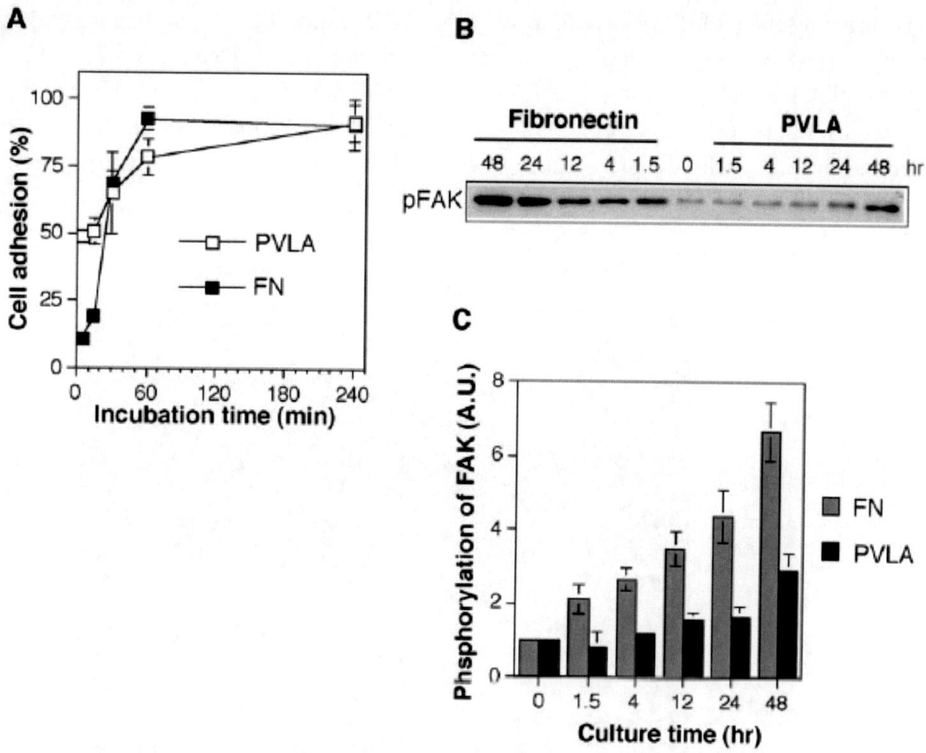

Figure 4. FAK is not phosphorylated in hepatocytes adhered onto a PVLA matrix. A: Kinetics of hepatocyte adhesion: hepatocytes were allowed to adhere for an indicated time to PS dishes coated with 100 μg/ml PVLA or 10 μg/ml FN. The percent of adhered cells was measured by protein assay. All values were normalized to the percentage of maximum adhesion, which was 100%. B: Western blot analysis of FAK phosphorylation: hepatocytes were allowed to adhere for an indicated time. C: The mean intensity of the band was quantified as described in Section 2; the values were determined from experiments performed separately at least three times. All values were normalized to the ratio of phosphorylated FAK to total FAK, which was 1. Source: From Ref. [85].

5.2. Epidermal Growth Factor (EGF) Signaling

During liver regeneration quiescent hepatocytes undergo one or two rounds of replication and they return to a nonproliferative state [86]. Growth factors regulate this process by providing both stimulatory and inhibitory signals for cell proliferation. Epidermal growth factor receptor (EGFR) mediates EGF signals [87]. It is highly expressed in hepatocytes, and EGF is potent mitogen for hepatocytes in culture [88]. EGF is produced by the Brunners glands of the duodenum, and made available for the liver by the portal circulation. Reduction of EGF plasma levels by sialoadenectomy has been shown to decrease the hepatic regenerative response [89]. Shortly after partial hepatectomy, the EGF plasma level is elevated two- and three-fold [90], suggesting that activation of EGFR is likely to be involved in G1 phase progression during liver regenerative growth [89]. Also, DNA synthesis is affected by EGF stimulation, as well as ECM, in hepatocytes [91]. Kim et al. examined EGF signaling of hepatocytes cultured on PVLA [92]. They suggested that hepatocytes cultured on PVLA expressed only tryptophan 2,3-oxygenase (TO) as a differentiated hepatocyte marker [93] in the presence of EGF whereas they lost TO expression on collagen within 48h regardless of EGF stimulation as shown in figure 6A. Also, LY 294002 as a phosphatidylinostitol (PI) 3-kinase specific inhibitor [94] significantly reduced TO expression as shown in figure 6B, suggesting that EGF-induced PI 3-kinase influences hepatocyte differentiation and cell division during diminished and increased integrin signaling, respectively.

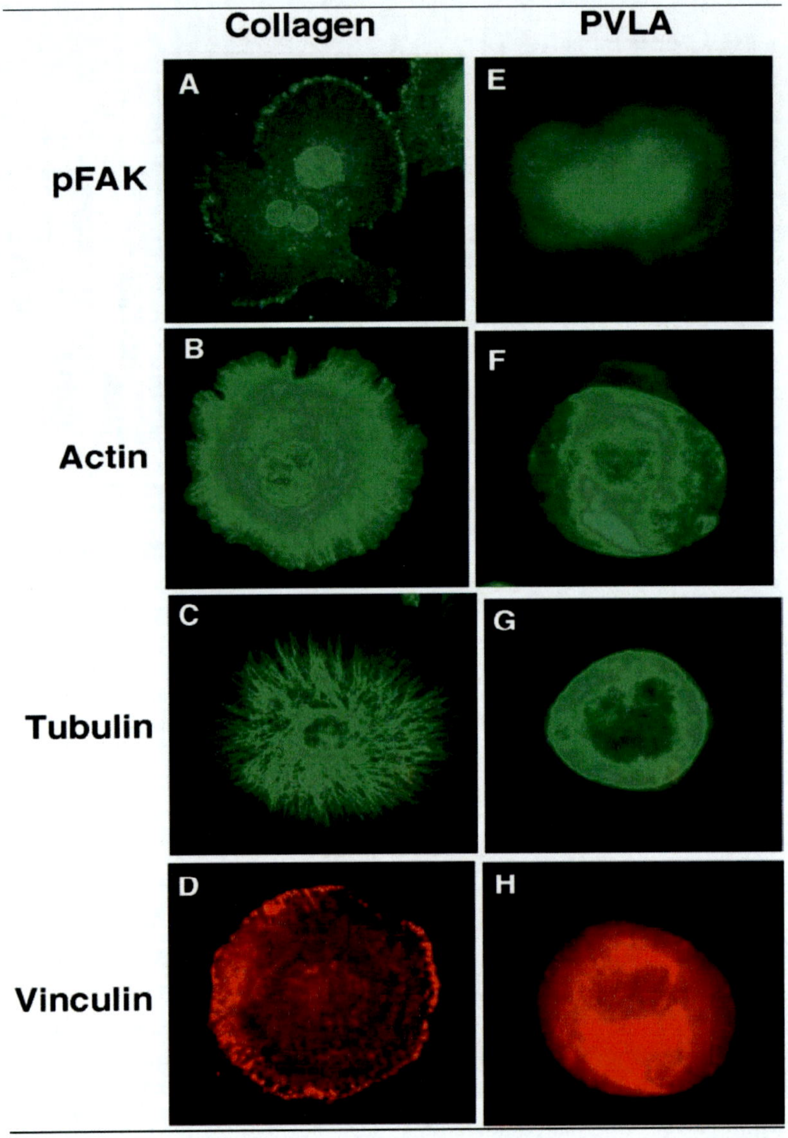

Figure 5. Focal adhesion and cytoskeleton proteins are not reorganized in hepatocytes adhered to the PVLA matrix. Hepatocytes were cultured for 24 h on cover glasses coated with 10 μg/ml type I collagen (A-D) or 100 μg/ml PVLA (E-H). The hepatocytes were stained for the detection of pFAK (A,E), F-actin (B,F), tubulin (C,G), and vinculin (D,H). Source: From Ref. [85].

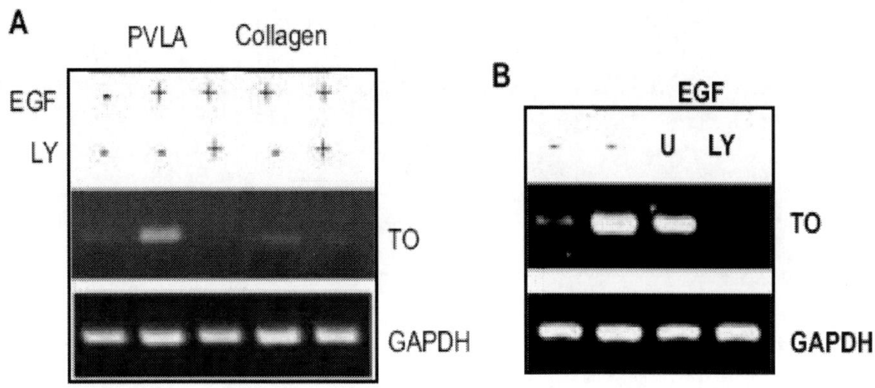

Figure 6. RT-PCR analysis of TO depending on EGF stimulation. Hepatocytes were cultured on collagen or PVLA for 4 h in WE base, and then media was replaced with new media containing 50 ng/mL EGF. For inhibitory experiments, the hepatocytes that had cultured for 4 h in WE base were treated with LY200094 (25 μM) or U0126 (20 μM) for 30min before EGF stimulation. Total RNA was extracted from hepatocytes 48 h after EGF stimulation. Source: From Ref. [94].

5.3. Survival Signaling

Hepatocytes require cell-ECM interaction for their survival [95] and detachment of the cells from ECMs induces cell death with the activation of Fas signaling [96]. Hoshiba et al. studied phosphorylated level of Akt as a survival signal molecule of hepatocytes on PVLA as the non-integrin-recognizable matrix and collagen or fibronectin as the integrin-recognizable matrix [97]. The results showed that the Akt was weakly phosphorylated in the hepatocytes cultured on PVLA or PLL matrix and agarose gel as a non-adhesive matrix whereas the Akt was highly phosphorylated in the hepatocytes cultured on type I collagen and fibronectin as shown in figures 7 A and 7 B, suggesting that Akt enhances hepatocyte survival in the adherent culture of hepatocytes on integrin-recognizable matrices but not non-integrin-recognizable ones. They also reported that hepatocyte viability decreased in the presence of wortmannin as a specific inhibitor of PI 3-kinase activation on type I collagen whereas it was not changed on PVLA and PLL matrices as shown in figure 7 C.

6. Summary

In this review, we have presented summary of the published works on design of biomimetic glycopolymers for liver tissue engineering. The galactose-carrying natural and synthetic polymers among the artificial ECMs can guide hepatocyte behaviors such as adhesion, morphology, proliferation, migration, differentiation and survival through receptor-mediated interaction between ASGPR of hepatocytes and galactose-carrying biomimetic glycopolymers. Also, galactose-density, topology of ECM, deposited natural ECM, coculture as well as cell source affected hepatocyte behaviors. Precise design of galactose-carrying biomimetic glycopolymers as synthetic ECMs will contribute to liver tissue engineering as a new form of bioartificial liver-assist devices or regeneration of liver-tissue substitutes.

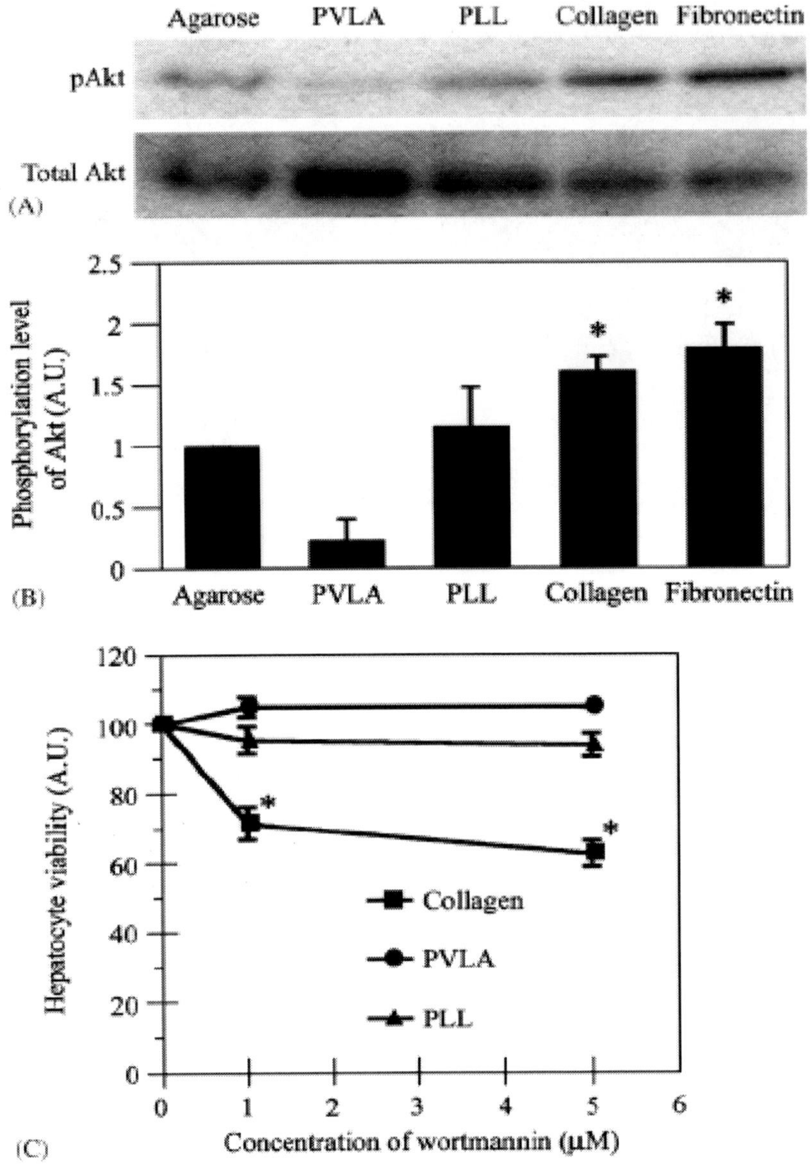

Figure 7. Activation of Akt rescues hepatocytes from anoikis in the adherent culture on integrin-recognizable matrices. (A) Western blot analysis of phosphorylated Akt after 24 h of culture. pAkt is phosphorylated Akt. Agarose and collagen are agarose gel and type I collagen, respectively. (B) The mean intensity of the band was quantified using Scion Image. Data shown are means±SE (n = 3, except PLL). *p<0.01 vs agarose gel. (C) Effect of wortmannin on hepatocyte viability after 24 h. After 3 h of incubation, hepatocytes were treated with wortmannin at indicated concentrations. Hepatocyte viability was measured by MTT assay. Black circle: PVLA, black square: type I collagen, and black triangle: PLL. Data shown are means±SE (n = 3). *p<0.01 vs. the viability in the absence of inhibitor. Source: From Ref. [97].

REFERENCES

[1] Drotleff S., Lungwitz U., Breunig M., Dennis A., Blunk T., Tessmar J., Göpferich A. (2004). Biomimetic polymers in pharmaceutical and biomedical sciences. *Eur. J. Pharm. Biopharm,* 58, 385-407.

[2] Cima L.G. (1994). Polymer sustrates for controlled biological interactions. *J. Cell Biochem,* 56, 155-161.

[3] Blunk T., Göpferich A., Tessmar J. (2003). Special issue biomimetic polymers. *Biomaterials,* 24, 4335-4335.

[4] Tessmar J., Mikos A., Göpferich A. (2003). The use of poly(ethylene glycol)-block-poly(lactic acid) derived copolymers for the rapid creation of biomimetic surfaces. *Biometerials,* 24, 4475-4486.

[5] Ruoslahti E., Pierschbacher M. D. (1987). New perspectives in cell adhesion: RGD and integrins. *Science,* 238, 491-497.

[6] Mann B. K., Schmedlen R. H., West J. L. (2001). Tethered-TGF-[beta] increases extracellular matrix production of vascular smooth muscle cells. *Biomaterials,* 22, 439-444.

[7] Kuhl P. R., Griffith-Cima L. G. (1996). Tethered epidermal growth factor as a paradigm for growth factor-induced stimulation from the solid phase. *Nature Med,* 2, 1022-1027.

[8] Vacanti J. P., Langer R. (1999). Tissue engineering: the design and fabrication of living replacement devices for surgical reconstruction and transplantation. *Lancet,* 354 (suppl I), 32-34.

[9] Shin H., Jo S., Mikos A. G. (2003). Biomimetic materials for tissue engineering. *Biomaterials,* 24, 4353-4364.

[10] Langer R., Vacanti J. P. (1993). Tissue engineering. *Science,* 260, 920-926.

[11] Lee K. Y., Mooney D. J. (2001). Hydrogels for tissue engineering. *Chem. Rev,* 101, 1869-1879.

[12] Putnam A. J., Mooney D. J. (1996). Tissue engineering using synthetic extracellular matrices. *Nature Med,* 2, 824-826.

[13] Hubbell J. A. (1999). Bioactive biomaterials. *Curr. Opin. Biotechnol,* 10, 123-129.

[14] Zakim D., Boyer T. D., eds. *Hepatology. A Textbook of Liver Disease* (Vol. 1). Philadelphia: W. B. Saunders Col, 1990.

[15] Davis M. W., Vacanti J. P. (1996). Toward development of an implantable tissue engineered liver. *Biomaterials,* 17, 365-372.

[16] Pricer W. E., Ashwell G. (1971). The binding of desialylated glycoproteins by plasma membranes of rat liver, *J. Biol. Chem,* 246, 4825-4833.

[17] Neufeld E., Ashewell G. (1979). Carbohydrate recognizing systems for receptor-mediated pinocytosis. In *Biochemistry of Glycoprotein and Proteoglycans.* W. Lennarz, ed. Plenum Press, New York, 241-266.

[18] Wall D. A., Wilson G., Hubbard A. L. (1980). The galactose-specific recognition system of mammalian liver: the route of ligand internalization in rat hepatocytes, *Cell,* 21, 79-93.

[19] Ashwell G., Harford J. (1982). Carbohydrate-specific receptors of the liver. *Annu. Rev. Biochem,* 51, 531-544.

[20] Drickamer K., Mamon J. F., Binns G., Leung J. O. (1984). Primary structure of the rat

liver asialoglycoproten receptor. *J. Biol. Chem,* 259, 770-778.

[21] Haynes P. A., Oka J.A., Weigel P. H. (1994). The rat hepatic lectin 1 subunit of the rat asialoglycoprotein receptor is a phosphoprotein and contains phosphotyrosine. *J. Biol. Chem,* 269, 33146-33151.

[22] Chan C., Berthiaume F., Nath B. D., Tilles A. W., Toner M., Yarmuch M. L. (2004). Hepatic tissue engineering for adjunct and temporary liver support critical technologies. *Liver Transpl,* 10, 1331-1342.

[23] Hansen L. K., Wilhelm J., Fassett J. T. (2006). Regulation of hepatocyte cell cycle progression and differentiation by type I collagen structure. *Curr. Top Dev. Biol,* 72, 205-236.

[24] Cho C. S., Seo S. J., Park I. K., Kim S.H., Kim T. H., Hoshiba T., Harada I., Akaike T. (2006). Galactose-carrying polymers as extracellular matrices for liver tissue engineering. *Biomaterials,* 27, 576-585.

[25] Rosso F., Marino G., Giordano A., Barbarishi M., Parmeggiani D., Barbarisi A. (2005). Smart materials as scaffolds for tissue engineering. *J. Cell Physiol,* 203, 465-470.

[26] Drury J. L., Mooney D. J. (2003). Hydrogels for tissue engineering: scaffold design variables and applications. *Biomaterials,* 24, 4337-4351.

[27] Yang J., Goto M., Ise H., Cho C. S., Akaike T. (2002). Galactosylated alginate as a scaffold for hepatocytes entrapment. *Biomaterials,* 23, 471-479.

[28] Park I. K., Yang J., Jeong H. J., Bom H. S., Harada I., Akaike T., Kim S. I., Cho C. S. (2003). Galactosylated chitosan as a synthetic extracellular matrix for hepatocytes attachment. *Biomaterials,* 24, 2331-2337.

[29] Yang J., Chung T. W., Nagaoka M., Goto M., Cho C. S. Akaike T. (2001). Hepatocyte-specific porous polymer-scaffolds of alginate/galactosylated chitosan sponge for liver-tissue engineering. *Biotech. Lett,* 23, 1385-1389.

[30] Chung T. W., Yang J., Akaike T., Cho K. Y., Nah J. W., Kim S. I., Cho C. S. (2002). Preparation of alginate/galactosylated chitosan scaffold for hepatocyte attachment. *Biomaterials,* 23, 2827-2834.

[31] Guo X. L., Yang K. S., Hyun J. Y., Kim W. S. Lee D. H., Min K. E., Park L. S., Seo K. H., Kim Y. I., Cho C. S., Kang I. K. (2003). Moropology and metabolism of Ba-alginate-encapsulated hepatocytes with galactosylated chitosan and poly(vinyl alcohol) as extracellular matrices. *J. Biomater. Sci. Polym. Ed,* 14, 551-565.

[32] Yamatoya K., Shirakawa M. (2003). Xyloglucan: structure, rheological properties, biological functions and enzymatic modification. *Curr. Trends Polym. Sci,* 8, 27-71.

[33] Seo S. J., Park I. K., Yoo M. K., Shirakawa M., Akaike T., Cho C. S. (2004). Xyloglucan as a synthetic extracellular matrix for hepatocyte attachment. *J. Biomater. Sci. Polymer Edn,* 15, 1375-1387.

[34] Seo S. J., Akaike T., Choi Y. J., Shirakawa M., Kang I. K. Cho C. S. (2005). Alginate microcapsules prepared with xyloglucan as a synthetic extracellular matrix for hepatocyte attachment. *Biomaterials,* 26, 3607-3615.

[35] Neumann P. M., Zur B., Ehrenreich Y. (1981). Gelatin-based sprayable foam as skin substitute. *J. Biomed. Mater Res,* 15, 9-18.

[36] Hong S. R., Lee Y. M., Akaike T. (2003). Evaluation of a galactose-carrying gelatin sponge for hepatocytes culture and transplantation. *J. Biomed. Mater Res. A,* 67, 733-741.

[37] Zhou C., Confalonieri F., Medina N., Zivanovic Y., Esnault C., Yang T., Jacquet M.,

Janin J., Duguet M., Perasso R., Li Z. (2000). Fine organization of *Bombyx mori* fibroin heavy chain gene. *Nucl. Acids Res,* 28, 2413-2419.

[38] Gotoh Y., Niimi S., Hayakawa T., Miyashita T. (2004). Preparation of lactose-silk fibroin conjugates and their application as a scaffold for hepatocyte attachment. *Biomaterials,* 25, 1131-1140.

[39] Weigel P. H., Schmell E., Lee Y. C., Roseman S. (1978). Specific adhesion of rat hepatocytes to beta-galactosides linked to polyacrylamide gels, *J. Biol. Chem.* 253, 330-333.

[40] Weigel P. H. (1980). Rat hepatocytes bind to synthetic galactoside surfaces via a patch of asialoglycoprotein receptors. *J. Cell Biol,* 87, 855-861.

[41] Kobayashi A., Kobayashi K., Akaike T., (1992). Control of adhesion and detachment of parenchymal liver cells using lactose-carrying polystyrene as substratum. *J. Biomater. Sci. Polym. Ed,* 3, 499-508.

[42] Tobe S., Takei Y., Kobayashi K., Akaike T. (1992). Receptor-mediated formation of multilayer aggregates of primary cultured adult rat hepatocytes on lactose-substituted polystyrene. *Biochem. Biophys. Res. Commun,* 184, 225-230.

[43] Kobayashi A., Goto M., Kobayashi K., Akaike T.(1994). Receptor-mediated regulation of differentiation and proliferation of hepatocytes by synthetic polymer model of asialoglycoprotein. *J. Biomat. Sci. Polym. Ed,* 6,325-342.

[44] Kobayashi K., Kobayashi A., Akaike T. Culturing hepatocytes on lactose-carrying polystyrene layer via asialoglycoprotein receptor-mediated interactions. In Lee Y. C., Lee R. T., Editors. *Methods in Enzymology: Neoglycoconjugates part B Biomedical Applications,* Vol. 247, San Diego, California: Academic Press: 1994. p 409-418.

[45] Takei R., Seo S. J., Cho C.S., Okahata Y., Akaike T. (2005). Adsorption behaviors of poly[*N*-P-vinylbenzyl-4-O-β-D-galactopyranosyl-(1→4)-D-gluconamide] by quarz-crystal microbalance. *Colloids Surf. B Biointerfaces,* 42,137-140.

[46] Lopina S. T., Wu G., Merrill E. W., Griffith-Cima L. (1996). Hepatocyte culture on carbohydrate-modified star polyethylene oxide hydrogels. *Biomaterials,* 17, 559-569.

[47] Donati I., Gamini A., Vetere A., Campa C., Paoletti S. (2002). Synthesis, characterization, and preliminary biological study of glycoconjugates of poly (styrene-co-maleic acid). *Biomacromolecules,* 3, 805-812.

[48] Kang I. K., Moon J. S., Jeon H. M., Meng W., Kim Y. I., Kim S. (2005). Morphology and metabolism of Ba-alginate encapsulated hepatocytes with galactosylated poly (allyl amine) and poly (vinyl alcohol) as extracellular matrices. *J. Mater Sci. Mater Med,* 16, 533-539.

[49] Yoon J. J., Nam Y. S., Kim J. H., Park T. G. (2002). Surface immobilization of galactose onto aliphatic biodegradable polymers for hepatocyte culture. *Biotechnol. Bioeng,* 78,1-10.

[50] Ying L., Y C., Zhou R. X., Leong K. W., Mao H. Q., Kang E. T., Neoh K. G. (2003). Immobilization of galactose ligands on acrylic acid-graft-copolymerized poly(ethylene terephthalate) film and its application to hepatocyte culture. *Biomacromolecules,* 4, 157-165.

[51] Kang I. K., Lee D. W., Lee S. K. Akaike T. (2003). Grafting of lactose-carrying styrene onto polystyrene dishes using plasma glow discharge and their interaction with hepatocytes. *J. Mater Sci. Mater Med,* 14, 611-616.

[52] Yin C., Liao K., Mao H. Q, Leong K. W., Zhuo R. X., Chan V. (2003). Adhesion

contact dynamics of HepG2 cells on galactose-immobilized substrates. *Biomaterials,* 24, 837-850.

[53] Baenziger J. U. Fiete D. (1980). Galactose and N-acetylgalactosamine-specific endocytosis of glycopeptides by isolated rat hepatocytes. *Cell,* 22, 611-620.

[54] Akaike T., Kobayashi A., Kobayashi K. New functionality materials. In: Tsuruta T., Doyama M., Seno M. Eds. *Synthesis and Function Control of Biofuntionality,* Vol. B, Elsevier Science Pub. B. V., London, 1993, pp173.

[55] Kim S.H., Kim J. H., Akaike T. (2003). Regulation of cell adhesion signaling by synthetic glycopolymer matrix in primary cultured hepatocytes. *FEBS Lett,* 553, 433-439.

[56] Ise H., Sugihara N., Negishi N., Nikaido T., Akaike T. (2001). Low asialogly- coprotein receptor expression as markers for highly proliferative potential hepatocytes. *Biochem. Biophys. Res. Commun,* 285, 172-182.

[57] Yin C., Ying L., Zhang P. C., Zhuo R. X., Kang E. T., Leong K. W. Mao H. Q. (2003). High density of immobilized galactose ligand enhances hepatocyte attachment and function. *J. Biomed. Mater Res. A,* 67, 1093-1104.

[58] Lu H. F., Lim W. S., Wang J., Tang Z. Q., Zhang P. C., Leong K. W., Chia S. M., Yu H., Mao H. Q. (2003). Galactosylated PVDF membrane promotes hepatocyte attachment and functional maintenance. *Biomaterials,* 4893-4903.

[59] Berthiaume F., Moghe P. V., Toner M., Yarmush M. L. (1996). Effect of extracellular matrix topology on cell structure, function, and physiological responsiveness: hepatocytes cultured in a sandwich configuration. *FASEB J,* 10, 1471-1484.

[60] Hansen L. K., Hsiao C. C., Friend J. R., Wu F. J., Bridge G. A., Remmel R. P., Cerra F. B., Hu W. S. (1998). Enhanced morphology and function in hepatocytes spheroids: a model of tissue self-assembly. *Tissue Eng,* 4, 65-74.

[61] Kato S., Kidoaki S., Matsuda T. (2004). Substrate-dependent cellular behavior of Swiss 3T3 fibroblasts and activation of Rho family during adhesion and spreading processes. *J. Biomed. Mater. Res. A,* 68, 314-324.

[62] Hoshiba T., Cho C. S., Murakawa A., Okahata Y., Akaike T. (2006). The effect of natural extracellular matrix deposited on synthetic polymers on cultured primary hepatocytes. *Biomaterials,* 27, 4519-4528.

[63] Bhatia S. N., Balis U. J., Yamush M. L., Toner M. (1999). Effect of cell-cell interactions in preservation of cellular phenotype: cocultivation of hepatocytes and nonparenchymal cells. *FASEB J,* 1883-1900.

[64] Seo S. J., Kim I. Y., Choi Y. J., Akaike T., Cho C. S. (2006). Enhanced liver functions of hepatocytes cocultured with NIH 3T3 in the alginate / galactosylated chitosan scaffold. *Biomaterials,* 27, 1487-1495.

[65] Kuri-Harcuch W., Mendoza-Figueroa T. (1989). Cultivation of adult rat hepatocytes on 3T3 cells: expression of various liver differentiated functions. *Differentiation,* 41, 148-157.

[66] Conner J., Vallet-Collom I., Daveau M., Delers F., Hiron M., Lebreton J. P., Guillouzo A. (1990). Acute-phase-response induction in rat hepatocytes co-cultured with rat liver epithelial cells. *Biochem. J,* 266, 683-688.

[67] Morin O., Normand C. (1986). Long-term maintenance of hepatocyte functional activity in co-culture: requirements for sinusoidal endothelial cells and dexamethasone. *J. Cell Physiol,* 129, 103-110.

[68] Ricalton-Banks L., Liew C., Bhandari R., Fry J., Shakesheff K. (2003). Long-term culture of functional liver tissue: three-dimensional coculture of primary hepatocytes and stellate cells. *Tissue Eng,* 401-410.

[69] Takeda M., Yamamoto M., Isoda K., Higashiyama S., Hirose M., Ohgushi H. Kawase M., Yagi K. (2005). Availability of bone marrow stromal cells in three-dimensional coculture with hepatocytes and transplantation into liver-damaged mice. *J. Biosci. Bioeng,* 100, 77-81.

[70] Zinchenko Y. S., Schrum L. W., Clemens M., Coger R. N. (2006). Hepatocyte and Kupffer cells co-cultured on micropatterned surfaces to optimize hepatocyte function. *Tissue Eng,* 12, 751-761.

[71] van Poll D., Sokmensuer C., Ahmad N., Tilles A. W., Berthiaume F., Toner M., Yarmush M. L. (2006). Elevated hepatocyte-specific functions in fetal rat hepatocytes co-cultured with adult rat hepatocytes. *Tissue Eng,* 12, 2965-2973.

[72] Lee K. W., Lee S. K., Joh J. W., Kim S. J., Lee B. B., Kim K. W., Lee K. U. (2004). Influence of pancreatic islets on spheroid formation and functions of hepatocytes in hepatocyte-pancreatic islet spheroid culture. *Tissue Eng,* 10, 965-977.

[73] Kulig K. M., Vacanti J. P. (2004). Hepatic tissue engineering. *Transpl. Immunol,* 12,303-310.

[74] Allen J. W., Bhatia S. N. (2002). Engineering liver therapies for the future. *Tissue Eng,* 8, 725-737.

[75] Thorgeirsson S. S. (1996). Hepatic stem cells in liver regeneration. *FASEB J,* 10,1249-1256.

[76] Theise N. D., Nimmakayalu M., Gardner R., Illei P. B., Morgan G., Teperman L., Henegariu O., Krause D. S. (2000). Liver from bone marrow in humans. *Hepatology,* 32,11-16.

[77] Petersen B. E., Bowen W. C., Patrene K. D., Mars W. M., Sullivan A. K., Murase N., Boqqs S. S., Greenberger J. S., Goff J. P. (1999). Bone marrow as a potential source of hepatic oval cells. *Science,* 284, 1168-1170.

[78] Ong S. Y., Dai H., Leong K. W. (2006). Hepatic differentiation potential of commercially available human mesenchymal stem cells. *Tissue Eng,* 12, 3477-3485.

[79] Sun Y., Xiao D., Zhang R. S., Cui G. H., Wang X. H. Chen X. G. (2007). Formation of human hepatocyte-like cell with different cellular phenotypes by human umbilical cord blood-derived cells in the human-rat chimeras. *Biochem. Biophys. Res. Commun,* 357, 1160-1165.

[80] Kobayashi N., Okitsu T., Nakaji S., Tanaka N. (2003). Hybrid bioartificial liver : establishing a reversibly immortalized human hepatocyte line and developing a bioartificial liver for practical use. *J. Artif. Organs.* 6, 236-244.

[81] Wang L., Sun J., Li L., Mears D., Horvat M., Sheil A. G. (1998). Comparison of porcine hepatocytes with human hepatoma (C3A) cells for use in a bioartificial liver support system. *Cell Tranpl,* 7, 459-468.

[82] Lorenti A., Barbich M., Hidalgo A., Hyon S. H., Sorroche P., Cuinle A., Schenone A., Chamoles N., Argibay P. (2001). Culture of porcine hepatocytes: the dogma of exogeneous matrix revisted. *Artif Organs,* 25, 546-550.

[83] McDonald J. A. (1998) Receptors for extracellular matrix components. *Curr. Opin. Cell. Biol,* 1, 985-999.

[84] Kimata T., Nagaki M., Ogiso T., Naiki T., Kato T., Moriwaki H. (2006). *Hepatology,*

44, 140-151.

[85] Kim S. H., Kim J. H., Akaike T. (2003). Regulation of cell adhesion signaling by synthetic glycopolymer matrix in primary cultured hepatocyte. *FEBS Lett,* 553, 433-439.

[86] Fausto N., Laird A. D., Webber E. M. (1995). Role of growth factors and cytokines in hepatic regeneration. *FASEB J.* 9, 1527-1536.

[87] Carpenter G., Cohen S. (1990). Epidermal growth factor. *J. Biol. Chem,* 265, 7709-7712.

[88] Huitfeldt H. S., Skarpen E., Lindeman B., Becher R., Thrane E. V., Schwarze P. E. (1996). Differential distribution of Met and epidermal growth factor receptor in normal and carcinogen-treated rat liver. *J. Histochem. Cytochem,* 44, 227-233.

[89] Skarpen E., Oksvold M. P., Grosvik H., Widnes C., Huitfeldt H.S. (2005). *J. Cell Physiol,* 202, 707-716.

[90] Noguchi S., Ohba Y., Oka T. (1991). Influence of epidermal growth factor on liver regeneration after partial hepatectomy in mice. *J. Endocrinol,* 128, 425-431.

[91] Alisi A., Spagnuolo S., Leoni S. (2003). Treatment with EGF increases the length of S-phase after partial hepatectomy in rat, changing the activities of cdks. *Cell Physiol. Biochem.* 13,239-248.

[92] Kim S. H., Akaike T. (2007). Epidermal growth factor signaling for matrix-dependent cell proliferation and differentiation in primary cultured hepatocytes. *Tissue Eng,* 13, 601-609.

[93] Inoue D., Yamamoto H., Nakamura T., Ichihara A., Okamoto H. (1989). Nicotinamide prolongs survival of primary cultured hepatocytes without involving loss of hepatocyte-specific functions. *J. Biol. Chem,* 264, 4747-4750.

[94] Sharma G. D., Ottino P., Bazan N. G., Bazan H. E. (2005). Epidermal and hepatocyte growth factors, but not keratinocyte growth factor, modulate protein kinase Calpha translocation to the plasma membrane through 15(S)-hydroxyeicosatetraenoic acid synthesis. *J. Biol. Chem,* 280, 7919-7924.

[95] Smets F. N., Chen Y., Wang L. J., Soriano H. E. (2002). Loss of cell anchorage triggers apoptosis (anoikis) in primary mouse hepatocytes. *Mol. Genet. Metab,* 75, 344-352.

[96] Rosen K., Shi W., Calabretta B., Filmus J. (2002). Cell detachment triggers p38 mitogen-activated protein kinase-dependent overexpression of Fas ligand. A novel mechanism of Anoikis of intestinal epithelial cells. *J. Biol. Chem,* 277, 46123-46130.

[97] Hoshiba T., Nagahara H., Cho C. S., Tagawa Y., Akaike T. (2007). Primary hepatocyte survival on non-integrin-recognizable matrices without the activation of Akt signaling. *Biomaterials,* 28, 1093-1104.

In: Tissue Engineering: Roles, Materials and Applications ISBN: 978-1-60456-293-4
Editors: S. J. Barnes, L. P. Harris, pp. 119-140

Chapter 6

CRUCIAL ROLE OF SCAFFOLD IN TISSUE ENGINEERING

Yoshito Ikada
Faculty of Medicine; Nara Medical University
Shijo-cho 840, Kashihara-shi, Nara634-8521, Japan

ABSTRACT

Almost three decades have passed since tissue engineering started for the repair of diseased skin tissues of patients. Numerous studies have focused on the cells that are required for tissue engineering. Indeed, it is impossible to regenerate a tissue without cells. It should be, however, stressed that scaffold also plays a key role in tissue engineering, especially when tissue engineering aims at regenerating large, three-dimensional tissues. In this case, not only the cells in the tissue, but also the associated extracellular matrix has been diseased or lost. An important function of the scaffold is to temporarily substitute the lost extracellular matrix that had assisted the proliferation, differentiation, and functions of the cells that had presented in the pre-diseased tissue.

The purpose of this review article is to demonstrate how critical a scaffold is for the success of preclinical and clinical applications of tissue engineering. The resorbable biomaterial that has been most often selected in the tissue engineering studies described in this article is an L-lactide(LA)-ε-caprolactone(CL) copolymer with a molar LA/CL ratio of 50/50. This copolymer exhibits a bioabsorption kinetics ranging between that of poly(glycolide) (PGA) and that of poly(L-lactide) (PLLA). If a porous scaffold fabricated from this copolymer is to be used for *in situ* regeneration of large tissues, the scaffold should be reinforced with PGA or PLLA fabrics. The reinforced scaffold has proven very suitable, in particular, for the *in situ* cardiovascular tissue engineering using bone marrow cells.

Keyword: tissue engineering, scaffold, lactide-ε-caprolactone copolymer, P(LA/CL), bone marrow cells, *in situ* tissue engineering, cardiovascular tissues, and alveolar bone.

I. INTRODUCTION

It was in 1993 that Langer and Vacanti presented the principle and method of tissue engineering[1]. Since that time, tremendous amounts of investigations have been performed by scientists and engineers associated with biomaterials, cell and gene technologies, and clinical medicine as well as in biomedical industries[2]. As a result the tissue engineering has gained popularity in the biomedical area. However, few tissues created by the assist of tissue engineering have been used for the repair of damaged, lost, or missing tissues of patients in clinical settings, despite the high enthusiasm in the tissue engineering research. The tissue engineering-based tissues that have entered relatively widespread clinical use include only skin (both epidermal and dermal tissues), periodontal tissues, bone in the field of oral surgery and orthopedics, and articular cartilage.

A number of reasons may be mentioned to explain why clinical applications of tissue engineering have been still in its infancy although almost two decades have passed since the advent of tissue engineering paradigm. Cells, scaffolds, and biomolecules have been considered as the basic ingredients for the success of tissue development. Each of them should meet the minimum requirements so as to effectively serve in clinical settings. The source of cells is the most crucial for the successful tissue engineering applications, but it is not yet in general agreement which tissue provides the most cost-effective progenitor or stem cells among bone marrow, fat, umbilical cord blood, muscle, and embryo. Probably, it depends on the tissue to be regenerated. A concern on malignant tissue formation remains in the use of stem cells. Scaffolds that provide a temporary, structural support for the cells to accommodate and guide their growth in the three-dimensional space into a specific tissue should be biocompatible, bioabsorbable, porous, high enough in mechanical strength, and compatible with existing clinical modalities, but non-bioabsorbable polymers such as poly(ε -caprolactone) have been still studied as a candidate of scaffolds that should be bioabsorbed, and most of the scaffolds manufactured by electrospinning have too small pore size for cells to infiltrate and ingrow. Biomolecules, especially growth factors such as BMPs, bFGF, and VEGF are still expensive and often difficult to obtain for scientists in academy. Any widely useful technologies available for the sustained, local release of these biomacromolecules have not yet been established.

In addition to these technical and scientific issues, tissue engineering involves ethical, legal, and industrial problems. As the cells to be used for clinical tissue engineering are of human origin, it is conceivable that ethical issues may come up in regard to the cell source, in addition to safety issues such as pathogen transfer. Regulations on tissue-engineered products must be stricter than those on non-viable artificial organs and biomaterials. Recently, the number of studies undertaken using large animal models has greatly decreased despite of their necessity as the preclinical study. This may be not only because of strict regulation and high cost, but also because of the lack of proper scaffolds. Regulatory authority will not give approval to any tissue engineering-based tissues, unless they have proven to be more effective for patient treatments than any existing therapeutic means. To provide clear evidence for higher effectiveness in patient treatments is not an easy task at all. It is apparent that medical industry hesitates to bring tissue-engineered products into business, if the potential market capacity for them is too small to bring a profit to the industry. Despite R and D investment exceeding \$ 4.5 billion, no tissue-engineered products have yet been commercially successful and the capital value of publicly traded tissue engineering corporations continues to decrease[3]. It is further mentioned that tissue-engineered products tend to be of variable quality and are more expensive than those by conventional, existing therapies[4].

This review article focuses on the scaffold among several factors influencing tissue engineering applications, since the author has been long engaged in the usage of scaffolds in clinical practice. The bioabsorbable biomaterial that has most often been employed for the fabrication of scaffolds used in the studies described below is poly(L-lactide- ε -caprolactone)(P(LA/CL)) with an LA/CL molar ratio of 50/50. This copolymer exhibits a bioabsorption rate ranging between poly(glycolide)(PGA) and poly(L-lactide)(PLLA), as presented in figure 1. This copolymer is a flexible and crystalline polymer that is capable of yielding not only tough plastics but also strong fibers. A sponge-like scaffold can be manufactured upon freeze-drying of dioxane solution of P(LA/CL) with different pore sizes and porosities depending on the solution concentration. This sponge is applicable as a scaffold for small animal models without any modification, but needs reinforcement with absorbable fibers when this P(LA/CL) sponge is to be used as a scaffold for large animal models or humans, because in these cases the scaffold must be firmly fixed with suture to the adjacent host tissue. This scaffold degrades by hydrosis in PBS without calcium, loses half of its tensile strength in 3 weeks, and completely by 12 weeks. In the studies described below, the sponge-like P(LA/CL) scaffold without or with fiber reinforcement will mostly be implanted in the form of sheet or tube immediately after attachment of seeded cells on the scaffold. This strategy is called *in situ* (or *in vivo*) tissue engineering that relies on the body's innate ability to heal without bioreactor and media for cell culture, in marked contrast with *ex vivo* (or *in vitro)* tissue engineering.

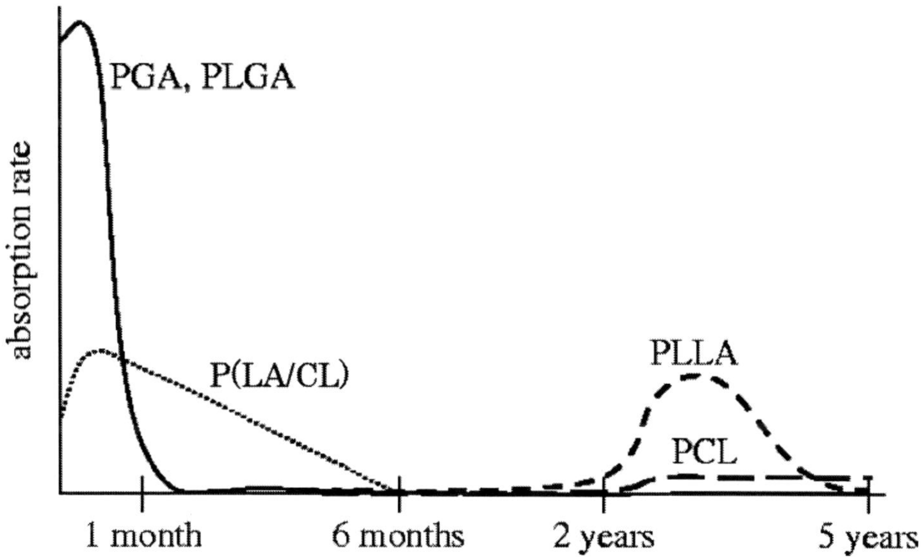

Figure 1. Schematic for the absorption rate of bioabsorbable polyesters. PGA, polyglycolide; PLGA, lactide-glycolide copolymer; P(LA/CA), lactide –ε-caprolactone copolymer; PLLA, poly-L-lactide; and PCL, poly-ε-caprolactone.

II. Examples of Preclinical and Clinical Trials of Tissue Engineering

1. Myocardial Tissue

Children with truncus arteriosus require a conduit from the right ventricle to the pulmonary artery and patients with tetralogy of Fallot may require a transannular patch. Currently, the conduits or patches are usually created from Dacron (polyethylene terephthalate), Gore-Tex (polytetrafluoroethylene)(PTFE), glutaraldehyde-treated bovine pericardium, or glutaraldehyde-treated homografts. These materials are not viable, do not grow with the children, and do not provide pulsatile pulmonary flow.

The optimal cardiac graft for the repair of congenital heart defects will be composed of autologous cells and will grow with the child. Ozawa *et al.* compared a P(LA/CL) sponge reinforced with knitted PLLA fabric (PCLA) to a gelatin sheet (GEL) and a non-woven PGA fabric (PGA)[5]. *In vitro* studies showed that the DNA content increased significantly in all patches between 1 and 3 weeks after seeding with smooth muscle cells (SMCs). Histology and staining SMCs for anti-α-smooth muscle actin (αSMA) revealed better growth of cells in the interstices of the grafts with GEL and PCLA than the PGA scaffold. *In vivo* studies demonstrated that seeded SMCs survived at least 8 weeks after the patch implantation in all groups. As shown in figure 2, PCLA scaffolds were replaced by more cells with larger αSMA-positive areas and by more extracellular matrix with larger elastin-positive areas than with GEL and PGA. As shown in figure 3, the PCLA patch did not thin and expand significantly, while the GEL and PGA patches thinned and expanded. All grafts had complete endothelialization on the endocardial surface.

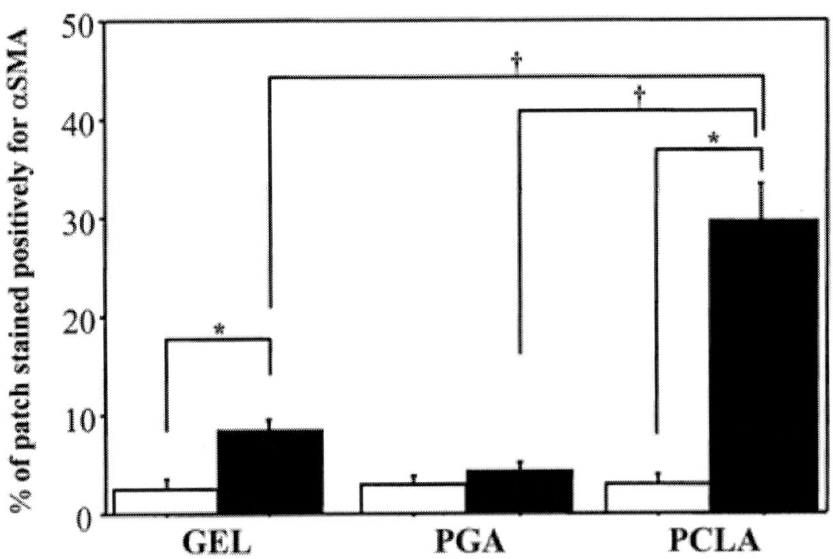

Figure 2. Percentage of the αSMA positive areas in the cell-seeded (■) and non-seeded (□) GEL, PGA, and PCLA scaffolds being used to repair the right ventricular outflow tract for 8 weeks. *$p<0.01$, †$p<0.001$ among the 3 cell-seeded groups. $n=5$, 5 for each biomaterial.

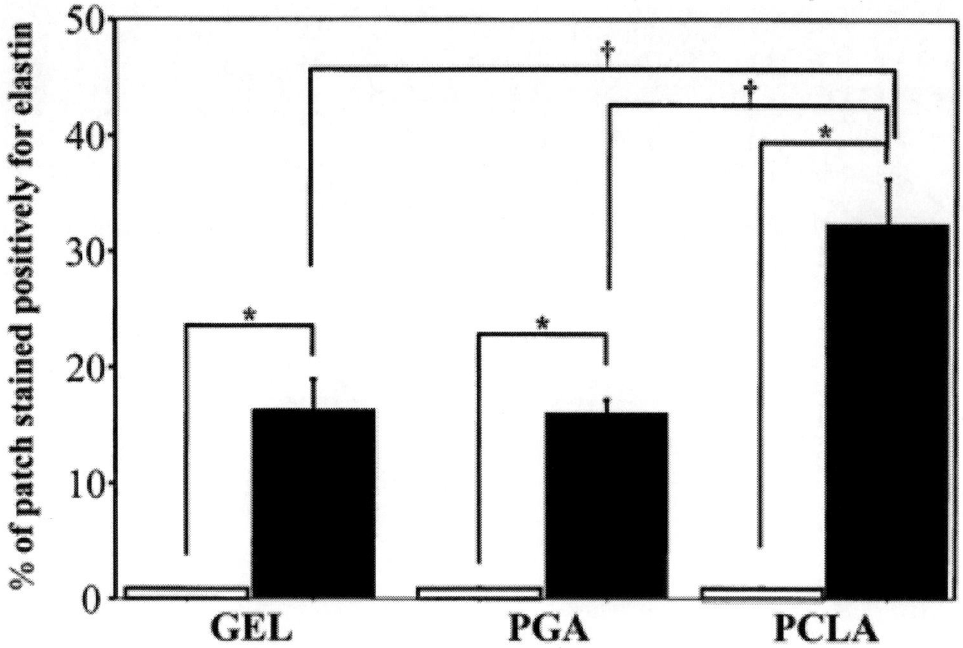

Figure 3. Percentage of the elastin-positive area in the patch section of cell-seeded (■) and non-seeded (□) GEL, PGA, and PCLA after being used to repair the right ventricular outflow tract for 8 weeks. *p<0.01,†p<0.001 among the 3 cell-seeded groups.

Since the PCLA scaffold had an important property that the seeded cells have sufficient time to form a tissue in the myocardial milieu and for new blood vessels to perfuse the tissue before the supporting outer material was completely biodegraded, Ozawa *et al.* further studied PCLA with SMC, because the cells were easily cultured, prevented ventricular remodeling in a myocardial infarct model, induced angiogenesis[6], formed an elastin-rich extracellular matrix[7,8], and were unlikely to cause a myocardial arrhythmia[9]. In this 5-month study they reported the morphological and histological changes of an implanted, cultured autologous SMC-seeded PCLA patch used to repair a surgically created, right ventricular outflow tract (RVOT) defect in an adult rat. The SMCs seeded in the patch were in a synthetic phenotype prior to implantation. At 8 weeks after implantation the spongy component was absorbed and replaced by cells and extracellular matrices. The number of capillaries in the grafts also increased (p < 0.001) between 8 and 22 weeks (figure 4). The fibrous PLLA component of the PCLA scaffold remained and the 22-week grafts maintained their thickness and surface area in the RVOT. The SMCs prior to implantation were in a synthetic phenotype and developed *in vivo* into a more contractile phenotype. By 8 weeks the patches were endothelialized on their endocardial surface. After 22 weeks of implantation, the phenotype of the SMCs in the scaffold changed slowly from a synthetic to a more contractile phenotype. Elastin fibers were colocalized with the SMCs and the elastin sites were included within the SMC distribution. Collagen was observed throughout the patches and no inflammatory reaction indicative of immunorejection was detected. At 22 weeks after implantation the outer fibrous components of PLLA of the PCLA patch remained.

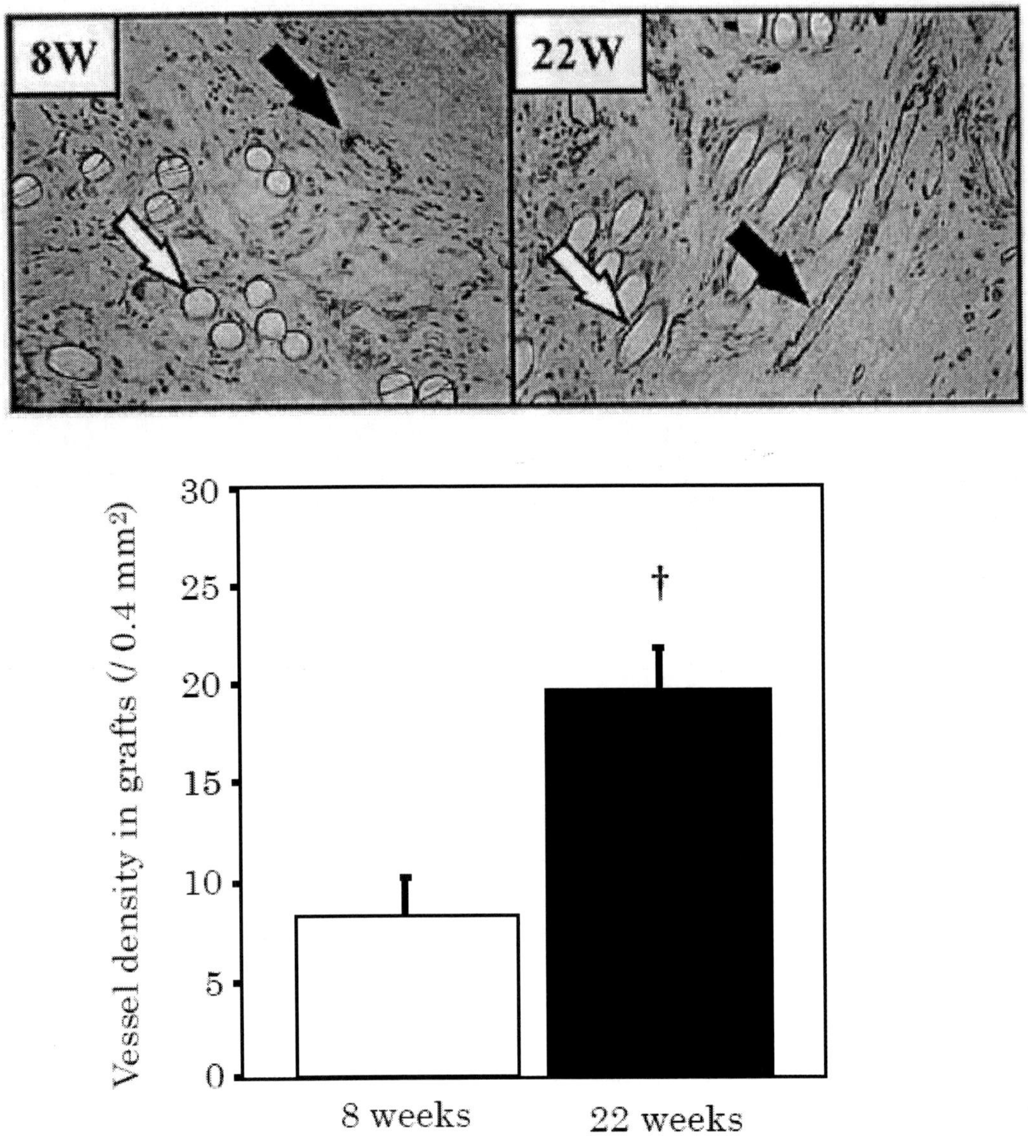

Figure 4. (Top)Photomicrographs of smooth muscle cell-seeded patches stained for factor VIII to identify the endothelial cells of blood vessels (200×). The PCLA scaffold did not stain (white arrows). The endothelial cells of the blood vessels (black arrows) stained brown in color. (Bottom) The blood vessel density of the 22-week patch increased († $p<0.0001$) compared with the 8-week patch.

2. Pericardial Tissue

Patients with congenital heart defects require re-sternotomy during reoperation and staged-operation. Re-sternotomy and dissections are risks of the reoperations as mediastinal adhesion formation may result in the iatrogenic injury of the heart, artery, and veins. Moreover, mediastinal adhesion makes surgeons difficult to identify anatomical feature, prolongs the operating time for dissection and might requie more blood transfusion. Closure of pericardium is one of the strategies to prevent patients and surgeons from these disadvantages; however, the pericardial closure is mostly impossible as it sometimes causes cardiac tamponade or late constriction after surgery. Moreover, in a second or more re-sternotomy cases, there remains no pericardium available for the chest closure.

Several attempts have been made to resolve the problem related to pericardial adhesions after cardiac surgery by using different types of pericardial substitutes, including expanded PTFE(ePTFE) patches, silicone rubber, and glutaraldehyde-treated bovine or porcine pericardium xenografts. Recently, many authors reported that bioabsorbable patches effectively prevented the formation of pericardial adhesion in their experimental studies. However, none of these materials has yet been routinely applicable in the clinical setting, as solid clinical evidence on their beneficial effects is absent. Therefore, Yoshioka *et al.* created a gelatin sheet reinforced with PGA (figure 5) and evaluated the biomechanical property of the sheet and the preventive effect on pericardial adhesion after pericardial replacement in a canine model[10]. The mechanical property was assessed by measuring tension of suture pull-out at first break test. Fifteen dogs underwent partial pericardial replacement with the gelatin sheet through left thoracotomy. The PGA-reinforced sheet exhibited tenfold higher tension of disruption at the suturing margin compared to the gelatin sheet without reinforcement (619 ± 141 versus 62 ±7 gf, $p < 0.001$), and demonstrated equivalent strength to that of clinically available ePTFE membrane. During rethoracotomy, adhesions between the epicardium and the pericardial substitutes were moderate at the 4-week interval and the sheet was resorbed completely after 12 weeks postoperatively. The degree of adhesion was quantitatively classified from 0 to 3, according to the adhesion score used by Heydorn and colleagues[11]: 0 = no adhesion; 1 = adhesion that could be readily separated by finger dissection; 2 = adhesion of intermediate strength; and 3 = adhesion which necessitated sharp dissection. The result is given in table 1, together with the result of inflammation, which was graded into 4 scales on histologic assessment. Inflammatory reaction significantly decreased from the 4-week interval to the 24-week interval after the pericardial replacement ($p < 0.05$).

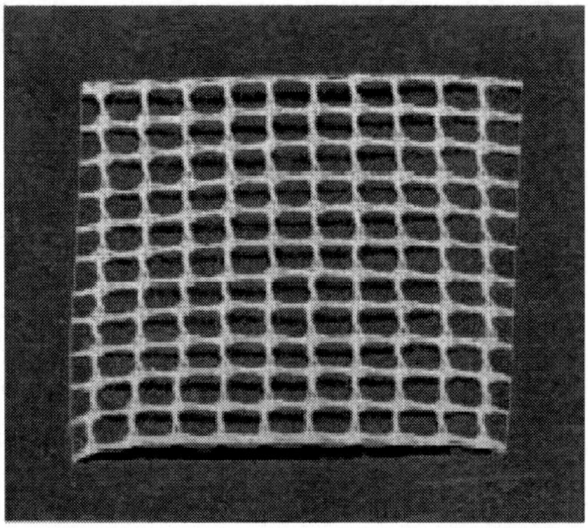

Figure 5. External appearance of the gelatin sheet with polyglycolic acid incorporated into the sheet as a latticework. Each square is 2.5 × 2.5 mm in size. The sheet was adequately malleable for surgical manipulation and it could be cut to a desired shape.

Table 1. Changes in adhesion and inflammatory reaction score around a gelatin sheet reinforced with polyglycolic acid

	2 weeks (n = 1)	4 weeks (n = 3)	12 weeks (n = 5)	24 weeks (n = 6)
Adhesion score	1	1 ± 1.7	0 ± 0.0	0.5 ± 0.6
Inflammatory reaction score	2	2 ± 0.0	1.6 ± 0.6	0.3 ± 0.5[a,b]

[a] $p < 0.05$ versus 4 weeks. [b] $p < 0.05$ versus 12 weeks.

Matsumura *et al.* reported the result of clinical trial performed using the gelatin sheet reinforced with PGA[12]. From February 2003 to July 2004, 5 patients, including 3 males and 2 females, who were anticipated for further repair of congenital defects or for repeated cardiac surgery had received the anti-adhesive sheet. There was neither mortality in this trial nor morbidity such as prolonged inflammatory response, infectious complication, signs of tamponade or early repeated thoracotomy due to the material. As shown in table 2, re-sternotomy took 19 to 34 minutes from the beginning of the operation to the sternum reopening. In each case, re-sternotomy was easily performed and there were no complications attributed to its use. All the doctors who joined the operation scored the anti-adhesive effectiveness as "Good". The impression of adhesiveness where the gelatin-sheet was used, other sites in the mediastinum, the time of operation until sternum reopened and cardiopulmonary bypass had established are summarized in table 2.

Table 2. Asssessment of the adhesion at re-sternotomy

	Degree of adhesion at re-sternotomy		Time	
Case	Gelatin sheet	Other site in mediastinum	op start to re-sternotomy (min)	re-sternotomy to CPB*(min)
1	slightly	severe	20	261
2	slightly	severe	34	196
3	slightly	moderate	26	265
4	slightly	severe	24	176
5	slightly	moderate	19	160

3. Vascular Tissue

It was shown in 2003 that bone marrow cells (BMCs) seeded onto a biodegradable scaffold contribute to the histogenesis of tissue-engineered vascular autografts (TEVAs) in canine *in vivo* models[13]. Histologic examinations revealed that seeded BMCs proliferated and were differentiated in TEVAs after implantation and led to endothelialization, SMC generation, and growth factor production. However, as no reports have described the function

* CPB : cardiopulmonary bypass

of endothelial cells or the mechanical properties and characteristics of TEVAs, Matsumura *et al.* evaluated the endothelial function and mechanical strength of TEVAs constructed with autologous mononuclear bone marrow cells (MN-BMCs) and a P(LA/CL) scaffold using a canine inferior vena cava (IVC) model(figure 6)[14]. Neither dilation nor ruptures was observed during the implantation study. At 1 and 3 months, TEVAs were significantly weaker than IVCs, although TEVAs gradually increased in strength. There were no statistical differences in strength among IVCs of dog (shown as a control) and 6- and 12-month TEVAs (figure 7). The stiffness of TEVAs gradually increased with time after implantation. There was no statistical difference in wall thickness between native IVC (shown as a control) and 6-month TEVAs (figure 8). To test the inflammatory response in TEVAs, expressions of mRNA for dog CD4 (254 base pairs) (figure 9A) and immunoexpression against CD4 in the specimens were examined. The ratio of CD4/s17 mRNA and the number of CD4-positive cells in the specimen significantly decreased as scaffolds partially and randomly degraded with time (figure 9B and C).

After the successful results of the supplementary examination in a dog IVC replacement model, Shin'oka *et al.* reconstructed peripheral pulmonary in a 4-year-old girl with the patient's own venous cells[15]. After that, 3 patients underwent tissue-engineered graft implantation with cultured autologous venous cells. However, since cell culturing was time-consuming and xenoserum had to be used, they began to use BMCs, readily available on the day of surgery, as a cell source. To assess the safety and feasibility of this technique, they reported the midterm clinical result of tissue-engineered vascular autografts seeded with autologous BMCs in 42 patients since September 2001[16]. A 5-mL/kg specimen of bone marrow was aspirated with the patient under general anesthesia before skin incision. A P(LA/CL) tube serving as a scaffold for the cells was the same as described above (figure 10). Twenty-three tissue-engineered conduits for extracardiac total cavopulmonary connection (TCPC) and 19 tissue-engineered patches were used for the repair of congenital heart defects. The patients'ages ranged from 1 to 24 years (table 3). All patients underwent a catheterization study, computed tomographic scan, or both, for evaluation after operation. The patients received anticoagulation therapy for 3 to 6 months after surgery.

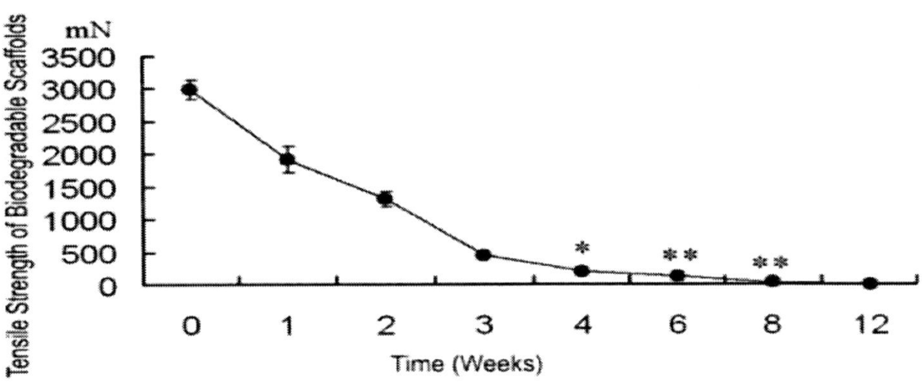

Figure 6. Tensile strength of P(LA/CL) scaffold *in vitro*. The scaffolds used in this study diminished in sequentially within 1 month. Data represent mean + standard error of 5 samples at each time point. *$p<.01$, **$p<.001$ vs. week 0.

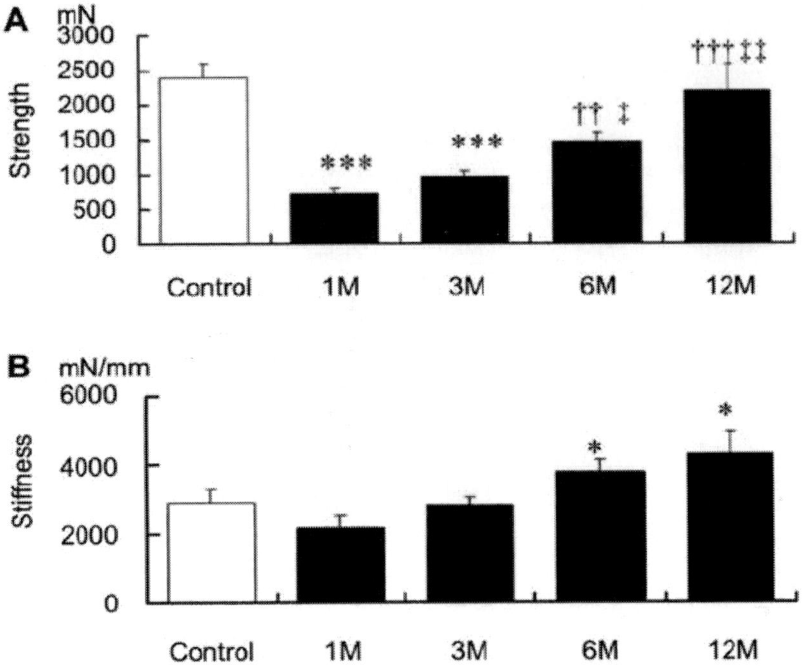

Figure 7. Increase in mechanical properties of tissue-engineered vascular autografts Dunnett's post hoc analysis (TEVAs). (A and B) Tensile strength (mN) and stiffness (mN/mm) of TEVAs (1, 3, 6, and 12 months after implantation). Kruskal-Wallis test; $p<.001$. ***$p<.001$ vs. inferior vena cava, ††$p<.01$ vs. TEVAs at 1 month, †††$p<.001$ vs. TEVAs at 1 month, ‡$p<.05$ vs. TEVAs at 3 months, ‡‡$p<.01$ vs TEVAs at 3 months. (B) Stiffness/width of TEVAs increased with time. Stiffness/width was calculated as stiffness (mN/mm) = elastic modules (mN/mm^2) × wall thickness (mm). Analysis of variance (ANOVA) $p<.01$. *$p<0.05$ vs. TEVAs at 1 month. Data from dog inferior vena cava are shown as control. Data represent mean ± standard error.

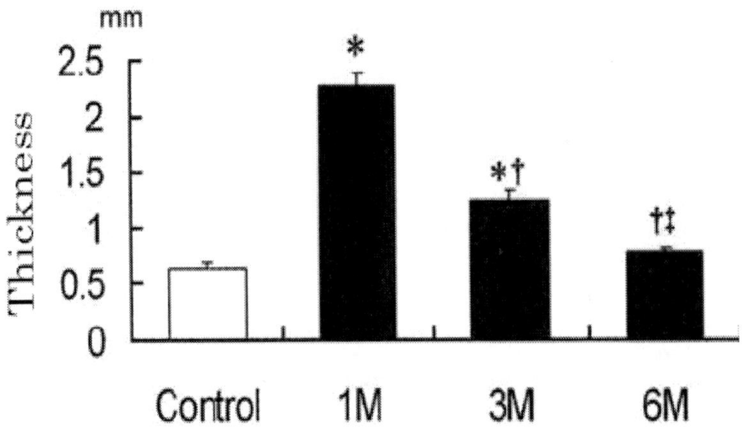

Figure 8. Wall thickness of tissue-engineered vascular autografts (TEVAs). Wall thickness of TEVAs was measured using specimens stained with Victoria blue (n=5). Kruskal-Wallis test; $p<.001$. Data represent mean ± standard error. *$p<.001$ vs. control, †$p<.001$ vs. TEVAs at 1 month, ‡$p<.01$ vs. TEVAs at 3 months.

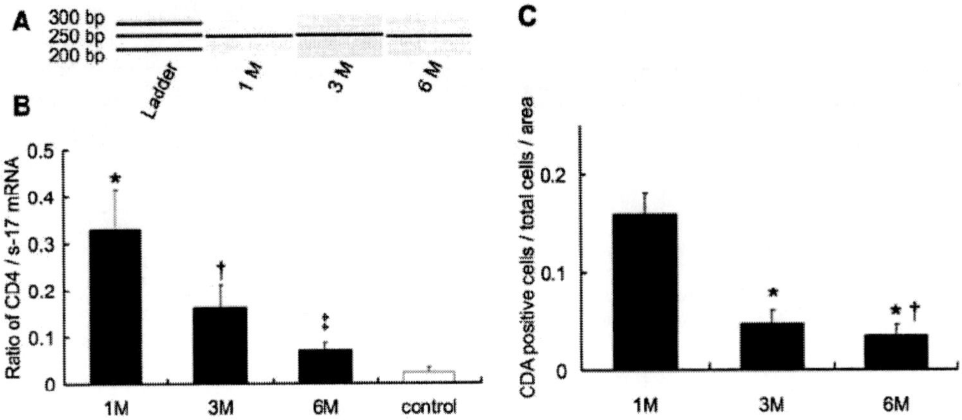

Figure 9. Inflammatory cells and calcium deposition in tissue-engineered vascular autografts (TEVAs). (A) Expression level of mRNA for dog CD4 (254 base pairs). (B) Ratio of CD4/s17mRNA significantly decreased with time. Data from the inferior vena cava neighboring TEVAs are shown as control. Kruskal-Wallis test; $p<.001$. *$p<.001$ vs. control, †$p<.05$ vs. control, ‡$p<.05$ vs. TEVAs at 1 month. (C) Changes in the number of CD4-positive cells in the specimens. Data represent mean±standard error. Kruskal-Wallis test; $p<.001$. *$p<.001$ vs. TEVAs at 1 month, †$p<.05$ vs. TEVAs at 3 months.

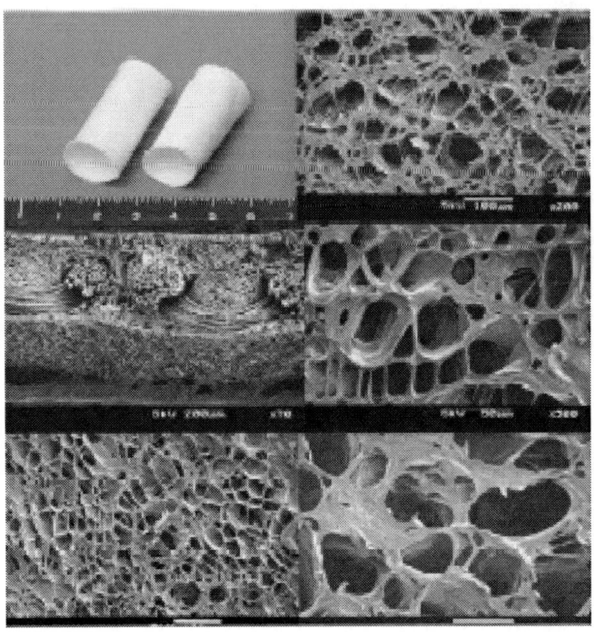

Figure 10. Macroscopic finding of biodegradable scaffolds and scanning electromicroscopic findings of polymer scaffolds. Upper left, Macroscopic finding; 18 mm in diameter. A copolymer sponge was reinforced with a PGA mesh. Bars represent 200 μm (middle left), 100 μm (lower left, upper right), and 50 μm (middle right, lower right).

Table 3. Characteristics of patients (female/male ratio 24/18)

Patient	Age (y)	Body weight (kg)	Chief diagnosis	Previous operations: Radical operations	Previous operations: Palliative operations	Operative procedure with tissue-engineered materials	Scaffold diameter (mm)	Components of biodegradable scaffold
Tissue-engineered sheets								
24	6	16.5	TF, PAPVD, valvular PS	ICR		RVOTR		P(LA/CL) + PLLA
25	10	36	D-TGA (I), RV outflow obstruction, branch PS	ASO, PA plasty		PAP		P(LA/CL) + PLLA
26	24	44	TF, PA, conduit stenosis	Rastelli	Waterston	RVOTR		P(LA/CL) + PLLA
27	9	29	SRV, DIRV, DORV, branch PS	Fontan	mBTS	PAP		P(LA/CL) + PLLA
28	7	16.1	TF, branch PS	ICR	mBTS	PAP		P(LA/CL) + PLLA
29	1	13.5	IAA (A), branch and main PA stenosis, SVC occlusion	ICR, arch repair		PAP and SVC Patch		P(LA/CL) + PLLA
30	3	13.8	Polysplenia, SRV, DORV		mBTS	TCPC (LT)		P(LA/CL) + PLLA
31	3	14	Truncus arteriosus communicans (I), conduit stenosis	Rastelli (×2)		PAP		P(LA/CL) + PLLA
32	11	36.5	Congenital aortic regurgitation (III)			RVOTR, Ross		P(LA/CL) + PLLA
33	2	8.5	Asplenia, SRV, DORV			TCPC (LT)		P(LA/CL) + PLLA
34	8	19	DORV, VSD, nonconfluent PA		mBTS (×2), CS + PAP	PAP		P(LA/CL) + PLLA
35	5	14.5	TF, branch PS	ICR	oBTS, mBTS, palliative RVOTR	RVOTR		P(LA/CL) + PGA
36	7	19.5	Polysplenia, L-DORV, PAA, bilateral branch PS	Double-switch operation	mBTS	PAP		P(LA/CL) + PGA
37	10	17.2	Polysplenia, SRV, left AVVA, right AVVR (II)		BDG	TCPC (LT)		P(LA/CL) + PGA
38	17	64	D-TGA (I), supravalvular PS, AR (II)	ASO		RVOTR		P(LA/CL) + PGA
39	3	13.8	TAC (II), conduit stenosis, PAPVR	Rastelli		RVOTR		P(LA/CL) + PLLA
40	9	42.5	BWG, PS	Takeuchi, right PA angioplasty		PAP		P(LA/CL) + PGA
41	8	20.5	TGA (II), CoA, LMT occlusion, left PA stenosis	ASO		PAP		P(LA/CL) + PGA
42	6	19.5	D-TGA (I), SVAS, left PA stenosis	ASO		PAP		P(LA/CL) + PGA

AVSD, Atrioventricular septal defect; *RV*, right ventricle; *mBTS*, modified Blalock-Taussig shunt; *P(LA/CL)*, copolymer of L-lactide and ε-caprolactone; *SRV*, single right ventricle; *DORV*, double-outlet right ventricle; *TAPVC*, total anomalous pulmonary venous connection; *PAA*, pulmonary artery atresia; *MS*, mitral stenosis; *oBTS*, original Blalock-Taussig shunt; *CS*, central shunt; *PA*, pulmonary artery; *TA*, tricuspid atresia; *AVVA*, atrioventricular valve atresia; *PAB*, pulmonary arterial banding; *PDA*, patent ductus arteriosus; *BDG*, bidirectional Glenn shunt; *ASD*, atrial septal defect; *TVP*, tricuspid valve plasty; *SLV*, single left ventricle; *CAVVR*, common atrioventricular valve regurgitation; *CAVVP*, common atrioventricular valve plasty; *HLHS*, hypoplastic left heart syndrome; *MA*, mitral atresia; *IAA*, interruption of aortic arch; *PPA*, pure pulmonary atresia; *DILV*, double-inlet left ventricle; *LV*, left ventricle; *VSD*, ventricular septal defect; *PS*, pulmonary stenosis; *cAVSD*, complete atrioventricular septal defect; *CAVV*, common atrioventricular valve; *CA*, common atrium; *RA*, right atrium; *AFL*, atrial flutter; *af*, atrial fibrillation; *APC*, atrial premature contraction; *DIRV*, double-inlet right ventricle; *PLSVC*, persistent left superior vena cava; *SVC*, superior vena cava; *TF*, tetralogy of Fallot; *PAPVD*, partial anomalous pulmonary vein drainage; *ICR*, intracardiac rerouting; *RVOTR*, right ventricular outflow tract reconstruction; *D-TGA*, dextrotransposition of the great arteries; *ASO*, arterial switch operation; *PAP*, pulmonary arterial patch angioplasty; *LT*, lateral tunnel method; *AVVR*, atrioventricular valve regurgitation; *AR*, aortic regurgitation; *TAC*, truncus arteriosus; *PAPVR*, partial anomalous pulmonary vein return; *BWG*, Bland-White-Garland syndrome; *CoA*, coarctation; *LMT*, left main trunk of coronary artery; *SVAS*, supravalvular aortic stenosis.

Mean follow-up after surgery was 490 ± 276 days. There were no complications such as thrombosis, stenosis, or obstruction of the tissue-engineered autografts. As shown in figure 11, the maximal trans-scctional area was calculated and compared with the implanted size in the TCPC group. In patients with azygous connection (in which only hepatic flow goes up through the tissue-engineered graft), the larger grafts implanted tended to decrease in diameter with time. Depending on the blood flow, some grafts decreased in diameter and some increased, as shown in figure 12. This seems to be some sort of environmental adaptation of the tissue-engineered grafts.

One late death at 3 months after TCPC was noted in a patient with hypoplastic left heart syndrome; this was unrelated to the tissue-engineered graft function. There was no evidence of aneurysm formation or calcification on cineangiography or computed tomography. All tube grafts were patent and the diameter of the tube graft increased with time (110 ± 7% of the implanted size) ,suggesting that these vascular structure may have the potential for growth, repair, and remodeling, and provide an important alternative to the use of prosthetic materials in the field of pediatric cardiovascular surgery.

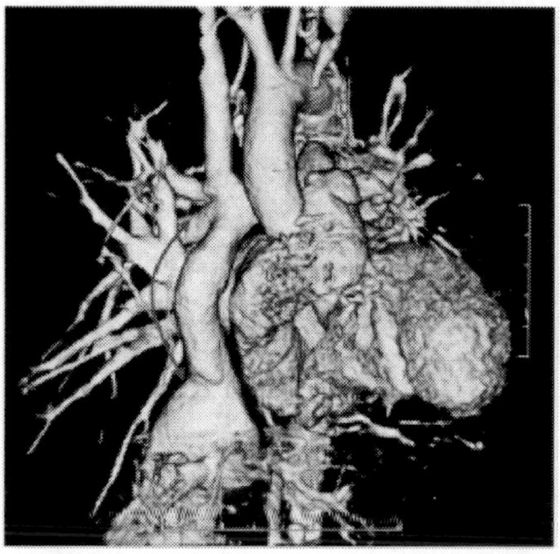

Figure 11. Three-dimensional computed tomographic image 12 months after implantation of TCPC graft (patient 20). An arrow indicates the location of tissue-engineered conduit.

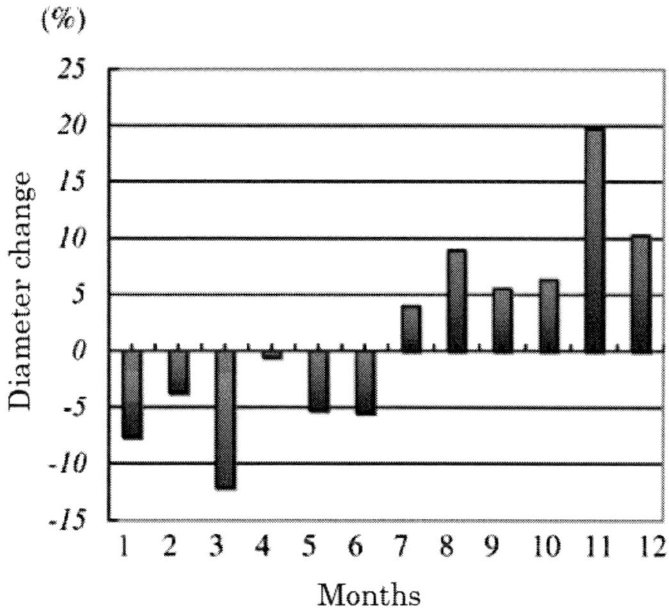

Figure 12. Percentage changes in cross-sectional area of tissue-engineered grafts 1 year after operation.

4. Small Intestine

There are some histological investigations on regenerating intestine for surgical treatment of short bowel syndrome, but few reports detail the regeneration of gut endocrine cells (ECCs) and the nerve system of the regenerated tissue-engineered intestine[17,18]. Nakase *et al.* attempted to regenerate a large size of small intestinal tissue and evaluated the regeneration of the ECCs and nerve system of autologous tissue-engineered small intestine[19]. The implantation effect using a large number of SMCs for the regeneration of small intestine was also investigated.

Two types of scaffolds with different cell densities were fabricated. For scaffold preparation, collagen sponge with a pore size of 70 to 110 μm and a pore volume fraction of 80 to 95% was used and SMCs were cultured in the collagen sponge. The SMC density of collagen solution was 5×10^6 cells/cm^3 [low density (LD) group] and 1×10^8 cells/cm^3 [high density (HD) group]. Both scaffolds were implanted into defects of isolated ileum in a canine model, as shown in figure 13. The area of engineered small intestine in the HD group was four times larger than that in the LD group, although that was smaller in size than the original size of the defect. There were no significant changes in the thickness of regenerated smooth muscle layer (SML) in the LD and HD groups. The number of ECCs gradually increased after implantation. At 18 weeks of regeneration, the number of ECCs reached a level comparable to that of normal mucosa. The nerve fibers extended to the center of the graft area and were observed in the regenerated SML and villi at 24 weeks.

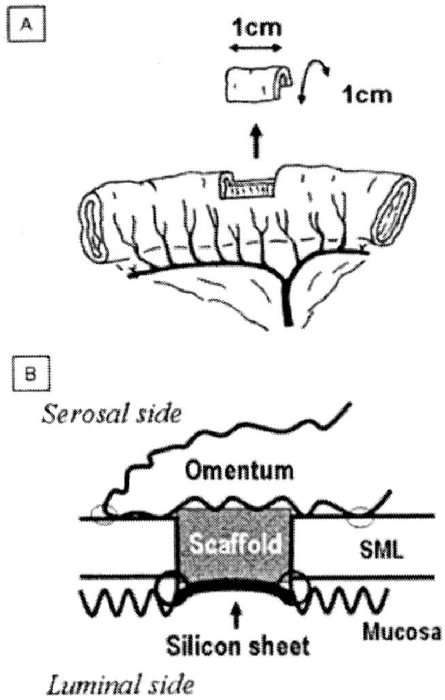

Figure 13. Schema of implantation procedure. (A) The defect was created at the middle of isolated ileal loop. The defect size was 1 cm × 1 cm. (B) The defect was patched with silicone sheets using 6-0 Vicryl suture. Collagen scaffolds were placed below the silicone sheet and covered on the serosal aspect with omentum.

5. Bile Duct

In the current treatment for an extra-hepatic bile duct affected by cancer or stenosis, the affected portion is resected and the hilar bile duct is anastomosed to the small intestine. In many cases, however, the post-operative course is complicated by retrograde infections via the intestine or stenosis at the anastomosis[20], requiring reanastomosis. There have been few reports on attempts to create bile ducts[21-23]. Miyazawa *et al.* fabricated a bile duct organoid unit (BDOU) by seeding autologous BMCs onto a P(LA/CL) scaffold and then transplanted the BDOU into pigs to investigate two of its features: functional potential as a replacement for the native bile duct and morphology after transplantation[24].

The scaffold tube measured 5 mm in diameter and 0.7 mm in thickness. Eighteen hybrid pigs were implanted with BDOU. Twelve of the transplanted BDOUs had been seeded with autologous BMCs in advance. Six animals, the controls, were grafted with the scaffold alone with no BMCs seeded. The common bile duct was cut, the hepatic cut end of the native common bile duct was anastomosed to the BDOU, and the other end was anastomosed to the duodenum, as shown in figure 14. The controls underwent a similar operation. The neo-bile duct was removed at pre-determined time points and investigated histologically. All 18 recipient pigs survived until their sacrifice at 6 weeks, 10 weeks or 6 months. Histological examination revealed incomplete epithelialization of the neo-bile duct at 6 and 10 weeks after transplantation. At 6 months, the organoid exhibited a morphology almost identical to that of the native common bile duct. No differences were found between the controls and the BMC seeded pigs. These results suggest that the tissue-engineered bile duct can serve as a substitute for the native one.

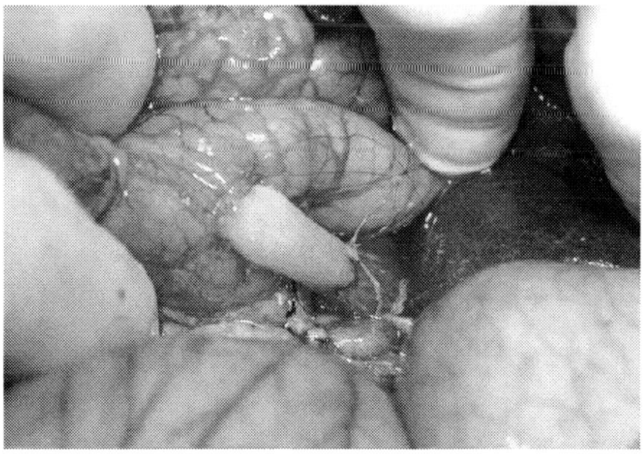

Figure 14. Bile duct organoid units. Bile duct organoid units were anastomosed to the native common bile duct at the one end and to the descending duodenum at the other to reconstruct the bile duct.

6. Auricular Cartilage

While the selection of articular chondrocytes had led to some progress in the tissue engineering of cartilage, the choice of such cells may not be ideal for all applications and other sites may yield additional benefits. In this context, cartilage is recognized as consisting of three classes or types of tissue—hyaline, elastic, and fibrocartilage—each distinct in its structure and biochemical nature, depending on site-specific functional adaptation. To date, little is known about the specific features of chondrocytes from these different sources or types of cartilage and their responses in tissue-engineered constructs designed to recapitulate the human auricle.

Isogai *et al.* investigated the chondrogenic potenial of chondrocytes harvested from the articular, nasoseptal, costal, and auricular cartilages in bovine donor animals[25]. Respective chondrocytes were seeded onto disc-shaped P(LA/CL) scaffolds. Cell-copolymer constructs were cultured and subsequently implanted in the subcutaneous space of athymic mice for up to 20 weeks. Cell culture followed up to 4 weeks showed varied chondrocyte proliferation from the tissue sources (figure 15), as did levels of type II collagen and aggrecan gene expression. For both genes, the highest expression was found for costal chondrocytes, followed by nasoseptal, articular, and auricular cells, as shown in figure 16. Retrieval of 20-week discs from mice revealed changes in construct dimensions with different chondrocytes. The greatest disc diameter was found for scaffolds seeded with auricular chondrocytes, followed by those with costal, nasoseptal, and articular cells. The greatest disc thickness was measured for the scaffolds containing costal chondrocytes, followed by those with nasoseptal, auricular, and articular cells. Retrieved copolymer alone was the smallest in diameter and thickness. Only auricular scaffolds developed elastic fibers after 20 weeks of implantation. Type II collagen and aggrecan were detected with different expression levels on quantitative RT-PCR of discs implanted for 20 weeks. These data demonstrate that bovine chondrocytes obtained from different cartilaginous sites in an animal may elicit distinct responses during their respective development of a tissue-engineered neocartilage.

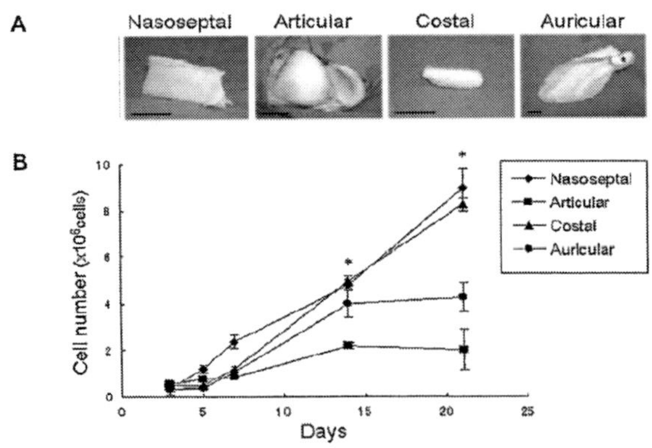

Figure 15. Gross morphology of the sites for bovine cartilage harvest used in this study (A) and the proliferation *in vitro* of chondrocytes isolated from nasoseptal, articular, costal, and auricular cartilage of newborn calves (B). Sources of cartilage were cleaned of adherent tissues (A). Cell numbers were counted on days 3, 5, 7, 14, and 21 of culture and values are expressed as mean ± SE for the number of samples ($n = 3$) at each time point. After day 7, nasoseptal and costal cells maintained higher levels of proliferation compared to auricular and articular chondrocytes. Articular chondrocytes were the least numerous of the cell types over the 21 days of culture. Significant differences ($*p < 0.05$) in cell numbers were found between articular chondrocytes and any other cell source at day 14 and between articular or auricular chondrocytes and either costal or nasoseptal cells at day 21. Bars, 2 cm (A).

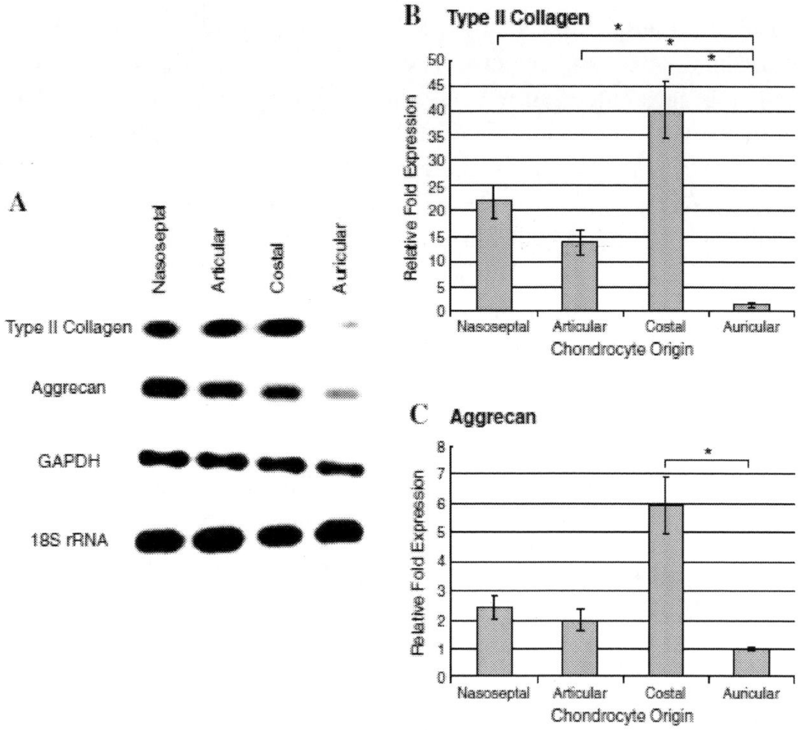

Figure 16. Qualitative (A) and quantitative (B and C) RT-PCR of bovine chondrocytes harvested at 28 days of culture. (A) The blot confirms the presence of type II collagen, aggrecan, GAPDH, and 18S rRNA in cultures developed from all four types of chondrocytes. All four genes were determined to be single bands at the correct molecular weight (not shown) when the amplicons were electrophoresed on agarose gels. Quantitative RT-PCR demonstrates relative abundance of gene levels for type II collagen (B) and aggrecan (C). Costal chondrocytes expressed relatively the greatest levels of type II collagen and aggrecan and auricular cells expressed the lowest levels. Significant differences in expression for type II collagen were found between auricular and all other chondrocyte cultures ($*p < 0.05$). Auricular cells were also significantly different in aggrecan expression levels compared to costal cells ($*p < 0.1$). No significant differences between the chondrocyte sources were noted for GAPDH ($p > 0.1$) (data not shown). Values of gene expression levels are expressed as mean ± SE for the number of independent cultures sampled ($n = 4$) at 28 days of culture.

7. Dura Mater

Water-tight closure of the dura mater at the end of an intradural neurosurgical procedure is essential for the surgical wound healing process. Theoretically, autografts are best for this purpose, but *in situ* harvesting sufficient periosteum or fascia for a graft is sometimes difficult. Sheets of PTFE have been used as an alternative material for dural repair. Long-term follow-up results covering the use of this dural substitute, however, have never been described.

To avoid these encountered complications, Yamada *et al.* developed a new bioabsorbable dural substitute composed of P(LA/CL) films and PGA non-woven fabric, as shown in figure 17. This composite sheet could be easily trimmed and sutured in a water-tight fashion by using several types of suturing thread at body temperature. The surface of the brain could be inspected through this transparent composite sheet even after dural closure. Results of a study showed that the dural defect was re-constructed by *de novo* autologous dura-like tissue and that the grafted dural substitute itself was absorbed within 6 months of transplantation[26,27]. The results of the clinical application of this dura mater substitute was reported by Yamada *et al.*[28]. The newly developed dural substitute was placed in 53 patients during neurosurgical procedures and the average follow-up period was 35.5 months. The handling properties and biocompatibility of this dural substitute were highly satisfactory and no significant complication was observed. In patients who underwent a second surgery performed more than 18 months after the initial operation, this dural substitute was found to have been replaced by an autologous collagenous tissue formed. Because of its bioabsorbability, chronic foreign body reactions to this dural substitute were negligible.

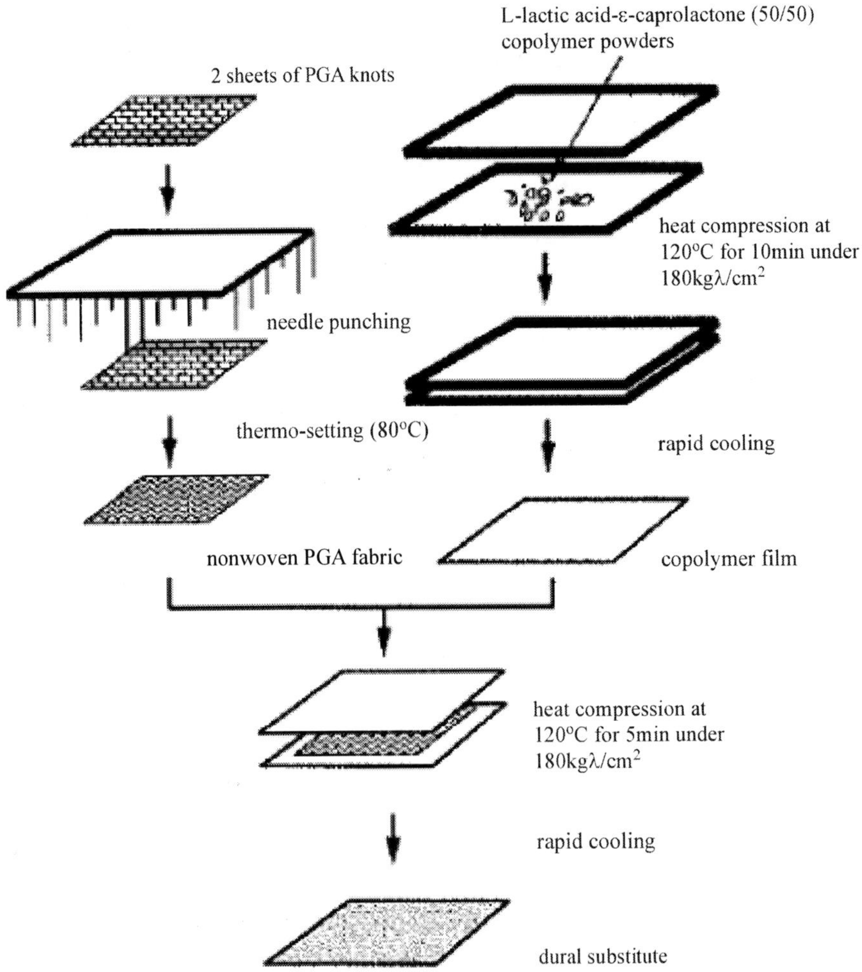

Figure 17. Schematic representation of fabrication of the composite sheet made of an L-lactic acid-ε-caprolactone (50% L-lactic acid, 50% ε-caprolactone) copolymer and PGA.

8. Alveolar Bone

Reconstruction of a thin or low-volume alveolar bone ridge is an important topic in the field of implant and prosthetic dentistry. Guided bone regeneration (GBR) has attracted much attention as a promising method, where a GBR membrane plays an essential role in preventing the ingrowth of soft tissue, and guides new bone formation into a desired shape. So far, titanium mesh, ePTFE, collagen, and several biodegradable polymers have been used as a membrane for GBR. However, non-absorbable membranes are often associated with complications including membrane exposure and infection, and even if there are no complications, they have to be removed by a second surgical procedure after bone formation. Although membranes made of collagen are cell-adhesive and absorbable, their mechanical strength is too poor to provide space for bone formation.

Kinoshita *et al.* evaluated the effects of combining a porous P(LA/CL)/β-TCP membrane and a bFGF incorporating-gelatin sponge on bone regeneration in mandibular ridges[29]. Four full-thickness saddle-type defects (10mm long x 5mm deep) were symmetrically created in both edentulous mandibular alveolar ridges of 6 beagles, as shown in figure 18. The dome-shaped membrane was secured to each defect site, and a gelatin sponge containing 200*μg* of bFGF was implanted on the left side of each defect (experimental group). Only the membranes (control group) were secured to the defect sites on the right. As given in figure 19, soft X-ray examination revealed an increase in newbone volume in the experimental group 6 months postoperatively. pQCT examination showed that immature bone density was higher in the experimental group. Furthermore, micro-CT images revealed well-formed new bone along the original contour of the dome-shaped membrane in the experimental group. Histologically, inflammatory infiltration in the surrounding tissue of the membrane was slight. These results suggested that combining the P(LA/CL)/ β-TCP membrane and the bFGF-gelatin sponge was promising for alveolar ridge reconstruction.

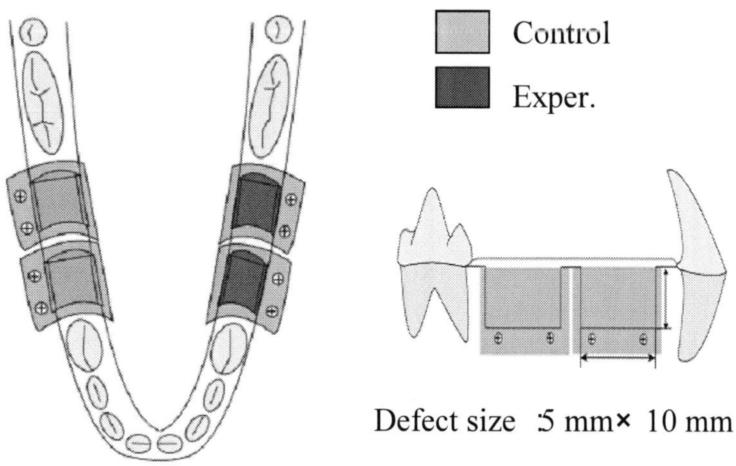

Figure 18. Schema of experimental procedure. P(LA/CL)/β-TCP membrane adapted to the bone-defect site and a 20 *mg* acidic hydrogel sponge containing 200 *μ*m of bFGF (bFGF-gelatin sponge) was implanted in each of the left-side bone defects (the experimental group); then, each membrane was secured to the existent bone on the buccal site with two PLLA pins. Only the P(LA/CL)/β-TCP membranes were adapted and secured to the right-side bone defects (control group).

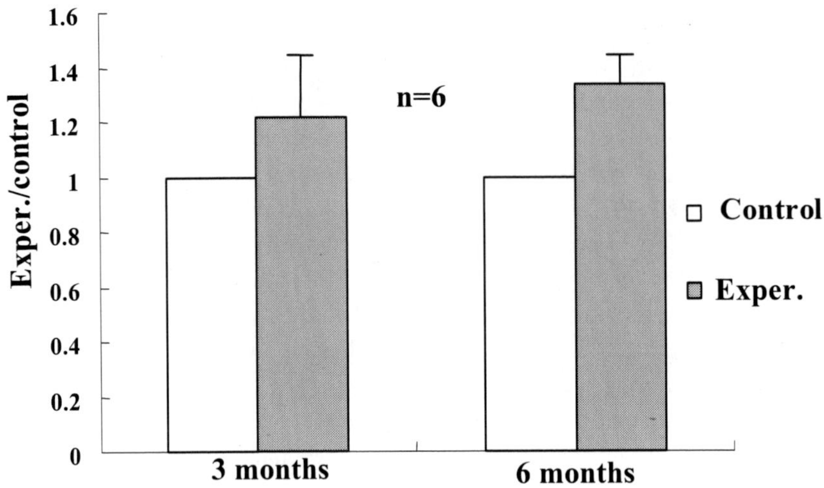

Figure 19. Volume of new bone 3 and 6 months postoperatively by micro-CT analysis. Y axis shows the relative volume of new bone.

III. Conclusion

As is well known, there are several approaches to tissue regeneration by the help of tissue engineering, in particular, *ex vivo* tissue engineering and *in situ* tissue engineering. Collectively, the representative examples of which were described above seem to suggest how simple, practical, economical, and promising the *in situ* tissue engineering is. Indeed, *ex vivo* tissue engineering facilitates development of large-scaled tissues like human pulmonary artery of 2 cm diameter and mandible of 7 cm size of patients[30]. In this system the scaffold is the key determinant factor. It is noteworthy that the scaffolds represented above are far from the natural extracellular matrix in terms of biological functions, but are capable of regenerating a variety of tissues. The minimum requisites for the scaffolds applicable to large animals and humans are briefly summarized in table 4. These points look very simple and readily satisfied, but majority of the scaffolds described in literature tend to ignore these requirements. This may account for the fact that most of the current studies on tissue engineering still remain in the stage, where small-sized tissues are the target of tissue engineering research with rat models. It seems too difficult to fabricate massive tissues by the laborious *ex vivo* tissue engineering strategy, since it has yet no means to produce vascularized tissues. Although cell biology has greatly advanced in these decades, the knowledge and information that is required for the creation of tissues outside the body is still too little for tissue engineers.

Now, it may be a good time to reevaluate with more interest the practical tissue engineering that has grown into the stage very close to human clinical applications that numerous patients have long been awaiting. The first paper published by Vacanti and his colleagues in 1988[31] was related to *in situ* tissue engineering, not to *ex vivo* tissue engineering.

Table 4. Requirements for the scaffolds to be applied for large animals and humans

1	should be fixed at the proper site in the body (*e.g.*, by suturing; gel is difficult to fix).
2	should be protected from compressive force if a soft scaffold is implanted (*e.g.*, by mesh cover).
3	should be processed or trimmed in the operation theater (porous β-TCP cannot be cut).
4	should be absorbed at the proper rate (PGA is absorbed too fast, whereas PLA remains too long).

References

[1] R.Langer and J.P.Vacanti, *Science,* 260, 920-926(1993).
[2] Y.Ikada, Tissue Engineering—Fundamentals and Applications—, Academic Press, 2006.
[3] M.J.Lysaght and A.L.Hazlehurst, *Tissue Eng.,* 10, 309-320(2004).
[4] C.H.Evance, G.D.Palmer, A.Pascher *et al., Tissue Eng.,* 13, 1987-1993(2007).
[5] T.Ozawa, D.A.Mickle, R.D.Weisel *et al., Circulation,* 106, 176-182(2002).
[6] R.K.Li, Z.Q.Jia, R.D.Weisel *et al., J. Mol. Cell. Cardiol.,* 31, 513-522(1999).
[7] T.Ozawa, D.A.Mickle, R.D.Weisel *et al., Circulation,* 106, 176-182(2002).
[8] M.Watanabe, T.Shin'oka, S.Tohyama *et al., Tissue Eng.,* 7, 429-439(2001).
[9] T.Ozawa, D.A.Mickle, R.D.Weisel *et al., Cell Transplant.,* 13, 169-177(2004).
[10] I.Yoshioka, Y.Saiki, K.Sakuma *et al., Ann. Thorac. Surg.,* 84, 864-870(2007).
[11] W.H.Heydorn, J.S.Daniel, and C.E.Wade, *J. Thorac. Cardiovasc. Surg.,* 94, 291-296(1987).
[12] G.Matsumura, Y.Ikada, and T.Shin'oka, *Pediatric Cardiac Surg.,* submitted.
[13] G.Matsumura, S.Miyagawa-Tomita, T.Shin'oka *et al., Circulation,* 108, 1729-1734 (2003)
[14] G.Matumura, Y.Ishihara, S.Miyagawa-Tomita *et al., Tissue Eng.,* 12, 3075-3083 (2006).
[15] T.Shin'oka, Y.Imai, and Y.Ikada, *N. Engl. J. Med.,* 344, 532-533(2001).
[16] T.Shin'oka, G.Matsumura, N.Hibino *et al., J. Thorac. Cardiovasc. Surg.,* 129, 1330-1338 (2005).
[17] T.C.Grikscheit, A.Siddique, E.R.Ochoa *et al., Ann. Surg.,* 240, 748-754(2004).
[18] T.C.Grikscheit, E.R.Ochoa, A.Ramsanahie *et al., Ann. Surg.,* 238, 35-41(2003).
[19] Y.Nakase, T.Nakamura, S.Kin *et al., J. Surg. Research,* 137, 61-68(2007).
[20] B.J.Ammori, S.Joseph, M.Attia *et al., Int. J. Pancreatol.,* 28, 15-21(2000).
[21] K.Shimono and Y.Nose, *Artif. Organs.,* 19, 115-116(1995).
[22] J.Xu, S.Sun, Q.Zhang *et al., Chin. Med. J. Engl.,* 111, 86-87(1998).
[23] M.Rosen, J.Ponsky, R.Petras *et al., Surgery,* 132, 480-487(2002).
[24] M.Miyazawa, T.Torii, Y.Toshimitsu *et al, Am. J. Transplant.,* 5, 1541-1547(2005).
[25] N.Isogai, H.Kusuhara, Y.Ikada *et al., Tissue Eng.,* 12, 691-703(2006).
[26] S.Miyamoto, K.Yamada, I.Nagata *et al., J. Artif. Organs.,* 1, 10-14(1998).
[27] K.Yamada, S.Miyamoto, I.Nagata *et al., J. Neurosurg.,* 86, 1012-1017(1997).

[28] K.Yamada, S.Miyamoto, M.Takayama *et al.*, *J. Neurosurg.*, 96, 731-735(2002).
[29] Y.Kinoshita, M.Matsuo, K.Todoki *et al.*, *J. Oral Maxillofac. Surg.*, submitted.
[30] Y.Kinoshita, *JMAJ,* 47, 294-297(2004).
[31] J.P.Vacanti, M.A.Morse, W.M.Saltzman *et al.*, *J. Pediatr. Surg.*, 23, 3-9(1988).

In: Tissue Engineering: Roles, Materials and Applications ISBN: 978-1-60456-293-4
Editors: S. J. Barnes, L. P. Harris, pp. 141-189

Chapter 7

RAPID PROTOTYPING AS AN ELEGANT PRODUCTION TOOL FOR POLYMERIC TISSUE ENGINEERING SCAFFOLDS: A REVIEW

Tim Desmet, Etienne Schacht and Peter Dubruel *

Polymer Chemistry and Biomaterials Research Group,
Ghent University, Krijgslaan 281 (building S4), B-9000 Ghent, Belgium

ABSTRACT

In the current review, we aim to give an overview of different rapid prototyping (RP) methods which have been applied so far for the production of polymeric scaffolds for tissue engineering applications. After a general introduction on tissue engineering, scaffolds and biomaterials, both the well established RP techniques and the novel more recent developments in the field will be discussed. Comparison between RP and the classical techniques for producing scaffolds, each with their specific advantages and drawbacks will enable the reader to gain insight in the power of RP as an elegant scaffold production technique. Finally, an overview of some commonly used techniques for characterising the materials developed will be given.

1. INTRODUCTION

1.1. Tissue Engineering

In modern medicine the transplantation of living tissue is of major importance in the treatment of diseases and injuries. A distinction can be made between organ-transplantation, which is considered live-saving, and tissue-transplantation, which is considered live-

* Corresponding author: Prof. Dr. Peter Dubruel; Krijgslaan 281 (Building S4), B-9000 Ghent; e-mail: Peter.Dubruel@UGent.be / tel: 003292644466

enhancing. In this review focus will be mainly on tissue-transplantation. Tissue transplantation consists of the replacement of damaged tissue by functional, healthy tissue which is subsequently incorporated in the host and as such restores the biological function[1].

According to the source of the implanted tissue, it is possible to distinguish several types of transplantation. Autografting is the transplantation of tissue or organs from one location to another in the same patient[2, 3]. Examples include skin-grafts and vein-extraction. A particular example of autografting is the removal of tissue from a patient, followed by the treatment of the tissue or the patient before re-implantation. An example includes the application of stem-cell autografts. Allografting is the transplantation between organisms of the same species[2, 3]. Allografting typically includes the transplantation of a heart, liver of kidney from a donor to a patient. Xenografting is the transplantation of organs and tissues from one species to another [3].

Each of these strategies imposes particular problems. For both autografts and allografts additional procedures are required, the amount of available tissue is limited and the morphology is problematic. Infections, transfer of diseases[4], rejection of the tissue by the host and donor site morbidity are all possible complications related to transplantation. As a result of an increased life expectancy and the limitations due to the previously mentioned complications the demand for organs/tissues outnumbers the availability. The most recent annual report of the US department of Health and Human Services mentions a total number of 94,086 registrations for transplantation, from which 65,859 were organ type[5]. This shortage remains the most important obstacle preventing patients to benefit from organ transplantation[6].

Several approaches are proposed in order to provide a sufficient solution for this problem. These approaches include, amongst others, the increase of organ-donation[7], more efficient utilization of potential donors [8], surgical techniques[9] and artificial implants[10].

The past decades the latter approach has led to the development of tissue engineering (TE) as a new scientific field. Tissue engineering is an interdisciplinary field that applies the principles of engineering and the life sciences toward the development of biological substitutes that restore, maintain, or improve tissue function[10-13]. Tissue engineering integrates the advantages of tissue/organ transplantation and purely synthetic prostheses by combining cells with synthetic extracellular matrices (ECM)[14].

Three major strategies of tissue engineering can be distinguished[10, 14]. The first strategy involves the transplantation of cells or cell substitutes. This approach has several advantages but the implant is not capable of fulfilling a supportive function. Furthermore, cells can lose their function or can be immunologically rejected. The second strategy consists of a scaffold which is incorporated in the patient without any cells present on the implant. The tissue formation is due to cells already present in the body. These cells are stimulated to migrate, proliferate and differentiate into the scaffold. Often, directing and stimulation of cells is realised by means of biological factors such as growth factors. The purification and large-scale production of growth-factors, in combination with the development of appropriate methods to deliver these growth-factors are difficulties to overcome[15]. The third strategy involves the transplantation of cells on, or within matrices. These cells can be encased in a semipermeable barrier (closed systems)[10]. As a result the implanted cells can secrete functional proteins which can diffuse through the barrier. Higher molecular weight molecules and cells cannot diffuse through the membrane, thus avoiding immunological rejection. In open systems, cells are seeded onto synthetic extracellular matrices (scaffolds), incubated in a bioreactor and subsequently implanted in the body, where the cell-seeded material can integrate with the surrounding tissue. Immunological response may be minimized by using autologous cells and/or immunosuppressive drugs. In this paper focus will be on the latter approach.

This third strategy combines the advantages of autologous implants and the controlled properties of artificial fabricated ECM. Scientists attain to develop new autologous tissue with appropriate biochemical and mechanical properties in order to restore functional tissue. Sometimes the tissue engineering paradigm is described as to isolate specific cells through a biopsy from a patient, to grow them on a three-dimensional biomimetic scaffold under precisely controlled culture conditions, to deliver the construct to the desired site in the patient's body and to direct new tissue formation into the scaffold that can be degraded over time[16, 17]. The artificial *in vitro* fabrication of a complete functional organ such as kidneys and livers is considered to be achievable in the coming decade [12]. The goal of tissue engineering is to surpass the limitations of conventional treatments such as organ/tissue transplantation and biomaterial implantation[10]. This final goal of fabricating functional tissue imposes constraints on the scaffolds design, physicochemical and mechanical properties.

1.2. Scaffolds

From the previous discussion it is clear that a scaffold has to meet several requirements. It is not yet clear what defines the so-called "ideal" scaffold and what the most relevant properties for scaffold design would be[18, 19].

A tissue engineering scaffolds should serve three primary purposes: (1) they must define a space that will shape the regenerating tissue, (2) they must provide temporary function in a defect while tissue regenerates and (3) they must facilitate ingrowth of tissue, thus serve as an adhesion substrate, and possibly deliver cells, proteins and/or other biofactors such as DNA. [20, 21].

These purposes result in certain characteristics which should be considered when designing a scaffold[22]. First of all, if a scaffold has to define a certain space it is necessary to have a three-dimensional structure with a defined architecture, shape and volume[23]. The material has to be easily processable into a variety of shapes and sizes, including patient-specific scaffolds starting from a CT-scan or MRI-data[21, 24]. From the second purpose it is clear that the scaffold should have mechanical properties matching those of the native ECM in order to be able to fulfil a supportive function[23]. This aspect also determines that, for some applications, the scaffold should be biodegradable or bioresorbable. The biodegradation (or resorption) rate should match the cell/tissue growth *in vitro* and/or *in vivo*. Biomaterials with a controllable biodegradation rate are thus of great interest for tissue engineering applications. As a consequence of the third purpose the scaffold should have a suitable surface chemistry for cell attachment, proliferation and differentiation, if required. Furthermore, the degradation products of the surface and bulk material should be biocompatible, causing minimal inflammatory or immune responses[25]. The availability of functional groups in order to covalently couple or include biofactors in the scaffolds should be considered. Another consequence is that a highly porous, interconnective structure which allows flow transport of nutrients and metabolic waste is desirable. Pore-size should allow migration of cells through the scaffold without blocking the flow transport[20, 22, 26].

These requirements will impose certain constraints on both the design of the scaffold and the biomaterials and will therefore guide the design of a successful novel scaffold for a specific application. However, these different aspects which should be considered can lead to conflicting requirements. For example, a high internal surface area/volume ratio is essential in order to accommodate the large numbers of cells required to replace or restore tissue or organ function. A compromise between pore size and internal surface area is thus required, without losing mechanical strength. It is also clear that a successful scaffold should balance mechanical function with biofactor delivery[26]. Several macro and micro-structural properties critical for rapid cell growth were identified and will be discussed further on [27-29].

An overview of how the different requirements interfere in a complex, multidisciplinary way to lead to an intelligent 3D scaffold applicable in regenerative medicine is presented in Figure 1.

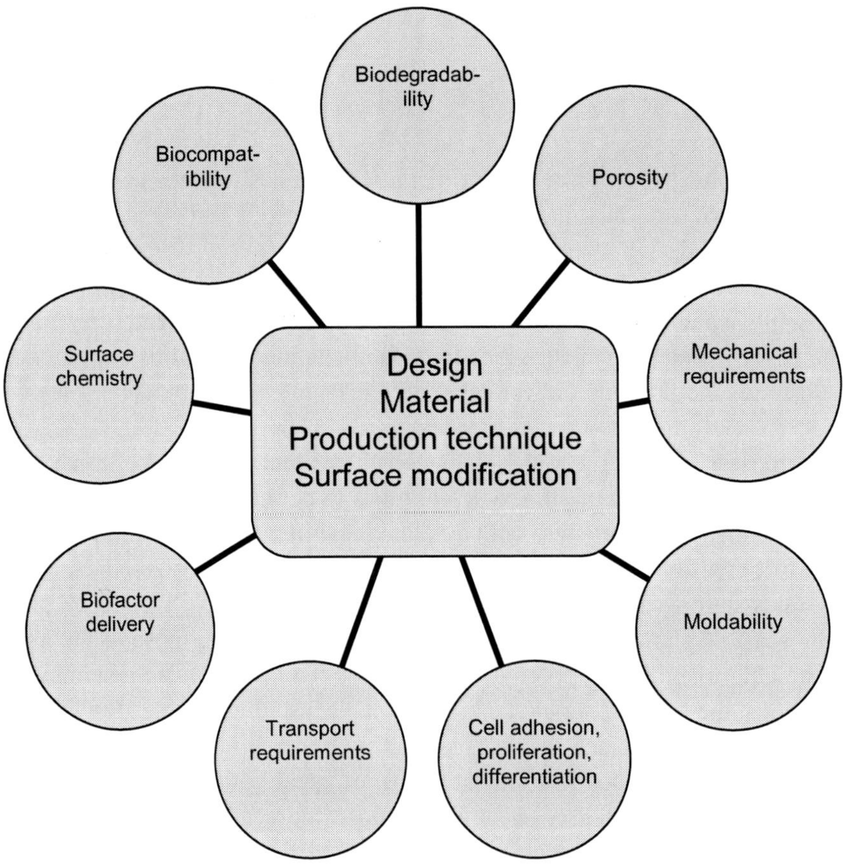

Figure 1. multiple requirements interfere in a complex, interdisciplinary manner and determine the constraints for the ideal scaffold for a specific application.

The choice of biomaterial and the design of the scaffold are primary concerns when creating a scaffold. The most ideal production technique will be mainly determined by the choice of material.

1.3. Biomaterials

1.3.1. Introduction

In this paper the focus will be on biodegradable and bioresorbable scaffolds, fabricated from biocompatible polymers. The shift from biostable, such as polyethylene, to biodegradable polymers for the fabrication of implants can be seen as a crucial paradigm shift in biomaterial science [26, 30].

Biomaterials have been defined as substances other than foods or drugs contained in therapeutic or diagnostic systems[31, 32]. Through history several attempts to replace tissue by compounds made of biomaterials are known. Early examples include the use of artificial eyes, ears, teeth and noses in ancient Egypt[33] or Aztec, Chinese and Roman culture [34]. The materials used include metals such as gold in dentistry, but also wooden teeth and glass eyes were found [32]. As medicine evolved implants became more sophisticated and the use of ceramics, metals and polymers for the fabrication of implants has become common practice[35]. A major concern is the development of methods that avoid risk, rejection or complications in the body [27, 36-38]. In order to achieve this, biocompatible materials, which do not have the potential to elicit an immunological, or clinically detectable primary or secondary foreign-body reaction, are of great interest. In general two strategies can be distinguished in order to achieve a biocompatible implant[39]. First of all, it is possible to apply a suitable biocompatible coating layer on the implant. Secondly a biocompatible bulk material can be used for the fabrication of the device. Biocompatibility is defined as the ability of a material to perform with an appropriate host response in a specific application[40].

The term “biodegradable polymer” was defined by Vert in 1994 as a solid polymeric material or device which breaks down due to macromolecular degradation with dispersion *in vivo* but no proof for elimination from the body[41]. This includes that the degradation of a biodegradable polymer is enzymatic. The term “bioresorption” is used to indicate that degradation is followed by a complete elimination of the by-products from the body[42]. It is necessary to investigate the consequences of the presence of these biodegradation products in the human body. As an example, the synthetic poly(lactic acid) (PLA) degrades to lactic acid by hydrolysis, which is a compound normally present in the metabolic pathways and thus non-toxic. However, the degradation of an implant can cause higher local concentrations and accumulation which can not easily be disposed, resulting in a local tissue inflammation[43] or enzymatic hydrolysis [44, 45]. The acidic by-products due to scaffold degradation were correlated with limited cell growth and angiogenesis by Sung et al. [46]. Earlier studies concerning the degradation of PLA and other aliphatic polyesters are available. The results indicate that inflammatory reactions should be considered when developing a scaffold [23, 25, 42, 47, 48]. It is thus important that sufficient vascularisation of the scaffold occurs, in order to expose the scaffold/cell construct to a neutral environment and thus limiting the foreign body reaction[49].

1.3.2. Natural Versus Synthetic Biomaterials

It is necessary to make a distinction between polymeric biomaterials from natural origin and these of synthetic origin. Natural polymers show an excellent biocompatibility. Furthermore they closely mimic native cellular environments, especially when the compounds are naturally present in ECM and are mostly easily degradable by an enzymatic or hydrolytic mechanism. They have unique mechanical properties but these properties can show batch-to-batch variations. Other concerns are a risk of viral infection (such as BSE for collagen), antigenecity, and unstable material supply[40].

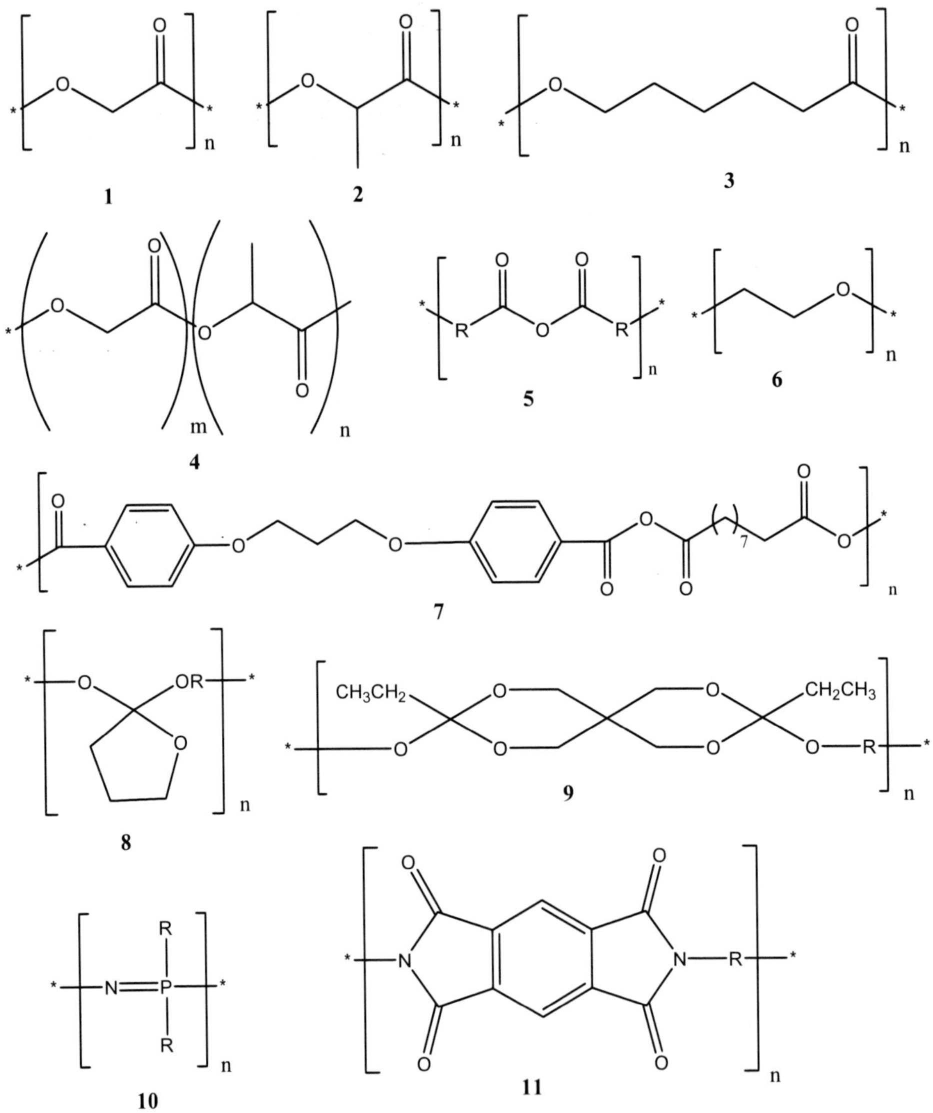

Figure 2. Some synthetic biomaterials : 1 poly(glycolic) acid (PGA), 2 poly(lactic acid) (PLA), 3 poly-ε-caprolactone (PCL), 4 poly(DL-lactide-co-glycolide) (PLGA), 5 polyanhydride, 6 poly(ethylene oxide) (PEO), 7 poly[(carboxy phenoxy)propane-sebacic acid] ,8 poly(ortho ester I), 9 poly(ortho ester II), 10 polyphosphazene, 11 polyimide.

Synthetic polymers do not show this variation in chemical and physical properties. They can be tailor-made, and thus match the exact properties required for a specific application. The properties of a specific class of synthetic polymers can be correlated with their composition, the fabrication method and modifications due to post-polymerization reactions[50]. As a consequence it is feasible to control the degradation rate of these polymers by varying their structure. This flexibility, as well as their unlimited availability and high reproducibility are important advantages compared to polymers from natural origin[25]. Investigations based on the use of synthetic biodegradable polymers have mainly focused on the polymers already approved by the food and drug administration (FDA)[49]. On the other hand, compared with natural polymers there is a higher chance of inflammation and foreign-body reaction for implants fabricated using synthetic polymers. Research on skin, cartilage, bone, vascular, nerve and liver regeneration have already been reported and seem promising approaches[40, 51]. One of the first commercially available systems includes Dermagraft, a skin prosthesis[52].

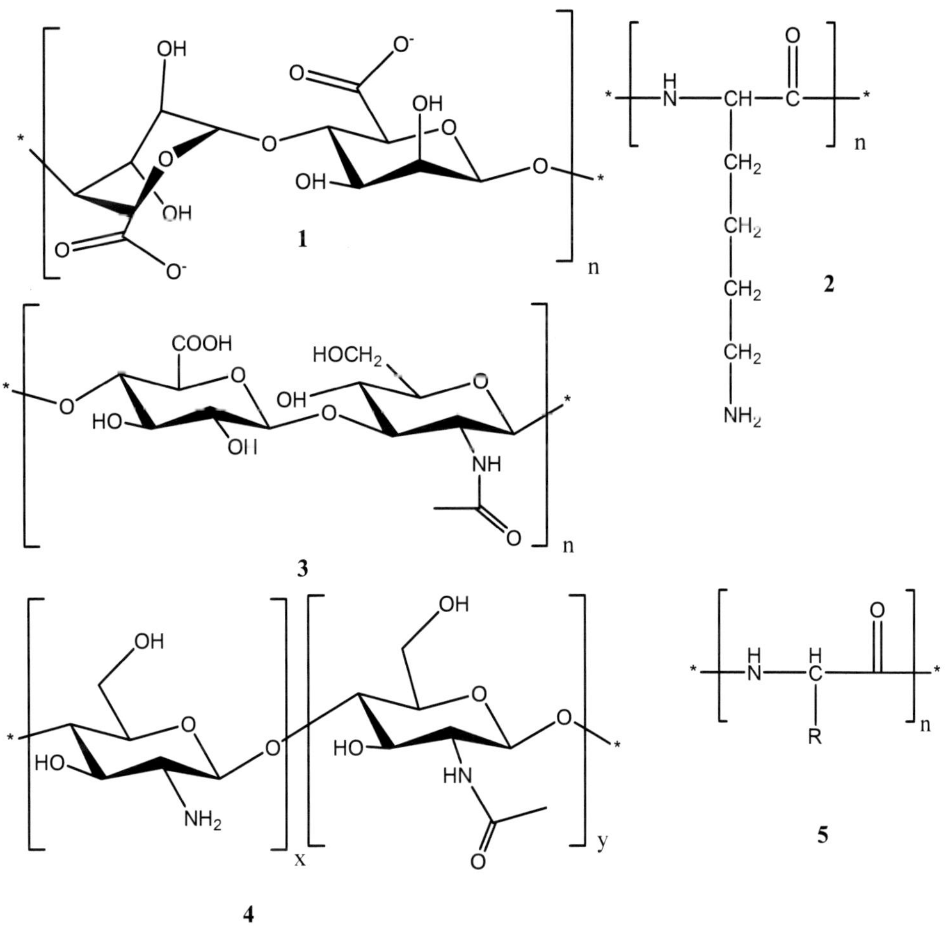

Figure 3. Some natural biomaterials 1 alginate, 2 poly(α-lysine), 3 hyaluronan, 4 chitosan, 5 poly(α-aminoacids).

1.3.3. Composites

It is clear that each biomaterial has unique advantages and limitations. Clinical studies and laboratory experiments evaluate these properties and the suitability of a material for a specific application[53]. When designing a scaffold the art lays in choosing the right biomaterial for the right application, or to identify a successful combination of biomaterials. The past decade a tendency towards the application of these composites, or hybrid materials, was observed. The combination of the high biocompatibility and biomimetic properties of natural polymers with the desired properties of synthetic polymers has led to numerous combinations of polymeric materials[54]. A general overview on the use of polymer-composite materials for biomedical applications is available [36] and more recent publications on the use of composites in tissue engineering indicate the potential of this strategy[53, 55-59]. Some hybrid materials were developed using a polymer with an inorganic filler such as tricalcium phosphate (TCP), hydroxyapatite (HA)[59-61] or basic salts[62]. Such a composite material can not only improve the biocompatibility but also enables the tailoring of the bulk degradation and resorption kinetics [57]. The potential use of poly(α-hydroxyacids) – silica sol-gels for replacement of cancellous bone has been studied by several groups. One of the main advantages is that by varying the molecular weight of the poly(α-hydroxyacid) and the silica content, the mechanical properties of these materials can easily be modified in order to match these of cancellous bone [63].

Collagen can be combined with other biopolymers or bioceramics to create a composite scaffold. Different authors reported increased mechanical strength and enhanced cell interactions of these scaffolds [56, 64, 65].

1.3.4. Biomimetic Materials

In order to achieve tissue formation the scaffold does not only have to provide a support for cell adhesion and proliferation but also has to stimulate the desired differentiation of cells and thus promote the secretion of new ECM constituents.

Generally, cells bind to the ECM through the receptors on the cell walls. One class of receptors are integrins, which bind selectively to specific binding sites such as arginine-glycine-aspartic acid (RGD) tripeptide found in cell adhesive proteins such as vitronectin, laminin and fibronectin. Beside attachment, these connections mediate several intracellular signals. These define mobility, cellular shape and regulate the cell cycle[66]. In the case of biomaterial-cell interactions, the receptor binding to the ligand that is presented from the biomimetic material determines the strength of the cell attachment, the cell migration rate and the extent of cytoskeletal organization formation[16].

It is required to understand how biomaterials interact with different kinds of cells and how they can be modified in order to guide cells. Considering the cell-ECM interaction two strategies can be proposed: immobilisation of complete proteins which contain the desired binding sites or the incorporation of the specific binding sites (e.g. RGD). Here, natural biomaterials seem to be advantageous. These materials posses receptor-binding ligands and are biologically recognizable. Still, the same concerns exist concerning pathogen transmission and immunogenicity. The previous years, several research groups have investigated how synthetic biomaterials and biomolecular approaches can be developed that recapitulate the elaborate biological recognition and signalling functions of the extracellular milieu[66]. The idea is to synthesize biomaterials equipped with molecular cues mimicking certain aspects of structure or function of natural extracellular microenvironments[66]. This modification of the biomaterial can be achieved by surface or bulk modification with one or more cell recognition molecules [16] as shortly outlined in the following paragraphs.

Bulk Modification

Bulk modification is a strategy whereby cell-signaling peptides are incorporated into the biomaterials and the resulting recognition sites are not only present on the surface but also in the bulk of the materials[16].This is mostly beneficial for applications focusing on injectable biomimetic materials. The encapsulation of cells within these materials is desirable for localised delivery by injection, thus avoiding a surgical procedure. Two main groups of injectable materials are hydrogels and microspheres.

An example includes the synthesis of alginate-based hydrogels covalently coupled with RGD spaced by a four amino acids sequences, Gly-Gly-Gly-Gly (GGGG) which was reported by Marler et al. [67]. The fine-tuning of the visco-elastic properties of these kinds of gels is necessary in order to achieve a gel which is injectable and able to occupy a defined space which matches the complex shape at the site of defect. Another example is the delivery of mesenchymal stem cells through injection of a peptide-based hydrogel [68]. Fittkau et al. reported that the desired response of specific cell types to tissue engineering scaffolds could be optimized through a combinatory approach to the use of biomimetic peptides[69]. This group studied the bulk modification of polyethyelene glycol (PEG) with RGD, YIGSR and PHSRN. Hydrogels modified with YIGSR or PHSRN showed only limited adhesion. When these peptides were combined with RGD, the YIGSR peptide was found to selectively enhance the migration of human microvascular endothelial cells by 25% compared to RGD modified materials. This indicates that the incorporation of multiple biomimetic factors can result in a more successful scaffold. Nevertheless, fine-tuning of the incorporated concentrations and the visco-elastic properties of the delivery system remain challenges.

The bulk incorporation of enzymatically degradable peptide sequences and/or functional groups in biomaterials offers opportunities to fine-tune the degradation rate. Mann et al. reported the development of a photo crosslinkable PEG hydrogels with protease sensitive peptide sequences[70].

Surface Modification

The specific nature of a surface determines how the living host tissue will interact with the implant. Both physical and chemical properties of the surface are of major importance. Physical factors include the wettability, surface energy and the micro-and nanotopographic properties such as roughness. From a chemical point of view, the presence of biological binding sites and the released products are critical.

The interaction of a surface with the surrounding tissue after implantation is a process which can be sub-devided in four different stages[71]. In the first stage a water shell is formed around the material in a couple of nanoseconds. Secondly, proteins will be adsorbed from the ECM. During the third stage, cells reach the surface and will interact with the implant through the protein covering. This complex process is dependant on several critical surface properties and is strongly influenced by biological molecules. This third stage occurs from as early as minutes after and up to days after implantation. In a final stage, the useful life of the implant, is the continuing development of the early implant stages. The duration of this stage can vary from days to several decades.

A cell interacts with its surrounding through protrusions, which are micron sized sheet-like structures composed of an actin filament mesh with, on the end, smaller hair-like protrusions, '*filopodia*'. These filopodia consist of long, thin actin filament bundles which act as feelers that sense the ECM and substrate surface. When the filopodia find a suitable binding site, such as the previous discussed lock and key integrin binding sites, a feedback signalling pathway within the cell allows more integrin receptors to be localised in that region of the cell[71].

When considering this complex interaction it is clear that the coating of a surface with a protein to which cells can bind, or the incorporation of the cell-binding sites derived from these proteins can significantly enhance adhesion. A surface can be functionalised by physical adsorption or chemical modification. In order to achieve a more stable cell adhesive stratum, it is desirable to bind biomolecules covalently to the surface[72]. The most important chemical and radiation induced techniques include hydrolysis [73-75], aminolysis [76, 77], UV-activation [78, 79], organosilanization[80] and plasma treatment [81-86].

Kallröt et al. reported a new solvent-free, single-step, non-destructive grafting procedure whereby the substrate is subjected to a vapor phase of a vinyl monomer and grafting is induced by photoinitiation using benzophenone[87, 88]. This technique was also used to enhance the *in vitro* degradation rate of poly-L-lactide (PLLA) by grafting of acryl amide, N-vinyl pyrrolidone, or acrylic acid [89].

Another emerging field is the modification of surfaces by means of highly energetic laser beams. Khorasani et al. used a pulsed CO_2 laser to modify the surface of a polydimethysiloxane (PDMS) membrane of 0.3 mm thickness[90]. This laser treatment introduced peroxide groups onto the PDMS surface, thus providing initiating sites for graft polymerisation of hydroxyethylmethacrylate (HEMA). The modified surface exhibited a reduced platelet adhesion[91]. Tiaw et al. used femtosecond and excimer lasers to perforate and modify the surface of poly-ε-caprolactone (PCL) membranes. Both laser types enhanced the wettability of the PCL membrane [92].

After activation of the surface, the immobilization of naturally derived macromolecules, small peptide moieties or cell-selective ligands derived from the cell adhesive proteins is feasible[72]. Both approaches will be outlined below.

The immobilization of proteins such as collagen or gelatin onto the surface of tissue engineering scaffolds in order to endow cell adhesive properties to the scaffold is reported extensively in literature[57, 75, 77, 81, 83, 93, 94]. Collagen is a natural protein, and a major structural component forming the natural ECM of connective tissues and organs[95]. The immobilization of collagen on PLLA, polyglycolide (PGA) and/or their copolymers poly(DL-lactide-*co*-glycolide) (PLGA) showed enhanced adhesion and spreading of fibroblasts[96] and chondrocytes [97]. Foo et al. reported the successful immobilization of collagen onto PCL membranes using a catalytic graft polymerization. Human umbilical vein endothelial cells (HUVECs) showed higher proliferation and differentiated cobblestone morphology on these substrates[93]. The major problem concerning the use of collagen is its immunogenicity and antigenicity. The use of gelatin can be considered a good alternative because of its absence of antigenicity and ease of handling at high concentrations[72]. Gelatin immobilized on polymers also shows enhanced attachment, proliferation and ECM deposition[98]. Choong et al. conducted a co-culture study of bone human bone marrow-derived fibroblasts (HBMFs) and human bone marrow endothelial cells (HBMEC) on modified PCL. The gelatine-coated and hydroxapatite coated substrates were superior for HBMEC attachment and proliferation, compared to hydrolyzed-only or untreated PCL substrates[74]. The co-culture of HBMFs with HBMECs was compared with the co-culture of skin fibroblasts with HBMECs. Enhanced HBMFs differentiation and proliferation was observed, compared to skin fibroblasts. Other natural polymers such as chitosan immobilised on PCL, were reported to enhance growth of human dermal fibroblasts[99].

Another example of biomimetic surface modification includes the incorporation of hyaluronic acid, a non-sulfated glycosaminoglycan (GAG), in order to promote wound-healing capabilities[100].

The immobilization of short chain peptides such as RGD in order to mimic the natural environment of cells was also studied by many research groups [44, 72, 76, 101-104]. The incorporation of RGD on the surface is reported to not only enhance cell adhesion and growth but also cases where differentiation and ECM production were stimulated were shown e.g. for osteoblasts[105]. Also the promotion of differentiation of bone marrow stem cells into osteoblast-like cells was found to be stimulated by incorporation of GRGDY peptide sequence in porous scaffolds[106]. Reyes et al. coated titanium implants with a GFOGER collagen-mimetic peptide. Their findings showed that osseointegration of the GFOGER-modified implants was significantly enhanced compared to titanium modified with full-length type I collagen[94]. These findings seem to indicate that not only cell-attachment can be controlled by the incorporation of short chain biofunctional peptides but also provide experimental evidence for the statement that the integrin-RGD interactions guide cells towards a final phenotype. This technology can be of great importance in creating successful scaffolds that will, besides supporting autologous stem cells, will also stimulate the differentiation. Research should enable us to control the exact concentration and spatial distribution of these short chain peptides through the scaffold in order to mimic more complex natural tissues.

Also certain sugars were studied e.g. the immobilization of galactose to achieve better hepatocyte attachment, viability and metabolic functions[107].

1.3.5. Delivery of Biofactors

Next to the fact that scaffolds should be highly biocompatible, exhibit desired mechanical properties and have the ability to mimic the natural ECM it is also desirable that implanted scaffolds should possess the ability to deliver different biofactors to the site of implantation in a controllable and sustained manner. These biofactors can be growth factors (or inhibition factors), cells, DNA, drugs or combinations.

Growth Factors

Growth factors can be incorporated by bulk encapsulation, specific or non-specific surface adsorption, and adding growth-factor containing microspheres encapsulating them. Research is focused on growth factors stimulating angiogenesis, bone and cartilage regeneration, wound repair, liver regeneration and nerve growth[51, 108]. It is critically important that the profile of growth factor delivery at the diseased treatment site should be local, low and sustained to induce localized growth of healthy tissue. Secondly, the combined delivery of multiple growth factors with complementary functions should be pursued[109]. Another concern is the bioactivity of the released growth factor[110].

As mentioned before, an important aspect which should be taken into account when creating a larger implant is the vascularisation in order to maintain a sufficient nutrient and waste product flow. Implants with insufficient diffusion (interconnectivity, pore size, etc.) were reported to sustain bone ingrowth only on the surface (500μm)[111]. It is thus of major importance to be able to stimulate endothelial cells to diffuse into the scaffold and form vascularised tissue, thus creating an environment in which cells can survive. Some of the most common angiogenic growth factors include vascular endothelial growth factor (VEGF), acidic or basic fibroblast growth factor (aFGF, bFGF), angiopoietin, and platelet-derived growth factor (PDGF) It is shown that dosage of these factors must be tightly controlled, as excess amounts of VEGF can cause severe vascular leakage and hypotension[109].

Bone and cartilage regeneration can be accelerated by localized delivery of several growth factors with complementary functions. As bone is a highly vascularised tissue, this problem is a complex compromise between several factors [112]. Oest et al. reported the co-delivery of BMP-2 and TFG-β3, combined with RGD-alginate hydrogel, to femoral defects within a polymer scaffold. The improvement of scaffold properties by incorporating growth factors significantly increased bone formation but still, only in a limited number of cases bone bridging of the defect region was established[113].

Coculture studies showed that not only the pro-angionic growth-factor delivery systems should be taken into account but also the influence of the growth-factor of different cell types could be a very useful technique in TE in order to achieve a certain pre-vascularization of the scaffold before implantation [114].

Cells

The delivery of cells is highly dependent on the cell adhesion, cell-seeding efficiency and viability of the cells in the scaffold. Seeding-efficiency remains a problem and several attempts were undertaken to seed cells in a more efficient manner. Zhou et al. reported a combined seeding method whereby bone marrow stromal cell sheets (BMSC) were wrapped around a pre-seeded composite scaffold, thus preventing the enclosed cells of migrating out of the scaffold and simultaneously providing cells on the outside of the scaffold. It was reported that the layered cell-sheets integrated well with the scaffold and remained viable [115]. Chen et al. also reported that cell-sheet approaches in combination with custom-shaped scaffolds have potential in producing bone-tissue[116]. Liao et al. used composite hyaluronic acid/collagen I hydrogels to increase the seeding-efficiency and cellular retention of porcine chondrocytes into poly(propylene fumarate) scaffolds[117]. Attempts were made by Schantz et al. to enhance cell seeding efficiency by means of fibrin glue [118]. Chimizu et al. reported the use of magnetic force and magnetite nanoparticles to increase the cell-seeding efficiency and demonstrated the potential use of this "Mag-seeding" for vascular tissue engineering[119].

Another strategy for increasing the cell-seeding efficiency is the direct incorporation of the cells in the biomaterial during the fabrication of the scaffold[49]. This strategy seems very promising but in order to be successful the complete fabrication process has to be compatible with living cells. This imposes new complications such as the use of solvents, higher temperatures and shear/stress limitations.

Genes

On the other hand, growth factor releasing scaffolds have some inherent limitations such as short half life of the protein, low bioavailability, enzymatic inactivation and a high cost of production[120]. Therefore, the use of gene transfer techniques in order to achieve the localized, long-term production of therapeutic levels of the desired growth factors in appropriate concentrations seems to be a possible alternative. Several gene transfer strategies, both non-viral and viral techniques, have been used in tissue engineering[121]. The use of several different gene delivery strategies is described in some reviews[121, 122]. Ko et al. reported composite biomaterials based on natural polymers with calcium phosphate in order to make the scaffolds mechanically strong and chemically suitable for use as a gene delivery vehicle[123]. Jang et al. reported a subtrate-mediated delivery strategy whereby polyethylenimine/DNA complexes were immobilized on PLGA scaffolds. The levels of transgene expression were similar to or greater than control studies based on bolus DNA delivery[124].

Drugs

Small molecular weight drugs can also be incorporated into biodegradable scaffolds in order to stimulate cells or prevent an inflammatory reaction. Prabu et al. reported the synthesis of a composite PCL/collagen scaffold which was subsequently loaded with amikacin and gentamycin in order to study their release profile [125]. Moroni et al. created scaffolds consisting of hollow poly[poly(ethylene oxide) terephtalate-co-(butylene) terephtalate] (PEO/PBT) fibers. These hollow cores were loaded with lysozyme in order to study the release properties of these scaffolds. A burst effect was detected over the first 24 hours for the release from hollow-fiber scaffolds, but when the loaded hollow fibers were sealed at both ends, a controlled release over a month was obtained[126, 127].

Concerning future prospects, it is generally accepted that the combined use of biomaterial scaffolds, drug delivery technology and gene therapy has great potential to provide improved tissue replacements[128].

1.3.6. Understanding Material-Cell Interactions

There is a limited knowledge of the specific interactions between materials and cells. The bioactivity of a material describes its positive effect or interaction with living tissue[71]. Bioactive materials elicit specific biological responses from the host[108]. In general, the development of new biomaterials has been an iterative process: polymers were rationally designed and then individually tested for their bioactive properties[129]. Until now, most of the studies addressing the cell-biomaterial interactions, use this trial-error approach, resulting in different experimental systems. Consequently it is not always clear to what extend the underlying biomaterial has a determining effect on the final phenotype of the seeded cells. This was demonstrated by McGuigan et al. in a recent comparison of different studies on the thrombogenicity of seeded endothelial cells[130]. The authors stated that future studies should set up more uniform experimental systems which should take in account the control over experimental variables. In addition, quantitative results should be reported in a comparable way.

Several emerging fields of science attempt to provide more evolved techniques for biomaterial evaluation, in order to enable scientists to develop diverse libraries of polymeric biomaterials and screen the biomaterial's interactions with specific cells in a fast, convenient manner[32, 129, 131]. Anderson et al. reported the construction of a polymeric microarray using a library of polyesters. A large amount of human mesenchymal stem cell-biomaterial composite interactions (> 3000) could be evaluated, thus leading to the fast selection of polymeric blends which support growth and attachment of cells. This emerging scientific field is often referred to as *materiomics*.

An overview of different studies of surface chemistry describing different functional groups and the varied cell responses from different kinds of cells is available[71].

Huebsch et al. recently reported the application of fluorescent resonance energy transfer (FRET) as a tool for probing molecular cell-biomaterial interactions in three-dimensions. This technique should enable scientists to quantify several cell-material interactions such as molecular changes in matrix proteins induced by interactions both with surfaces and cells, the number of bonds between integrins and their adhesion ligands and changes in the crosslink density of hydrogel synthetic ECM analogs[132]. The same group also used FRET as a non-invasive method to study the spatial organization of polymer chains in alginate hydrogels[133].

2. Scaffold Fabrication Techniques

2.1. Classical Versus Rapid-Prototyping

A scaffold can be fabricated by several techniques which allow the introduction of pores into the internal architecture.

2.1.1. Conventional Techniques

Classical techniques used compression/injection/melt moulding, fibre bonding, phase separation, solvent casting/particulate leaching, membrane lamination, gas foaming/high pressure processing, the lost mould method, emulsion freeze drying, and combinations of these techniques and machining techniques[39]. Non of these techniques was able to meet the constraints a more ideal scaffold is likely to fulfil. Especially the lack of mechanical strength, the limited porosity and the poor interconnectivity impose problems[134]. The precise variation of the external geometry can not be obtained with these techniques, thus making patient-specific implants hard to manufacture. Furthermore it is impossible to create a spatial distribution of the pores, or to create internal channels within the scaffold[111]. An overview of the conventional techniques with a short description of the basic principle and some references is presented in Table 1.

2.1.2. Rapid Prototyping

Solid freeform fabrication (SFF) is the general name for a number of techniques which produce solid objects by the specific, sequential delivery of energy and/or material. In literature these are referred to as layered manufacturing, additive fabrication, rapid prototyping (RP) rapid manufacturing (RM)[153] and rapid tooling (RT). RP, RM and RT respectively refer to a fast production of a prototype, a finished object or a tool. However, in literature, this terminology is not always used in a straightforward way.

Table 1. An Overview of Different Conventional Scaffold Fabrication Techniques

Technique	Principle	Main disadvantages	Reference
Solvent-casting particulate leaching	A uniform suspension of a polymer with salt particles of a specific diameter is prepared. After evaporation of the solvent the structure is immersed in water, leaching out the salt particles and thus creating a porous structure.	Limit to membranes of 3mm thick, not applicable for load-bearing tissue, solvent residue may be harmful.	[106]
Gas foaming	A gas (e.g. CO_2) is dissolved into a polymer under high pressure until saturation. When bringing back the pressure to atmospheric level the solubility of the gas decreases, leading to the formation of gas bubbles.	Mostly a non-porous surface, closed-pore structure inside the matrix.	[135]
Fibre networking	Scaffolds are produced from biodegradable fibres which were created through classic textile technology or electrospinning (ESP). Different routes, such as non-woven, knitting and fibre bonding have been reported. In the latter one a solution/heating process can be used in order to bind the fibers together, leading to scaffolds with better mechanical properties.	Not useful for load-bearing tissues, limit applications to other polymers. Difficult to control pore size and shape. In fibre bonding situations might occur where toxic solvents are used.	[136-141]
Phase separation	A biodegradable polymer is dissolved in a solvent The temperature is decreased, leading to a liquid-liquid phase separation and quenched to form a two-phase solid. Sublimation of the solvent leads to a porous biopolymer scaffold.	Scaffold morphology is difficult to control, solvent residue might be harmful.	[37]
Melt moulding	A mixture of solid biodegradable polymer and microspheres of a specific diameter (e.g. from gelatine) is brought into a mould. Temperature is increased above the glass transition temperature of the polymer, leading to the fusion of the polymer particles. After cooling the microspheres are leached out by immersion in water.	High temperature, labour-intensive process.	[142-144]
Emulsion freeze drying	A mixture of biodegradable polymer/solvent with water is homogenised to form a water-in-oil emulsion. After quenching and freeze-drying, a porous scaffold remains.	Limited thickness	[145]
Solution casting	A biodegradable polymer is dissolved in a solvent. Adding a non-solvent leads to the precipitation of the polymer.	Solvent residue might be harmful	[146, 147]
Freeze drying	A dispersion or solution of a biodegradable polymer is frozen, creating ice-cristals which force the polymer in the interstitial space, thereafter freeze-drying is used to yield a porous scaffold. Varying different parameters such as concentration, freezing rate, pH,.. leads to different pore-sizes. Chemical or physical crosslinking can be used to make the scaffolds less soluble.	Interconnectivity of pores, limited thickness.	[100, 148-152]

Rapid prototyping (RP) is the direct production of finished goods from a layer manufacturing device[39]. The different techniques using rapid prototyping were originally developed in order to create prototypes for industrial purposes. From a medical point of view, the use of medical models fabricated by rapid prototyping is of great value for the planning and simulation of oral, maxillofacial and orthopaedic surgery, or in forensics[154]. It is only the last decades that scientists showed interest in the possible application of rapid prototyping techniques in order to fabricate implantable scaffolds[155].

RP techniques are computer-aided techniques. This means that all the information is available in a 3D-design. Software will decompose the 3D-design or images of the object in 2D-layers. The RP machine will construct these 2D-layers one by one, thus reconstructing the 3D-object.

RP processes fabricate a finished object directly from this three-dimensional computer model. This model can be designed in CAD or, the 3D data can be retrieved from a CT-scan or MRI-scan. As the latter approach enables the automated, direct fabrication of scaffolds with a complex, patient-specific external geometry, this is the first major advantage of rapid prototyping techniques compared to classical methods.

The fabrication process is a layer-by-layer additive technique. Building a layer at a time in a automated, unattended, manner enables the construction of complex geometrical objects, with fully interconnected pores. The freedom of design is far better compared to conventional fabrication methods. Pore size can be precisely controlled and is only limited by the resolution of the applied technique, resulting in a highly designable microstructure where structural inhomogeneous pores can be easily created. The use of heterogeneous materials is another powerful advantage. The fabrication with RP is fast due to the high degree of automation and the limited number of process steps from image to finished device. For the production of a limited number of devices the cost-efficiency of RP techniques is superior to this of classical production methods[156].

Different RP technologies were used for the creation of *in vivo* devices. An overview of the different techniques with the basic principle is presented in Table 2 and is available in literature [21, 22, 27, 111, 141, 157-162]. Most authors make a distinction between direct manufacturing of a scaffold and indirect methods. The first technique uses rapid prototyping to directly create a scaffold from a biomaterial. The second approach uses rapid prototyping to create a sacrificial mould[163]. The biomaterial is cast into the mould and afterwards this mould is removed. This paper will mainly focus on direct manufacturing of scaffolds. Therefore, wax printing, a RP technique used to create wax sacrificial moulds will not be considered [164].

2.1.3. Different RP Techniques

It is possible to make a distinction between the different RP techniques as a function of the phase of the processed polymer. The polymer can be in a liquid phase; molten or dissolved, or can be in a solid state at the moment of processing.

Solid-phase processing implies that the polymer particles are fused by means of energy (selective laser sintering, SLS) or solvent addition (3D-printing, 3DP and Theriform); see Figure 4. A thin powder layer is deposited. Subsequently heat or binder-droplets are delivered on the programmed places, locally fusing the polymer particles together, resulting in a 2D-cross-section of the desired object. When this cross-section is finished the next layer of powder is deposited onto the first cross-section and so the process is repeated until the complete 3D-object is constructed.

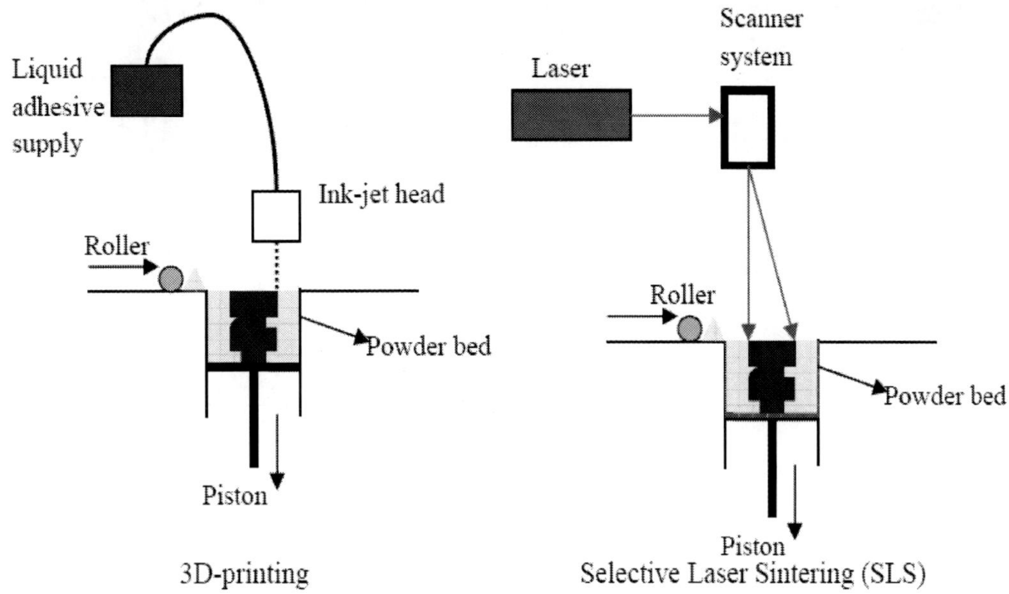

Figure 4. solid-phase RP strategies.

Liquid-phase processing of polymers can be based on several approaches; see Figure 5 . A first possibility is that a molten thermoplastic polymer can be extruded. Thin filaments are extruded in a software-controlled way, thus creating a 2D-cross-section. When a layer is finished the extrusion device moves up, or the construct goes down, and a second layer is constructed. This is the case for Fused Deposited Modelling (FDM) and analogues such as Three-dimensional fiber-deposition techniques (3DFDP), Precision extruding deposition (PED), and Precise extrusion manufacturing (PEM). An alternative for laser curing is "the masked lamb" approach which implies that the complete layer is illuminated, instead of using a point-to-point method. In literature, this method is rarely reported for use in tissue engineering, and will thus not be further considered.

Stereolitography apparatus (SLA) is the technique where energy is locally delivered to a thin layer of a solution of photopolymerisable polymers. These are crosslinked on those programmed places, thus creating a 2D-cross-section. A next layer of solution is deposited and the curing restarts, adding more and more layers until the 3D-object is finished. Micro-stereolitography and 2-photon polymerisation are similar techniques, but the latter is based on the reaction between monomers and initiator in a transparent medium.

It is also possible to dispense a polymer solution through a nozzle or a syringe, thus creating small filaments. This is the case for 3D bioplotting, pressure-assisted microsyringe (PAM), low-temperature deposition manufacturing (LDM) and Multi-nozzle deposition manufacturing (MDM).

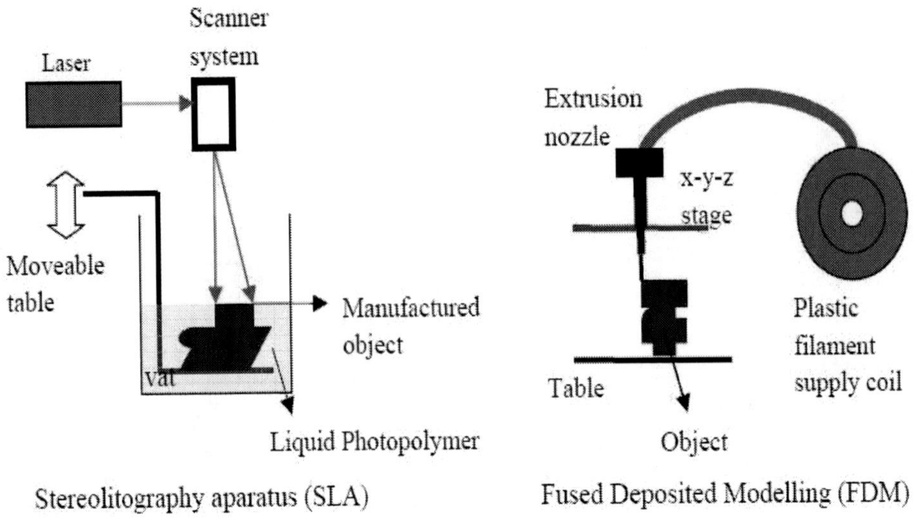

Figure 5. liquid-phase RP strategies.

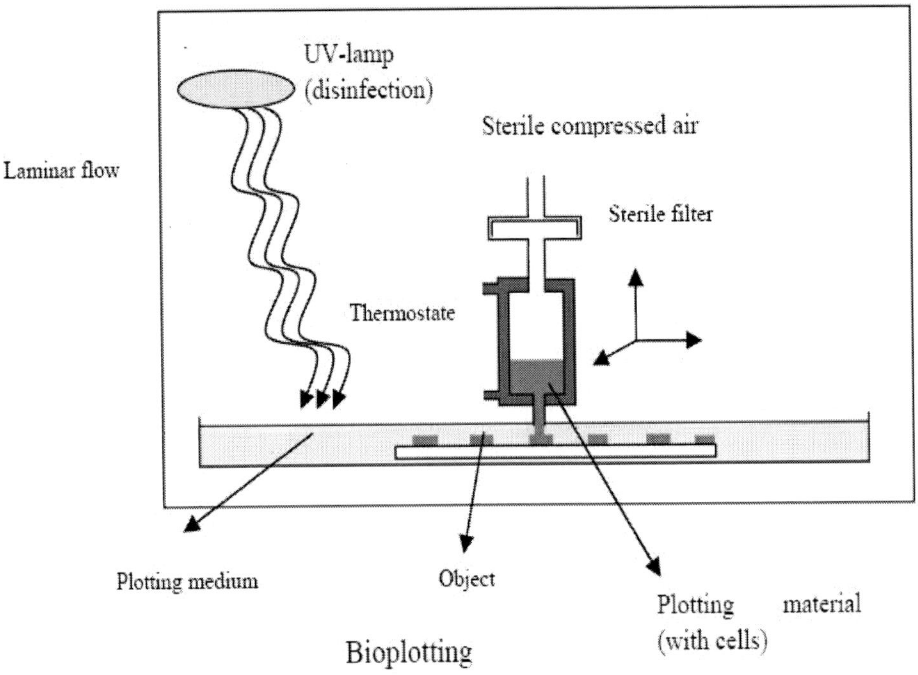

Figure 6. syringe-based RP processing.

Bioprinting is yet a somewhat different technique where cells or cell-aggregates, the so-called bio-ink, are directly dispensed onto a suitable substrate, the so-called biopaper. Commercially available printing-systems can be adapted for this purpose.

A schematic overview of these techniques is presented in Figure 7. The stronger and weaker points of each technique are summarised in Table 2 and Table 3.

The Laminated object manufacturing (LOM) and Multiphase jet solidification (MJS) are two other RP techniques worth mentioning but as their use for tissue engineering is rather limited they will not be considered. LOM is a patented sheet lamination technique. A CO_2 laser is used to cut a cross-section out of a feed material, e.g. paper doped with polymer, in a second stage these layers are bounded together by means of pressure and temperature[161, 165]. MJS is used for the creation of metal of ceramic parts out of low viscosity materials in powder of pellet form which are extruded through a jet in liquid form[158].

In the next paragraph we will discuss some of these liquid-phase techniques and the most significant improvements and results which were obtained the last decade.

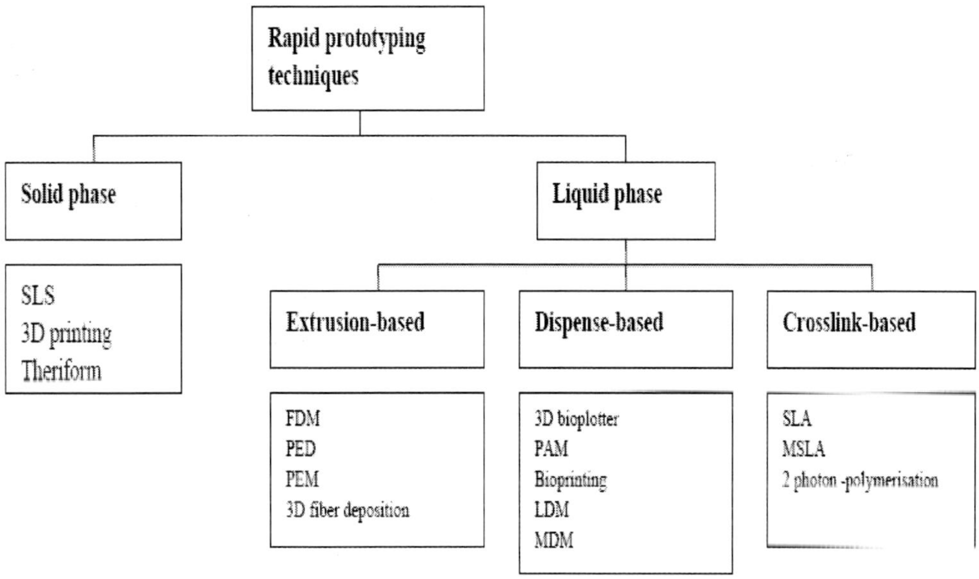

Figure 7. overview of different RP instruments.

2.2. Recent Advancements in Liquid-Phase RP for Tissue Engineering

As RP techniques originally were developed for industrial use no special features enabling them to work in sterile circumstances with biological, sometimes living, materials were considered in their original set-up. The past decade more and more systems were adapted to these circumstances and the application of biomaterials. We will briefly comment on some recent trends in this research area, mainly focusing on rapid prototyping technologies where a biomaterial is deposited in a controlled manner in order to directly fabricate a scaffold.

Table 2. overview of solid-phase RP-techniques

Rapid Prototyping Technique	Principle	Resolution	Advantages	Drawbacks	References
Selective Laser Sintering (SLS)	Solid parts of materials are locally fused together by means of laser energy. This process is repeated for each new layer of powder.	500 μm	Wide material choice, no support required, relatively cheap	Trapped materials in the interior might be difficult to remove, rough surface texture, curing step is required.	[166, 167]
3D-ink jet printing (3DP)	Solid parts of a powder are locally fused together by dispensing a solvent (e.g. water) on specific places in very low quantities. This process is repeated for each new layer of powder.	180 μm	Low temperatures, good control over pore-size and interconnectivity, no support required, large number of potential materials	Trapped materials might be difficult to remove, post-processing is required. When the binder is an organic solvent, residue might be harmful.	[168-173]
Theriform	Solid parts of a polymer powder are locally fused together by dispensing a solvent (e.g. water) on specific places in very low quantities by using ink jet printing technology. This process is repeated for each new layer of powder.	300 μm	Analogue to 3D-ink jet printing	Analogue to 3D-ink jet printing	[174-176]

Table 3. Overview of Liquid-Phase RP-Techniques

Rapid Prototyping Technique	Principle	Resolution	Advantages	Drawbacks	References
Fused Depostion Modeling (FDM)	A material is melted and extruded through a moveable heated nozzle forming thin filaments which are deposited and fuse together upon cooling.	250 µm	No trapped materials, good control of internal and external morphology, very little material waste.	High temperatures therefore direct cell-inclusion is not possible, difficult to insert a microscale porosity, limited choice in materials	[177]
Three-dimensional fiber-deposition technique (3D-FDP)	A molten material is dispensed from a heated stainless steel syringe by means of mechanical pressure.	250 µm	Same as FDM, input material in pellet form.	High temperatures, difficult to insert a microscale porosity	[178]
Precision extruding deposition (PED)	PED is analogue to FDM but in conventional FDM precursor filaments are required. The use of a mini-extruder system in PED allows the direct extrusion from granulates or pellets.	250 µm	Same as FDM, input material in pellet form.	High temperatures, difficult to insert a microscale porosity	[179]
Precise extrusion manufacturing (PEM)	PEM is analogue to FDM but a special sprayer is used. The sprayer consists of a cylindrical liquefier, allowing the material to get in a semi-liquid state. The semi-liquid is dispensed by compressed air.	200-500 µm	Same as FDM, input material in pellet form.	High temperatures, difficult to insert a microscale porosity	[180]
3D bioplotter	A versitale 3D dispense device which can deposit liquids ore pastes such as a hydrogel into a liquid plotting medium, which compensates the gravity and thus ensures structural integrity. Next to hydrogels it is also possible to extrude materials at higher temperatures.	250 µm	Compatible with many materials, biomolecules may be incorporated, scaffolds made of hydrogels offer opportunities for soft-tissue engineering. Versatile device.	Parts might exhibit low mechanical strength, low accuracy and resolution, rather slow processing times.	[181-184]
Pressure assisted microsyringe (PAM)	A microsyringe filled with polymer solution in a computer-controlled, three axis micropositioner uses compressed air to dispense the polymer solution through a glass capillary needle.	10-600 µm	Many materials can be used, biomolecules can be incorporated, very good resolution, not subject to heat.	Requires solvent; unable to incorporate particles which limits the use of composites and leaching-methods. Narrow range of printable viscosities.	[185, 186]

Table 3. (Continued).

Rapid Prototyping Technique	Principle	Resolution	Advantages	Drawbacks	References
Low-temperature deposition manufacturing (LDM)	A slurry of biodegradable materials is prepared and dispensed through a screw pump nozzle onto a refrigerated platform. The frozen scaffold is subsequently freeze-dried to remove solvent.	400 μm	Working at low temperatures offers the possibility to incorporate biomolecules.	The use of organic solvents can compromise the biocompatibility.	[187]
Multi-nozzle deposition manufacturing (MDM)	Four different nozzles can simultaneously deliver materials by dispensing liquids/slurry. This allows the construction of heterogeneous, complex structures and the direct incorporation of biofactors.	400 μm	Low temperature, high range of different materials can be used (alginate and PCL), biomolecules may be incorporated	The use of organic solvent might be necessary, as well as freeze-drying.	[38, 188]
Rapid prototyping robotic dispensing system (RPBOD)	The technology resembles the 3D-plotting presented by Landers et al. [182].	400-1000μm	Compatible with many materials such as hydrogels, biomolecules may be incorporated.	Difficult optimisation of process for specific material, freeze drying is required.	[189]
Robocasting	Technique based on the robotic deposition of colloidal pastes through extrusion apparatus.	100-1000 μm	Compatible with many materials, low temperatures, no organic solvents, biomolecules may be incorporated.	Precise control of suspension properties and fine-tuning of all the parameters involved is necessary.	[190, 191]
Stereolitography apparatus (SLA)	A laser (mostly UV) is used to cure a liquid-phase photopolymer.	250 μm	Compatible with many materials, no support needed, fast, no trapped material.	Only photocrosslinkable materials. Use of UV-light.	[168, 192, 193]
Micro-stereolitography (MSLA)	Idem as SLA but the focused laser beam is only a few μm's diameter.	A few microns	Same as SLA	Same as SLA	[194]
Two-photon polymerization	A transparent medium containing monomers and initiator molecules is polymerised by absorption of two photons through a so-called virtual state. Infra-red light is used.	< 100 nm	Involves IR, superior resolution leads to high control over external and internal morphology	Only transparent media, material must be photcrosslinkable and cytocompatible	[195, 196]

Table 3. (Continued).

Rapid Prototyping Technique	Principle	Resolution	Advantages	Drawbacks	References
Bioprinting	A layer by layer deposition of cells and/or cell aggregates into a 3D gel with sequential maturation into perfused and vascularised living tissue or organ.	100 μm	Precise localisation of cells and cell aggregates in the construct is possible. Printing a vascular tree is considered to overcome vascularisation problems.	Bioprinting is a recent technology, thus further development of the bio-ink, the bio-paper, bio-printer and bioreactor is necessary before tissues can be synthesized[12].	[12, 197, 198]

FDM is a more traditional RP technique which has been extensively studied at the National University of Singapore. This group focussed on bone-tissue engineering. A commercial available FDM device (Stratasys Inc.) was used to create scaffolds of PCL with different lay-down patterns and a honeycomb-like structure[177]. The compressive stiffness and stress-strain curves for different scaffolds, as well as preliminary cell-culture tests were performed and showed encouraging results,. This work demonstrated that FDM was a suitable technique for the design and fabrication of highly reproducible 3D scaffolds with a fully interconnected network. A high correlation between the scaffold porosity and the compressive properties was demonstrated, thus showing that mechanical properties can be modulated by varying the architecture of the scaffold[199]. As a proof of concept an extensive *in vivo* study was performed on white rabbits[118]. Bilateral parietal critical-size defects were created in the testing animals. Starting from CT-scans of these defects customized PCL-scaffolds were constructed and seeded with salvarial osteoblasts (group I) and mesenhymal progenitor cells (group II). Nonseeded PCL-scaffolds were implanted as a control group. Results indicated osseous tissue integration and functionally stable restoration of the calvarium. Furthermore, cell-seeded constructs measured about 60% more then unrepaired or unseeded scaffolds, indicating the advantage of seeding and culturing cells onto the scaffolds before implantation.

Rai et al. reported the use of PCL /tricalcium phosphate (PCL-TCP) scaffolds, Fibrin glue was seeded on the top of the scaffolds simultaneously with human recombinant bone morphogenetic protein-2 (rhBMP-2). The influence of this biological factor on the osteogenic expression of osteoblasts was demonstrated. Groups with rhBMP-2 showed an initial burst in cell-growth which was not dose-dependent. After 20 days a higher cell death was observed on these scaffolds, compared to rhBMP-2 free scaffolds[200]. It is remarkable that the the research in this group evolved from pure PCL scaffolds to PCL-composites loaded with biological factors.

FDM is a known technique but it is desirable to create predictive models, software tools and suitable biomaterials to fully exploit this technology in the application of tissue engineering[201].

Khalil et al. reported the use of a multi-nozzle deposition technique for scaffold creation[38]. This system incorporates four different nozzles systems: a pneumatic microvalve, piezoelectric nozzle, solenoid valve and a pressurised mini-extruder. A material delivery system of various arrangements was created to accustom each nozzle system. Each nozzle has his own advantages and disadavantages. This system has the potential to create heterogeneous scaffolds with a controlled spatial organisation, allowing the delivery of cells to specific places in the scaffold. Preliminary testing was mainly focused on alginate scaffolds for soft tissue engineering purposes[188]. A semi-empirical model was developed to predict the deposition flow rate and the deposition geometry. These models enabled the authors to optimise deposition conditions, and thus the fabrication of scaffold structures with specific pore sizes and strut diameters was possible. The process proved to be viable, as cells were able to survive for more than 21 days and had proliferated to six folds the initial number[188].

From the same group at the Drexel University a Precision Extruding Deposition (PED) technique was proposed by Wang et al.[179]. The technique is described in Table 3.

Scaffolds with a controlled pore size of 250 μm and designed structural orientations were directly fabricated from PCL pellets and preliminary biological experiments demonstrated the biocompatibility of the material. Further work on this technique was performed by Darling,

Sun et al. The use of micro-CT as a non-destructive technique to characterise these highly interconnected scaffolds was demonstrated [202].

Woodfield et al. described the use of a custom-designed fiber-deposition device[178]. The technology was named three-dimensional fiber-deposition technique (3D-FDP) and is comparable to FDM. Thermoplastic polymers, hydrogels and biomaterial pastes can be extruded from a heated syringe with nozzle onto a moveable x-y-z-stage. Their research mainly focused on cartilage and bone-engineering applications with PEOT/PBT scaffolds[203]. Malda et al. investigated the influence of the composition and the architecture of scaffolds on cartilage composition and on oxygen gradients[204, 205]. Results indicate that more hydrophilic scaffolds enhance cartilaginous tissue formation. Within this type of more hydrophilic scaffolds the ones with a more accessible pore volume and larger pore size were the better performing constructs. These effects were likely to have been mediated by differential serum protein adsorption and efficiency of nutrient/waste exchange[203]. The importance of nutrient/waste exchange was also studied by measuring the oxygen gradient in scaffolds formed by compression molding-particulate leaching and 3DF-scaffolds. Significantly more cells were present in the center of the 3DF-scaffolds, although similar gradients were observed regardless of scaffold type[204]. Woodfield et al. studied scaffolds with anisotropic pore architectures to influence the zonal organization of chondrocytes[206]. When scaffolds with comparable porosity and volume fraction available for cell attachment but some with homogeneously spaced pores and others with pore-size gradients (pore size from 200-1650 μm in diameter) were compared, it was observed that pore-size gradients promoted anisotropic cell distribution, similar to that in immature bovine cartilage. It was thus clearly demonstrated that the success of a scaffold as a tissue-engineering construct is highly dependant on the design. The authors also presented an interesting article concerning the influence of substrate composition and gas plasma treatment on the morphology and the attachment of chondrocytes to the PEOT/PBT material[86].

This technique was also used in order to fabricate titanium alloy scaffolds which were evaluated *in vivo*[207, 208].

Xiong et al. presented a new technology : "precise extrusion manufacturing" (PEM)[180]. This technique was used to create porous PLLA scaffolds for bone tissue engineering. The same group also presented a low-temperature deposition manufacturing (LDM) and a multi-nozzle deposition manufacturing (MDM) technique[187, 209]. This last technique was used to process PLLA/TCP composite into scaffolds which were subsequently implanted. Some of them were loaded with rhBMP, by spraying rhBMP particles suspended in water into the scaffold during the forming process. rhBMP was reported to be released gradually. Good biocompatibility and restoration of the defect was observed for scaffolds loaded with rhBMP[209].

The Bioplotter, an instrument invented in the year 2000 at the Freibrug Materials Research Centre is a commercially available plotting system (see Figure 6) which permits the creation of scaffolds from materials with a low mechanical strength. A plotting material (with cells) is dispensed from a pressurized cartridge in a computer-controlled manner, into a liquid medium with similar density. The cells can be cultures of the patient's own cells, thus creating an autologous implant. Landers et al. reported the creation of hydrogel scaffolds by plotting a 10 w% solution of gelatin in silicone oil at 40°C. An agar scaffold was created by dispensing a 5 w% agar solution in a 4w% gelatine medium at 70°C[183]. The fabrication of alginate hydrogel scaffolds was also reported by plotting a aqueous solution of sodium alginate, gelatine and EDTA into a aqueous $CaCl_2$ solution of 0,03 mol/l[184]. The alginate forms a gel when it is dispensed in the direct environement of the calcium ions. Another application was the creation of fibrin/alginic acid scaffolds. All these scaffolds showed a specific external shape and a designed internal pore structure with 100% interconnectivity. As an example the geometry of a nose was created. The controlled manufacturing of hydrogel scaffolds seeded with cells is considered a crucial technique in the field of soft tissue engineering.

Ang et al. used a similar technology, based on a pneumatic dispenser, to create chitosan and chitosan-hydroxyapatite scaffolds. With a preliminary study they demonstrated the potential of this RPBOD technique to create 3D scaffolds with regular and reproducible macropore architecture[189].

Cohen et al. recently reported a similar deposition process. A custom-built robotic platform was used to deposit alginate gels pre-seeded with chondrocytes into patient-specific implants.

Still, there is a great need for more basic research on hydrogel scaffolds in tissue engineering[183].

Moroni et al. adapted this Bioplotter-device together with Envisiontec in order to have it extruding thermoplastic polymers at high temperature (Tmax 230°C). This was a continuation of the work of Woodfield et al., which was described previously. In the 3D-FDP technique the table moves, on the other hand, in the case of a Bioplotter, the extrusion device moves in the x-y-z direction while the stage is fixed. Moroni et al. presented a mechanistic approach towards the design of scaffolds [127] by studying the influence of pores geometry and architecture on dynamic mechanical properties[210]. Different scaffold architectures were investigated by means of dynamic mechanical analysis (DMA). Results showed that viscoelastic properties could be modulated. Not only the porosity but also fibre deposition and orientation were crucial factors in modulating the scaffolds as shown in Figure 8. Subsequently the effect of porosity and chemical composition on dynamic stiffness was studied[211]. From these studies an experimental and numerical analysis was derived[212] . A mathematical model showed to be able to predict the dynamic stiffness of scaffolds varying in porosity, architecture and intrinsic physico-chemical properties. Further research included the fabrication of hollow struts (see Figure 9) which showed potential as drug-delivery systems for sustained release and the combination of RP fabricated scaffolds with electrospun fibers[127, 140, 213].

Yousefi et al. reported the use of a bioplotter device in order to create scaffolds from a PLLA paste. PLLA was dissolved in methyl ethyl keton and several scaffolds were fabricated in order to evaluate a new computational approach. This approach consists in identifying an optimum window for the porosity level and pore size of scaffolds. Both mechanical and physiological requirements are addressed with a numerical tool that predicts the stress field for the scaffold tissue construct at the site of implantation[214]. These kind of predictive models are valuable tools in order to construct scaffolds in a controlled manner, tailor-made for a specific application and are thus studied by different research groups[127, 162, 210, 212, 215, 216].

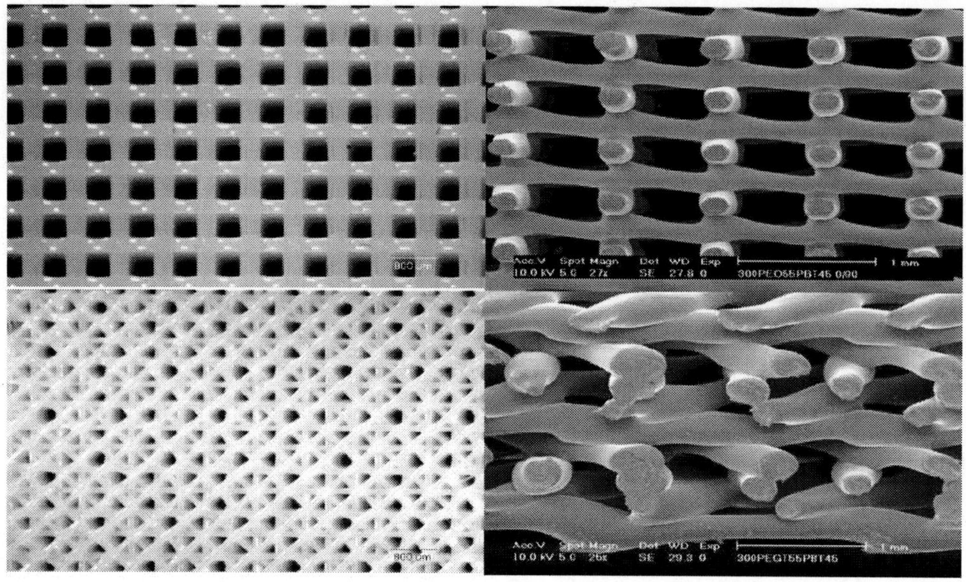

Figure 8. Optical Microscopy pictures of scaffolds surface (left) and SEM pictures of cross-sections (right). (upper) 0-90° pattern, (bottom) 0-45° pattern of PEOT/PBT scaffolds produced with an adapted Bioplotter device. Reprinted with the kind permission of Moroni et al. and Elsevier Limited.

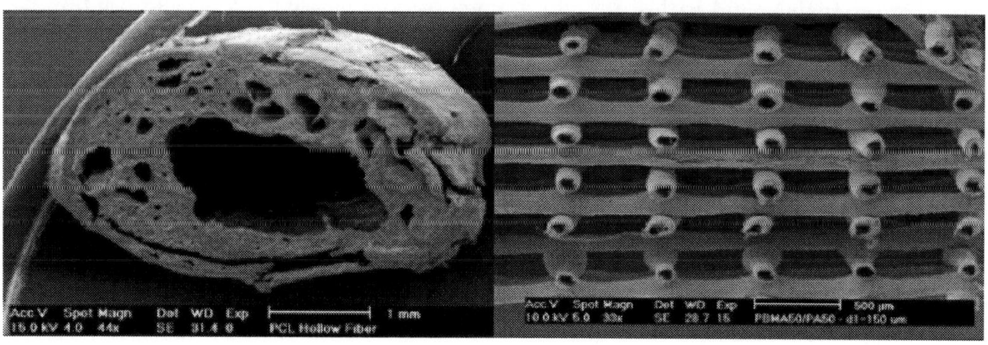

Figure 9. SEM micrographs of hollow fibrous scaffolds with different hollow cavity diameter and shell thickness. Reprinted with the kind permission of Moroni et al. and Liebert Pub.

A computer-controlled three-axis micropositioner is used to create scaffolds from a filaments extruded from a pressurized microsyringe by Vozzi et al.[185, 186]. Filaments of PLGA with a resolution up to 10 μm could be achieved. This high resolution is one of the main advantages of this so-called pressure assisted microsyringe (PAM) system[217]. Scaffolds made from PLLA-PCL blends were evaluated for their possible applications in cardiovascular tissue engineering[218]. First results suggest that the scaffolds could be used for engineering small diameter blood vessels without risk of occlusion. This research was further developed in order to address the important problem of creating larger, vascularised implants by Bianchi et al. [219]. The endothelial cells were transfected with an adenoviral vector carrying the gene for human tissue kallikrien. Furthermore, the co-culture of endothelial cells with fibroblasts showed encouraging results. The authors recently stated that biochemical, structural and mechanical cues all play a synergetic role in creating a successful scaffold[220]. This statement seems to be generally accepted within the scientific community.

A two-nozzle extrusion apparatus was used to deposit a slurry of alumina-poly(vinyl alcohol) in order to produce large-scale components by Cesarano et al.[190]. A customized HA lattice scaffold was robocasted based on a CT-scan and fitted excellent into the damaged human mandible [221]. This was a proof of concept exercise; the scaffold was not implanted into the patient. Dellinger et al. reported the use of robocasting to create HA scaffolds which were evaluated both *in vitro* and *in vivo*[191].

Stereolithography is a more conventional technique but recent investigations showed a potential in using this technique to fabricate moulds[222]. These sacrificial moulds can be used to fabricate scaffolds with a specific architecture. It is possible to combine this indirect fabrication methodology with more classic techniques such as particulate leaching, in order to introduce nano-porosity. This was also demonstrated for moulds fabricated by wax printing[223]. Micro-stereolithography is a similar technique but has a higher resolution. This recent technique was also used to create moulds[194]. Two-photon polymerization is a recent technique which allows very high resolution to be obtained. As this technique is quit recent no reports on *in vivo* experiments or extensive studies on scaffolds were performed yet, to the best of our knowledge. It is reported that the objects made by two-photon polymerization have a larger range of sizes, shapes, and materials than possible with conventional microfabrication techniques[224]. This supports the believe that this technique possesses great potential in the field of tissue engineering[195].

Another promising field is this of "bioprinting" or "organ printing". Organ printing was defined as a rapid prototyping computer-aided 3D printing technology, based on using layer by layer deposition of cell and/or cell aggregates into a 3D gel with sequential maturation of the printed construct into perfused and vascularised living tissue or organ[197]. The basic premises of this emerging field are somewhat different then the once of conventional tissue engineering[198]. This approach emphasizes the self-organisational properties of cells and tissue and believes that the fusion of developmental biology with tissue engineering will lead to unprecedented achievements[197]. The technology consists of a bioprinter, which can be a modified commercial printer[225], which can deliver biological materials ("the bio-ink) in a controlled, biocompatible environment, the "bio-paper". For the latter, simple, in situ cross-linkable hydrogels were developed that mimic the natural ECM as the bio-paper into which cells or cell aggregates can be printed[226]. The previously discussed Bioplotter can thus serve as a bioprinter. The fusion of endothelial cell aggregates printed on collagen was demonstrated by Boland et al. [227, 228] and aggregates of Chinese hamster ovary cells were successfully printed with motor neuron cells[229]. Mostly, in bioprinting, the synthetic polymer scaffold from more conventional RP-approaches is substituted by the ECM scaffold produced from the cultured cells. This overcomes the inability of more conventional RP techniques to precisely place cells or cell aggregates into a printed scaffold. A precise local delivery of different kinds of cells could possibly solve the vascularisation problem[230]. Matsusaki et al. recently reported a hydrogel template approach to create tissue from mouse fibroblast cells and ECM [231]. Anyhow, cell printing is a tool still in development[232]. Some issues such as sterility and viability of printed mammalian cells and fundamental understanding of the communication between cells and of the response of cells to their matrix remain challenging topics.

3. Characterization of Polymer Scaffolds

As the field of biomaterial science and tissue engineering using RP is a more recent development, characterization methods are still being developed, fine-tuned and adapted. Both *in vitro* and *in vivo* biological experiments are of major importance, but beforehand a good mechanical, structural and chemical analysis of the scaffolds is required.

As previously outlined, the constraints set on a scaffold are derived from the paradigm that a synthetic scaffold should mimic the host tissue to the highest possible extent. In order to evaluate if the scaffold really meets these constraints, a full characterization is essential.

The interest is dual. First, characterization of scaffolds enables researchers to evaluate the advantages and drawbacks of the fabrication method(s) applied. Secondly, the architecture of a scaffold will influence the mechanical properties and the biological performance. It is thus crucial to understand how these characteristics are correlated. A brief overview of some of the most important scaffold evaluation techniques is given below.

3.1. Structural Characterization: An Overview of the Most Common Techniques

In order to sustain living cells, sufficient diffusion of nutrients and waste products within a scaffold is necessary. The efficiency of these processes is determined by the pore interconnectivity, the porosity and the pore-size. A structural characterization should thus include an evaluation of the porosity, the pore-sizes, the interconnectivity, the surface to volume ratio, anisotropy, the cross-sectional area and the permeability[233]. Langer et al. recommended 90% porosity, but for some applications (e.g. load-bearing applications) this is not feasible due to the lack of mechanical strength. It is thus required to find a compromise between the porosity and the mechanical strength[20]. Generally, scaffolds should be highly porous and have fully interconnected pores with sufficient pore-size in order to show a sufficient permeability [234].

It is possible to make a purely theoretical estimation of the porosity based on a volume or mass measurement of the scaffold[199, 234]. In the unit cube method, the linear measurements of the scaffold cube volume and the volume of polymer material is calculated from the lay-down pattern. This method is based on several estimations and is unfortunately not very accurate.

When information on the strut diameter and pore size obtained from Scanning Electron Microscopy (SEM) is used for calculation of the porosity, more reliable results can be obtained[199]. SEM is one of the most commonly used techniques for scaffold characterization [235]. Different dimensions can be measured and evaluated by SEM and these can be correlated with processing parameters[178, 199] such as the nozzle used. The programmed lay-down pattern and the uniformity of the pores can be verified by SEM as well as the interconnection of the pores. Other theoretical estimations of porosity include the Archimedes method[164] and the liquid displacement method[236].

Porosity can also be determined by experimental means such as mercury porosimetry. This method is based on mercury intrusion into the scaffold with incremental increase in mercury pressure. The technique is applicable for pore sizes ranging from 0.0018-400 μm and closed pores are neglected [234].

Gas pycnometry is based on the volume measurement of the scaffold material by detecting the pressure when scaffold specimen displaces gas[199]. This technique does not take the closed pores into account and the measurement of linear dimensions of the scaffold is not sufficiently accurate. Furthermore, no volatile solvents can be present.

Gas adsorption is a known technique to determine the complete surface of a scaffold and the total pore volume. The sample is placed in an evacuated chamber where a small amount of the adsorbate gas is introduced. Measurements of pressure and volume of the chamber allow the calculation of the surface area from the Bruanauer, Emmett and Teller (BET) theory and the Langmuir model. When the pores are completely saturated with gas molecules, the total pore volume can be calculated[234]. This technique can address the nanoporosity of a scaffold but does not provide any further information on the scaffold's architecture.

An experimental technique to determine pore size is flow porosimetry[234]. This technique does not provide any further information.

More recently, micro computed tomography (μCT) showed great potential in analysing scaffolds both *in vitro* and *in vivo*[237]. This fully non-destructive technique allows direct quantification of the micro-structural architecture of scaffolds in three-dimensions[238]. μCT imaging is based on the attenuation of X-rays in a 2D-cross section. The reduced intensity of emergent X-rays is detected. Using an algorithm, the 2D section is reconstructed pixel by pixel[234]. It might be required to use a staining-reagent such as OsO4 or Lipiodol[237]. Zehbe et al. used synchrotron μCT in order to map the cell-distribution within scaffolds. The attached cells were labelled with a combination of Au-lysine and silver enhancer[239]. For further reading, we refer to different recent publications in the field[202, 233, 234, 237, 238, 240]. The main advantage of μCT is it's 3D nature allowing to analyze the entire scaffold. In that respect, it is noteworthy that, in a comparative study, mechanical data correlated well with μCT data on newly formed bone, but only to a lower extent with the SEM data obtained[237].

3.2. Mechanical Characterization

For load-bearing scaffolds, it is crucial to determine their elastic properties. Different possible techniques such as Dynamic Mechanical Analysis (DMA), compression testing, tensile tests and bending tests [144, 241] can be performed. The most suitable technique depends on the target application.

Mechanical analysis can be performed at different stages of the research. It can be a strategy to correlate scaffold architecture to it's mechanical strength and thus create predictive models. These models make it possible to design a certain scaffold in function of a targeted mechanical strength and porosity. Secondly, mechanical analysis can be performed as a quality-control for the produced scaffold. Subsequently, mechanical analysis can be performed at different stages of *in vivo* experiments, in order to evaluate whether the degradation kinetics of the polymer match the tissue formation.

It is crucial to perform measurements in an environment which is comparable to the site-of implantation [210], or to obtain transferrable experiments. This implies that experiments should be performed in wet conditions. The mechanical analysis should also be performed on the natural tissue in order to compare the scaffolds properties with the target-tissue.

Kaufman et al. reported a nanomechanical analysis of bone tissue engineering scaffolds using an atomic force microscope (AFM)[242]. Nanoindentation allows several mechanical measurements on small samples of each material in a library.

3.3. Biological Characterization : *In Vitro – In Vivo*

Scaffold-cell constructs can first be cultured *in vitro* in a bioreactor or they can be directly implanted *in vivo*. In order to evaluate cell-viability, cell distribution, proliferation and differentiation of cells many techniques are available. Most importantly SEM, confocal microscopy and cell metabolism assays are used. Endres et al. analysed the secretion of osteogenic marker molecules, type I procollagen and osteocalcin by semiquantitative immunocytochemistry assays[61]. (Epi-) fluorescent microscopy also provides a powerful tool to study tissue-formation in scaffolds[208].

A recent development is the use of fluorescence resonance energy transfer (FRET) in order to study material-cell interactions in three-dimensions[132]. It has been used to quantify several parameters of the cell-material interface relevant to cell response, including molecular changes in matrix proteins induced by interactions both with surfaces and cells, the number of bonds between integrins and their adhesion ligands and changes in the crosslink density of hydrogel synthetic extracellular matrix analogues. The application of FRET in scaffold-based tissue-engineering has the potential to provide us deeper understanding of the fundamental interactions at the surface of the scaffold.

Another emerging technology is the use of real-time PCR for absolute quantification of gene expression (aqPCR) instead of reverse transcription- polymerase chain reaction (rt-PCR) [243]. Leong et al. demonstrated that aqPCR was a better method to quantify gene expression then end-point PCR[243].

Preliminary studies can be performed in vitro by culturing cells on the scaffolds in a bioreactor[244]. These devices were developed in recent years in order to create a more biomimetic environment for cell culture. In perfusion flow bioreactors, metabolic waste is constantly removed and nutrients are transported towards the cells[107]. This type of bioreactors allows the combination of 3D culture and perfusion flow and thus mimics the physiological environment. Stephens et al. recently created such a perfusion flow bioreactor with an upright microscope[245]. This system allows the real-time imaging of the cell/material interactions. This might be very helpful in gaining deeper insight in the influence of surface properties, the type of biomaterial, the presence of flow-induced shear and scaffold architecture on cell response and growth of new tissues.

In vivo experiments are performed on the most interesting tissue engineered constructs. As an important step towards clinical application, research work in large animal models is essential[246]. Nevertheless, the results obtained from *in vivo* studies cannot directly be extrapolated to their clinical performance. The existing assays, both *in vitro* and *in vivo*, need improvements in order to be predictive for the clinical application.

Histological appearances of implants can be studied after several weeks of implantation. Not only the biocompatibility of the scaffold is thus evaluated but also the tissue formation can be assessed in a histological manner[247]. The applications of RP technologies in scaffold fabrication are wide and varied, however, only a small number have reached the clinical arena[30]. More recently, the group at the National University of Singapore[30], as well as Cao et al.[246], have undertaken several studies utilizing a PCL/CaP composite scaffold in large animals

3.4. Chemical Characterization: Trends and Limitations

As surface modified constructs show great potential in tissue engineering, techniques for the chemical and structural characterization of the scaffolds should be considered. Mostly, surface modified scaffolds are hard to assess in a quantitative manner. The amount of immobilised or grafted molecules is very low compared to the bulk material. Only a couple of surface characterization techniques are available and many side-effects might occur, for example with dyes.

Techniques such as X-ray Photoelectron Spectroscopy (XPS) and Fourier transform infrared (FT-IR) spectroscopy are commonly used for chemical surface characterization[81, 88]. XPS, however, is a relative technique. IR can only be applied if a significant difference can be observed between the original substrate and the modified substrate. Furthermore the staining of specific chemical groups present on the surface, and quantitative or semi-quantitative determination of the concentration is common practise[80]. Recently, Fourier Transform Infrared Imaging (FT-IR imaging) has been used by different researchers in order to chemically map a surface and to obtain an idea of the distribution of chemical reactive groups[248, 249]. Spatial distribution of surface roughness and chemical groups is considered of major interest in order to guide cells to a desired final faith[78, 79, 250]. Hatakeyama *et al.* reported the biofunctional patterning of thermo-responsive surfaces for spatiotemporally controlled cell adhesion, growth and thermally induced detachment[251]. The quantification of co-grafted polymers was performed by attenuated total reflectance-FT-IR. The site-selective immobilisation of the growth factor insulin was confirmed by fluorescence microscopy.

Other techniques such as Time-of-Flight Secondary Ion Mass Spectrometry (Tof-SIMS) are common but a complete overview is beyond the intention of this paper.

Many interesting techniques and creative solutions for the problem of chemical quantification of scaffolds and surfaces are presented. Still, in our opinion this remains a challenging area of research. In our opinion the correlation of chemical characteristics with *in vivo* or bioreactor results will lead to a deeper understanding of cell-biomaterial interaction.

4. Future Prospects

The future of biomaterials science and tissue engineering will greatly depend on the way in which different scientific disciplines (materials scientists, biologists, engineers, medical doctors, ...) will work together to solve the existing multi-disciplinary questions. Every answer raises a dozen of new and intriguing questions to be solved. Studies indicate that the ideal biomaterial does not exist and that depending on the application, certain composite materials which combine the interesting properties of different material classes might be the best available solution. Furthermore, the combined use of different production techniques might be required for developing optimal structures. However, critical issues that still require in-depth study for all biomaterials studied include amongst others, the optimisation of procedures for applying and analysing coatings on different types of 3D porous biomaterials, enhancing the seeding efficiency on scaffolds, studying to what extent the application of co-culture systems on biomaterials will still require the addition of growth factors and cytokines to the scaffold and a better understanding and characterization of biomaterial-cell interactions at the cellular level. Finally, we would also like to mention that, in our opinion, certain standards should be developed both for the material development and material analysis techniques. This will be the only way in which test results obtained in different laboratories can be efficiently and successfully compared and correlated to *in vivo* and bioreactor studies. This will result in a more rapid advancement of the field. Without any doubt, rapid prototyping techniques will be among the techniques of choice for producing porous scaffolds with a controllable pore size, pore distribution and overall porosity.

REFERENCES

[1] Lalan S, Pomerantseva I, Vacanti JP. Tissue engineering and its potential impact on surgery. *World Journal of Surgery.* 2001 Nov;25(11):1458-1466.

[2] Laurencin CT, Ambrosio AMA, Borden MD, Cooper JA. Tissue engineering: Orthopedic applications. *Annual Review of Biomedical Engineering.* 1999;1:19-46.

[3] Guarino V, Causa F, Ambrosio L. Bioactive scaffolds for bone and ligament tissue. *Expert Review of Medical Devices.* 2007 May;4(3):405-418.

[4] Chung HW, Kim KH, Kim JG, Lee SY, Yoon YH. Retinal complications in patients with solid organ or bone marrow transplantations. *Transplantation.* 2007 Mar 27;83(6):694-699.

[5] http://www.ustransplant.org/annual_reports/current/103_dh.htm.

[6] Punch JD, Hayes DH, LaPorte FB, McBride V, Seely MS. Organ donation and utilization in the United States, 1996-2005. *American Journal of Transplantation.* 2007;7(5):1327-1338.

[7] Bardell T, Hunter DJW, Kent WDT, Jain MK. Do medical students have the knowledge needed to maximize organ donation rates? *Canadian Journal of Surgery.* 2003 Dec;46(6):453-457.

[8] Al Sebayel MIM, Khalaf HA. The efficiency of utilization of potential donors for organ transplantation in Saudi Arabia: A pilot study\. *Transplantation Proceedings.* 2004 Sep;36(7):1881-1882.

[9] Rogiers X, Malago M, Gawad K, Jauch KW, Olausson M, Knoefel WT, et al. In situ splitting of cadaveric livers - The ultimate expansion of a limited donor pool. *Annals of Surgery.* 1996 Sep;224(3):331-339.

[10] Langer R, Vacanti JP. Tissue Engineering. *Science.* 1993 May 14;260(5110):920-926.

[11] Griffith LG, Naughton G. Tissue engineering - Current challenges and expanding opportunities. *Science.* 2002 Feb 8;295(5557):1009-+.

[12] Mironov V, Prestwich G, Forgacs G. Bioprinting living structures. *Journal of Materials Chemistry.* 2007;17(20):2054-2060.

[13] Bonassar LJ, Vacanti CA. Tissue engineering: The first decade and beyond. *Journal of Cellular Biochemistry.* 1998:297-+.

[14] Putnam AJ, Mooney DJ. Tissue engineering using synthetic extracellular matrices. *Nature Medicine.* 1996 Jul;2(7):824-826.

[15] www.becon1.nih.gov/becon_2001_final_report.doc.

[16] Shin H, Jo S, Mikos AG. Biomimetic materials for tissue engineering. *Biomaterials.* 2003 Nov; 24(24):4353-4364.

[17] Drury JL, Mooney DJ. Hydrogels for tissue engineering: scaffold design variables and applications. *Biomaterials.* 2003 Nov;24(24):4337-4351.

[18] Hutmacher DW, Sittinger M, Risbud MV. Scaffold-based tissue engineering: rationale for computer-aided design and solid free-form fabrication systems. *Trends in Biotechnology.* 2004 Jul;22(7):354-362.

[19] Butler DL, Goldstein SA, Guilak F. Functional tissue engineering: The role of biomechanics. *Journal of Biomechanical Engineering-Transactions of the Asme.* 2000 Dec;122(6):570-575.

[20] Hollister SJ, Maddox RD, Taboas JM. Optimal design and fabrication of scaffolds to mimic tissue properties and satisfy biological constraints. *Biomaterials.* 2002 Oct;23(20):4095-4103.

[21] Yeong WY, Chua CK, Leong KF, Chandrasekaran M. Rapid prototyping in tissue engineering: challenges and potential. *Trends in Biotechnology.* 2004 Dec;22(12):643-652.

[22] Hutmacher DW. Scaffold design and fabrication technologies for engineering tissues - state of the art and future perspectives. *Journal of Biomaterials Science-Polymer Edition.* 2001;12(1):107-124.

[23] Hutmacher DW. Scaffolds in tissue engineering bone and cartilage. *Biomaterials.* 2000 Dec;21(24):2529-2543.

[24] Hollister SJ, Levy RA, Chu TM, Halloran JW, Feinberg SE. An image-based approach for designing and manufacturing craniofacial scaffolds. *International Journal of Oral and Maxillofacial Surgery.* 2000 Feb;29(1):67-71.

[25] Peter SJ, Miller MJ, Yasko AW, Yaszemski MJ, Mikos AG. Polymer concepts in tissue engineering. *Journal of Biomedical Materials Research.* 1998 Win;43(4):422-427.

[26] Hollister SJ. Porous scaffold design for tissue engineering. *Nature Materials.* 2005 Jul;4(7):518-524.

[27] Leong KF, Cheah CM, Chua CK. Solid freeform fabrication of three-dimensional scaffolds for engineering replacement tissues and organs. *Biomaterials.* 2003 Jun;24(13):2363-2378.

[28] Freed LE, Grande DA, Lingbin Z, Emmanual J, Marquis JC, Langer R. Joint Resurfacing Using Allograft Chondrocytes and Synthetic Biodegradable Polymer Scaffolds. *Journal of Biomedical Materials Research.* 1994 Aug;28(8):891-899.

[29] Freed LE, Vunjaknovakovic G, Biron RJ, Eagles DB, Lesnoy DC, Barlow SK, et al. Biodegradable Polymer Scaffolds for Tissue Engineering. *Bio-Technology.* 1994 Jul;12(7):689-693.

[30] Hutmacher DW, Cool S. Concepts of scaffold-based tissue engineering-the rationale to use solid free-form fabrication techniques. *Journal of Cellular and Molecular Medicine.* 2007 Jul-Aug;11(4):654-669.

[31] Peppas NA, Langer R. New Challenges in Biomaterials. *Science.* 1994 Mar 25;263(5154):1715-1720.

[32] Langer R, Tirrell DA. Designing materials for biology and medicine. *Nature.* 2004 Apr 1;428(6982):487-492.

[33] Williams DF, Cunningham J. *Materials in clinical dentistry.* Oxford ; New York: Oxford University Press, 1979.

[34] Anderson DG, Burdick JA, Langer R. Materials science - Smart biomaterials. *Science.* 2004 Sep 24;305(5692):1923-1924.

[35] Ratner BD. *Biomaterials science : an introduction to materials in medicine.* 2nd ed ed. San Diego ; London: Elsevier Academic Press, 2004.

[36] Ramakrishna S, Mayer J, Wintermantel E, Leong KW. Biomedical applications of polymer-composite materials: a review. *Composites Science and Technology.* 2001 Jul;61(9):1189-1224.

[37] Chen VJ, Ma PX. Nano-fibrous poly(L-lactic acid) scaffolds with interconnected spherical macropores. *Biomaterials.* 2004 May;25(11):2065-2073.

[38] Khalil S, Nam J, Sun W. Multi-nozzle deposition for construction of 3D biopolymer tissue scaffolds. *Rapid Prototyping Journal.* 2005;11(1):9-17.

[39] Savalani MM, Harris RA. Layer manufacturing for in vivo devices. Proceedings of the Institution of Mechanical Engineers Part H-Journal of Engineering in Medicine. 2006 May;220(H4):505-520.

[40] Nair LS, Laurencin CT. Polymers as biomaterials for tissue engineering and controlled drug delivery. *Tissue Engineering I: Scaffold Systems for Tissue Engineering.* 2006;102:47-90.

[41] Vert M, Li SM, Spenlehauer G, Guerin P. Bioresorbability and Biocompatibility of Aliphatic Polyesters. *Journal of Materials Science-Materials in Medicine.* 1992 Nov;3(6):432-446.

[42] Vert M, Li SM, Garreau H. Attempts to Map the Structure and Degradation Characteristics of Aliphatic Polyesters Derived from Lactic and Glycolic Acids. *Journal of Biomaterials Science-Polymer Edition.* 1994;6(7):639-649.

[43] Bostman O, Hirvensalo E, Vainionpaa S, Makela A, Vihtonen K, Tormala P, et al. Ankle Fractures Treated Using Biodegradable Internal-Fixation. *Clinical Orthopaedics and Related Research.* 1989 Jan(238):195-203.

[44] Bacakova L, Filova E, Rypacek F, Svorcik V, Stary V. Cell adhesion on artificial materials for tissue engineering. *Physiological Research.* 2004;53:S35-S45.

[45] Fu K, Pack DW, Klibanov AM, Langer R. Visual evidence of acidic environment within degrading poly(lactic-co-glycolic acid) (PLGA) microspheres. *Pharmaceutical Research.* 2000 Jan;17(1):100-106.

[46] Sung HJ, Meredith C, Johnson C, Galis ZS. The effect of scaffold degradation rate on three-dimensional cell growth and angiogenesis. *Biomaterials.* 2004 Nov;25(26):5735-5742.

[47] Gilding DK, Reed AM. Biodegradable Polymers for Use in Surgery - Polyglycolic-Poly(Actic Acid) Homopolymers and Copolymers .1. *Polymer.* 1979;20(12):1459-1464.

[48] Reed AM, Gilding DK. Biodegradable Polymers for Use in Surgery - Poly(Glycolic)-Poly(Lactic Acid) Homo and Co-Polymers .2. Invitro Degradation. *Polymer.* 1981;22(4):494-498.

[49] Liu C, Xia Z, Czernuszka JT. Design and development of three-dimensional scaffolds for tissue engineering. *Chem. Eng. Res. Des.* 2007 Jul;85(A7):1051-1064.
[50] Sodergard A, Stolt M. Properties of lactic acid based polymers and their correlation with composition. *Progress in Polymer Science*. 2002 Aug;27(6):1123-1163.
[51] Seal BL, Otero TC, Panitch A. Polymeric biomaterials for tissue and organ regeneration. *Materials Science and Engineering R-Reports.* 2001 Oct 10;34(4-5):147-230.
[52] Mansbridge J. Commercial considerations in tissue engineering. *Journal of Anatomy.* 2006 Oct;209(4):527-532.
[53] Wang M. Developing bioactive composite materials for tissue replacement. *Biomaterials.* 2003 Jun;24(13):2133-2151.
[54] Hubbell JA. Biomaterials in Tissue Engineering. *Bio-Technology.* 1995 Jun;13(6):565-576.
[55] Olah L, Filipczak K, Jaegermann Z, Czigany T, Borbas L, Sosnowski S, et al. Synthesis, structural and mechanical properties of porous polymeric scaffolds for bone tissue regeneration based on neat poly(epsilon-caprolactone) calcium carbonate. *Polymers for Advanced Technologies.* 2006 Nov-Dec;17(11-12):889-897.
[56] Kikuchi M, Ikoma T, Itoh S, Matsumoto HN, Koyama Y, Takakuda K, et al. Biomimetic synthesis of bone-like nanocomposites using the self-organization mechanism of hydroxyapatite and collagen. *Composites Science and Technology.* 2004 May;64(6):819-825.
[57] Yefang Z, Hutmacher DW, Varawa SL, Meng LT. Human alveolar osteoblasts interactions with PCL-TCP composite modified with collagen type I. *Tissue Engineering.* 2006 Apr;12(4):1028-1028.
[58] Yefang Z, Hutmacher DW, Varawan SL, Meng LT. Comparison of Human alveolar osteoblasts cultured on polymer-ceramic composite scaffolds and tissue culture plates. *International Journal of Oral and Maxillofacial Surgery.* 2007 Feb;36(2):137-145.
[59] Chim H, Hutmacher DW, Chou AM, Oliveira AL, Reis RL, Lim TC, et al. A comparative analysis of scaffold material modifications for load-bearing applications in bone tissue engineering. *International Journal of Oral and Maxillofacial Surgery.* 2006 Oct;35(10):928-934.
[60] Nakamura T, Shimizu Y, Takimoto Y, Tsuda T, Li YH, Kiyotani T, et al. Biodegradation and tumorigenicity of implanted plates made from a copolymer of epsilon-caprolactone and L-lactide in rat. *Journal of Biomedical Materials Research.* 1998 Dec 15;42(4):475-484.
[61] Endres M, Hutmacher DW, Salgado AJ, Kaps C, Ringe J, Reis RL, et al. Osteogenic induction of human bone marrow-derived mesenchymal progenitor cells in novel synthetic polymer-hydrogel matrices. *Tissue Engineering.* 2003 Aug;9(4):689-702.
[62] Agrawal CM, Athanasiou KA. Technique to control pH in vicinity of biodegrading PLA-PGA implants. *Journal of Biomedical Materials Research.* 1997 Sum;38(2):105-114.
[63] Tian D, Blacher S, Dubois P, Jerome R. Biodegradable and biocompatible inorganic-organic hybrid materials - 2. Dynamic mechanical properties, structure and morphology. *Polymer.* 1998 Feb;39(4):855-864.

[64] Chen GP, Ushida T, Tateishi T. Development of biodegradable porous scaffolds for tissue engineering. *Materials Science and Engineering C-Biomimetic and Supramolecular Systems.* 2001 Nov 1;17(1-2):63-69.

[65] Chang MC, Tanaka J. XPS study for the microstructure development of hydroxyapatite-collagen nanocomposites cross-linked using glutaraldehyde. *Biomaterials.* 2002 Sep;23(18):3879-3885.

[66] Lutolf MP, Hubbell JA. Synthetic biomaterials as instructive extracellular microenvironments for morphogenesis in tissue engineering. *Nature Biotechnology.* 2005 Jan;23(1):47-55.

[67] Marler JJ, Guha A, Rowley J, Koka R, Mooney D, Upton J, et al. Soft-tissue augmentation with injectable alginate and syngeneic fibroblasts. *Plast. Reconstr. Surg.* 2000 May;105(6):2049-2058.

[68] Haines-Butterick LA, Rajagopal K, Lamm M, Pochan DJ, Schnieder JP. Controlling hydrogelation kinetics via peptide design for three-dimensional encapsulation and injectable delivery of cells. *Biopolymers.* 2007;88(4):518-518.

[69] Fittkau MH, Zilla P, Bezuidenhout D, Lutolf M, Human P, Hubbell JA, et al. The selective modulation of endothelial cell mobility on RGD peptide containing surfaces by YIGSR peptides. *Biomaterials.* 2005 Jan;26(2):167-174.

[70] Mann BK, Gobin AS, Tsai AT, Schmedlen RH, West JL. Smooth muscle cell growth in photopolymerized hydrogels with cell adhesive and proteolytically degradable domains: synthetic ECM analogs for tissue engineering. *Biomaterials.* 2001 Nov;22(22):3045-3051.

[71] Roach P, Eglin D, Rohde K, Perry CC. Modern biomaterials: a review-bulk properties and implications of surface modifications. *Journal of Materials Science-Materials in Medicine.* 2007 Jul;18(7):1263-1277.

[72] Chung HJ, Park TG. Surface engineered and drug releasing pre-fabricated scaffolds for tissue engineering. *Advanced Drug Delivery Reviews.* 2007 May 30;59(4-5):249-262.

[73] Sun H, Onneby S. Facile polyester surface functionalization via hydrolysis and cell-recognizing peptide attachment. *Polymer International.* 2006 Nov;55(11):1336-1340.

[74] Choong CSN, Hutmacher DW, Triffitt JT. Co-culture of bone marrow fibroblasts and endothelial cells on modified polycaprolactone substrates for enhanced potentials in bone tissue engineering. *Tissue Engineering.* 2006 Sep;12(9):2521-2531.

[75] Chong MSK, Lee CN, Teoh SH. Characterization of smooth muscle cells on poly(epsilon-caprolactone) films. *Materials Science and Engineering C-Biomimetic and Supramolecular Systems.* 2007 Mar;27(2):309-312.

[76] Zhu YB, Gao CY, Liu XY, He T, Shen JC. Immobilization of biomacromolecules onto aminolyzed poly(L-lactic acid) toward acceleration of endothelium regeneration. *Tissue Engineering.* 2004 Jan-Feb;10(1-2):53-61.

[77] Zhu YB, Gao CY, Liu XY, Shen JC. Surface modification of polycaprolactone membrane via aminolysis and biomacromolecule immobilization for promoting cytocompatibility of human endothelial cells. *Biomacromolecules.* 2002 Nov-Dec;3(6):1312-1319.

[78] Li B, Ma YX, Wang S, Moran PM. A technique for preparing protein gradients on polymeric surfaces: effects on PC12 pheochromocytoma cells. *Biomaterials.* 2005 May;26(13):1487-1495.

[79] Li B, Ma YX, Wang S, Moran PM. Influence of carboxyl group density on neuron cell attachment and differentiation behavior: Gradient-guided neurite outgrowth. *Biomaterials.* 2005 Aug;26(24):4956-4963.

[80] Goddard JM, Hotchkiss JH. Polymer surface modification for the attachment of bioactive compounds. *Progress in Polymer Science.* 2007 Jul;32(7):698-725.

[81] Cheng ZY, Teoh SH. Surface modification of ultra thin poly (epsilon-caprolactone) films using acrylic acid and collagen. *Biomaterials.* 2004 May;25(11):1991-2001.

[82] Chu PK, Chen JY, Wang LP, Huang N. Plasma-surface modification of biomaterials. *Materials Science and Engineering R-Reports.* 2002 Mar 29;36(5-6):143-206.

[83] Li YH, Huang YD. The study of collagen immobilization on polyurethane by oxygen plasma treatment to enhance cell adhesion and growth. *Surface and Coatings Technology.* 2007 Feb 26;201(9-11):5124-5127.

[84] Siow KS, Britcher L, Kumar S, Griesser HJ. Plasma methods for the generation of chemically reactive surfaces for biomolecule immobilization and cell colonization - A review. *Plasma Processes and Polymers.* 2006 Aug 15;3(6-7):392-418.

[85] Sodergard A. Perspectives on modification of aliphatic polyesters by radiation processing. *Journal of Bioactive and Compatible Polymers.* 2004 Nov;19(6):511-525.

[86] Woodfield TBF, Miot S, Martin I, van Blitterswijk CA, Riesle J. The regulation of expanded human nasal chondrocyte re-differentiation capacity by substrate composition and gas plasma surface modification. *Biomaterials.* 2006 Oct;27(7):1043-1053.

[87] Edlund U, Kallrot M, Albertsson AC. Single-step covalent functionalization of polylactide surfaces. *Journal of the American Chemical Society.* 2005 Jun 22;127(24):8865-8871.

[88] Kallrot M, Edlund U, Albertsson AC. Surface functionalization of degradable polymers by covalent grafting. *Biomaterials.* 2006 Mar;27(9):1788-1796.

[89] Kallrot M, Edlund U, Albertsson AC. Covalent grafting of poly(L-lactide) to tune the in vitro degradation rate. *Biomacromolecules.* 2007 Aug;8(8):2492-2496.

[90] Khorasani MT, Mirzadeh H, Sammes PG. Laser surface modification of polymers to improve biocompatibility: HEMA grafted PDMS, in vitro assay - III. *Radiat. Phys. Chem.* 1999 Aug;55(5-6):685-689.

[91] Khorasani MT, Mirzadeh H. In vitro blood compatibility of modified PDMS surfaces as superhydrophobic and superhydrophilic materials. *J. Appl. Polym. Sci.* 2004 Feb 5;91(3):2042-2047.

[92] Tiaw KS, Goh SW, Hong M, Wang Z, Lan B, Teoh SH. Laser surface modification of poly(epsilon-caprolactone) (PCL) membrane for tissue engineering applications. *Biomaterials.* 2005 Mar;26(7):763-769.

[93] Foo HL, Taniguchi A, Yu H, Okano T, Teoh SH. Catalytic surface modification of roll-milled poly(epsilon-caprolactone) biaxially stretched to ultra-thin dimension. *Materials Science and Engineering C-Biomimetic and Supramolecular Systems.* 2007 Mar;27(2):299-303.

[94] Reyes CD, Petrie TA, Burns KL, Schwartz Z, Garcia AJ. Biomolecular surface coating to enhance orthopaedic tissue healing and integration. *Biomaterials.* 2007 Jul;28(21):3228-3235.

[95] Riesle J, Hollander AP, Langer R, Freed LE, Vunjak-Novakovic G. Collagen in tissue-engineered cartilage: Types, structure, and crosslinks. *Journal of Cellular Biochemistry.* 1998 Dec 1;71(3):313-327.

[96] Yang J, Wan YQ, Yang JL, Bei JZ, Wang SG. Plasma-treated, collagen-anchored polylactone: Its cell affinity evaluation under shear or shear-free conditions. *Journal of Biomedical Materials Research Part.* A 2003 Dec 15;67A(4):1139-1147.

[97] Ma ZW, Gao CY, Gong YH, Shen JC. Cartilage tissue engineering PLLA scaffold with surface immobilized collagen and basic fibroblast growth factor. *Biomaterials.* 2005 Apr;26(11):1253-1259.

[98] Liu XH, Won YJ, Ma PX. Surface modification of interconnected porous scaffolds. *Journal of Biomedical Materials Research Part A.* 2005 Jul 1;74A(1):84-91.

[99] Chung TW, Wang YZ, Huang YY, Pan CI, Wang SS. Poly (epsilon-caprolactone) grafted with nano-structured chitosan enhances growth of human dermal fibroblasts (vol 30, pg 35, 2006). *Artif. Organs.* 2006 Feb;30(2):127-127.

[100] Mao JS, Liu HF, Yin YJ, Yao KD. The properties of chitosan-gelatin membranes and scaffolds modified with hyaluronic acid by different methods. *Biomaterials.* 2003 Apr;24(9):1621-1629.

[101] Marletta G, Ciapetti G, Satriano C, Pagani S, Baldini N. The effect of irradiation modification and RGD sequence adsorption on the response of human osteoblasts to polycaprolactone. *Biomaterials.* 2005 Aug;26(23):4793-4804.

[102] Santiago LY, Nowak RW, Rubin JP, Marra KG. Peptide-surface modification of poly(caprolactone) with laminin-derived sequences for adipose-derived stem cell applications. *Biomaterials.* 2006 May;27(15):2962-2969.

[103] Tugulu S, Silacci P, Stergiopulos N, Klok HA. RGD - Functionalized polymer brushes as substrates for the integrin specific adhesion of human umbilical vein endothelial cells. *Biomaterials.* 2007 Jun;28(16):2536-2546.

[104] Niu XF, Wang YL, Luo YF, Xin J, Li YG. Arg-gly-Asp (RGD) modified biomimetic polymeric materials. *J. Mater Sci. Technol.* 2005 Jul;21(4):571-576.

[105] Ho MH, Hou LT, Tu CY, Hsieh HJ, Lai JY, Chen WJ, et al. Promotion of cell affinity of porous PLLA scaffolds by immobilization of RGD peptides via plasma treatment. *Macromolecular Bioscience.* 2006 Jan 5;6(1):90-98.

[106] Yoon JJ, Song SH, Lee DS, Park TG. Immobilization of cell adhesive RGD peptide onto the surface of highly porous biodegradable polymer scaffolds fabricated by a gas foaming/salt leaching method. *Biomaterials.* 2004 Nov;25(25):5613-5620.

[107] Park TG. Perfusion culture of hepatocytes within galactose-derivatized biodegradable poly(lactide-co-glycolide) scaffolds prepared by gas foaming of effervescent salts. *Journal of Biomedical Materials Research.* 2002 Jan;59(1):127-135.

[108] Hubbell JA. Bioactive biomaterials. *Current Opinion in Biotechnology.* 1999 Apr;10(2):123-129.

[109] Zisch AH, Lutolf MP, Hubbell JA. Biopolymeric delivery matrices for angiogenic growth factors. *Cardiovascular Pathology.* 2003 Nov-Dec;12(6):295-310.

[110] Sheridan MH, Shea LD, Peters MC, Mooney DJ. Bioadsorbable polymer scaffolds for tissue engineering capable of sustained growth factor delivery. *Journal of Controlled Release.* 2000 Feb;64(1-3):91-102.

[111] Sachlos E, Czernuszka Jt. Making Tissue Engineering Scaffolds Work. Review On The Application Of Solid Freeform Fabrication Technology To The Production Of Tissue Engineering Scaffolds. *European Cells and Materials.* 2003;5:29 - 40.

[112] Lee SH, Shin H. Matrices and scaffolds for delivery of bioactive molecules in bone and cartilage tissue engineering. *Advanced Drug Delivery Reviews.* 2007 May 30;59(4-5):339-359.

[113] Oest ME, Dupont KM, Kong HJ, Mooney DJ, Guldberg RE. Quantitative assessment of scaffold and growth factor-mediated repair of critically sized bone defects. *Journal of Orthopaedic Research.* 2007 Jul;25(7):941-950.

[114] Unger RE, Sartoris A, Peters K, Motta A, Migliaresi C, Kunkel M, et al. Tissue-like self-assembly in cocultures of endothelial cells and osteoblasts and the formation of microcapillary-like structures on three-dimensional porous biomaterials. *Biomaterials.* 2007 Sep;28(27):3965-3976.

[115] Zhou YF, Chen FL, Ho ST, Woodruff MA, Lim TM, Hutmacher DW. Combined marrow stromal cell-sheet techniques and high-strength biodegradable composite scaffolds for engineered functional bone grafts. *Biomaterials.* 2007 Feb;28(5):814-824.

[116] Chen FL, Zhou YF, Barnabas ST, Woodruff MA, Hutmacher DW. Engineering tubular bone constructs. *Journal of Biomechanics.* 2007;40:S73-S79.

[117] Liao E, Yaszemski M, Krebsbach P, Hollister S. Tissue-engineered cartilage constructs using composite hyaluronic acid/collagen I hydrogels and designed poly(propylene fumarate) scaffolds. *Tissue Engineering.* 2007 Mar;13(3):537-550.

[118] Schantz JT, Hutmacher DW, Lam CXF, Brinkmann M, Wong KM, Lim TC, et al. Repair of calvarial defects with customised tissue-engineered bone grafts - II. Evaluation of cellular efficiency and efficacy in vivo. *Tissue Engineering.* 2003;9:S127-S139.

[119] Shimizu K, Ito A, Arinobe M, Murase Y, Iwata Y, Narita Y, et al. Effective cell-seeding technique using magnetite nanoparticles and magnetic force onto decellularized blood vessels for vascular tissue engineering. *J. Biosci. Bioeng.* 2007 May;103(5):472-478.

[120] Chen RR, Mooney DJ. Polymeric growth factor delivery strategies for tissue engineering. *Pharmaceutical Research.* 2003 Aug;20(8):1103-1112.

[121] Bleiziffer O, Eriksson E, Yao F, Horch RE, Kneser U. Gene transfer strategies in tissue engineering. *Journal of Cellular and Molecular Medicine.* 2007 Mar-Apr;11(2):206-223.

[122] Jang JH, Houchin TL, Shea LD. Gene delivery from polymer scaffolds for tissue engineering. *Expert Review of Medical Devices.* 2004 Sep;1(1):127-138.

[123] Ko HF, Sfeir C, Kumta PN. In situ synthesis and characterization of porous polymer-ceramic composites as scaffolds for gene delivery. *Materials Science and Engineering C-Biomimetic and Supramolecular Systems.* 2007 Apr;27(3):479-483.

[124] Jang JH, Bengali Z, Houchin TL, Shea LD. Surface adsorption of DNA to tissue engineering scaffolds for efficient gene delivery. *Journal of Biomedical Materials Research Part A.* 2006 Apr;77A(1):50-58.

[125] Prabu P, Dharmaraj N, Aryal S, Lee BM, Ramesh V, Kim HY. Preparation and drug release activity of scaffolds containing collagen and poly(caprolactone). *Journal of Biomedical Materials Research Part A.* 2006 Oct;79A(1):153-158.

[126] Moroni L, Hendriks JAA, Schotel R, De Wijn JR, Van Blitterswijk CA. Design of biphasic polymeric 3-dimensional fiber deposited scaffolds for cartilage tissue engineering applications. *Tissue Engineering.* 2007 Feb;13(2):361-371.

[127] Moroni L. A Mechanistic Approach to Design Smart Scaffolds for Tissue Engineering [PhD thesis]. Enschede: Twente; 2006.

[128] Bonadio J. Tissue engineering via local gene delivery. *J. Mol. Med-Jmm.* 2000;78(6):303-311.

[129] Anderson DG, Putnam D, Lavik EB, Mahmood TA, Langer R. Biomaterial microarrays: rapid, microscale screening of polymer-cell interaction. *Biomaterials.* 2005 Aug;26(23):4892-4897.

[130] McGuigan AP, Sefton MV. The influence of biomaterials on endothelial cell thrombogenicity. *Biomaterials.* 2007 Jun;28(16):2547-2571.

[131] Anderson DG, Levenberg S, Langer R. Nanoliter-scale synthesis of arrayed biomaterials and application to human embryonic stem cells. *Nature Biotechnology.* 2004 Jul;22(7):863-866.

[132] Huebsch ND, Mooney DJ. Fluorescent resonance energy transfer: A tool for probing molecular cell-biomaterial interactions in three dimensions. *Biomaterials.* 2007 May;28(15):2424-2437.

[133] Kong HJ, Kim CJ, Huebsch N, Weitz D, Mooney DJ. Noninvasive probing of the spatial organization of polymer chains in hydrogels using fluorescence resonance energy transfer (FRET). *Journal of the American Chemical Society.* 2007 Apr 18;129(15):4518-+.

[134] Yang SF, Leong KF, Du ZH, Chua CK. The design of scaffolds for use in tissue engineering. Part 1. Traditional factors. *Tissue Engineering.* 2001 Dec;7(6):679-689.

[135] Harris LD, Kim BS, Mooney DJ. Open pore biodegradable matrices formed with gas foaming. *Journal of Biomedical Materials Research.* 1998 Dec 5;42(3):396-402.

[136] Mikos AG, Bao Y, Cima LG, Ingber DE, Vacanti JP, Langer R. Preparation of Poly(Glycolic Acid) Bonded Fiber Structures for Cell Attachment and Transplantation. *Journal of Biomedical Materials Research.* 1993 Feb;27(2):183-189.

[137] Cima LG, Vacanti JP, Vacanti C, Ingber D, Mooney D, Langer R. Tissue Engineering by Cell Transplantation Using Degradable Polymer Substrates. *Journal of Biomechanical Engineering-Transactions of the Asme.* 1991 May;113(2):143-151.

[138] Greiner A, Wendorff JH. Electrospinning: A fascinating method for the preparation of ultrathin fibres. *Angew Chem. Int. Edit.* 2007;46(30):5670-5703.

[139] Kim GH, Han H, Park JH, Kim WD. An applicable electrospinning process for fabricating a mechanically improved nanofiber mat. *Polymer Engineering and Science.* 2007 May;47(5):707-712.

[140] Moroni L, Licht R, de Boer J, de Wijn JR, van Blitterswijk CA. Fiber diameter and texture of electrospun PEOT/PBT scaffolds influence human mesenchymal stem cell proliferation and morphology, and the release of incorporated compounds. *Biomaterials.* 2006 Oct;27(28):4911-4922.

[141] Weigel T, Schinkel G, Lendlein A. Design and preparation of polymeric scaffolds for tissue engineering. *Expert Review of Medical Devices.* 2006 Nov;3(6):835-851.

[142] Thomson RC, Yaszemski MJ, Powers JM, Mikos AG. Fabrication of Biodegradable Polymer Scaffolds to Engineer Trabecular Bone. *Journal of Biomaterials Science-Polymer Edition.* 1995;7(1):23-38.

[143] Thomson RC, Yaszemski MJ, Powers JM, Mikos AG. Hydroxyapatite fiber reinforced poly(alpha-hydroxy ester) foams for bone regeneration. Biomaterials 1998 Nov;19(21):1935-1943.

[144] Gomes ME, Ribeiro AS, Malafaya PB, Reis RL, Cunha AM. A new approach based on injection moulding to produce biodegradable starch-based polymeric scaffolds: morphology, mechanical and degradation behaviour. *Biomaterials.* 2001 May;22(9):883-889.

[145] Whang K, Thomas CH, Healy KE, Nuber G. A Novel Method to Fabricate Bioabsorbable Scaffolds. *Polymer.* 1995 Feb;36(4):837-842.

[146] Schmitz JP, Hollinger JO. A Preliminary-Study of the Osteogenic Potential of a Biodegradable Alloplastic-Osteoinductive Alloimplant. *Clinical Orthopaedics and Related Research.* 1988 Dec(237):245-255.

[147] Reuber M, Yu LS, Kolff WJ. Effect of Processing Temperature on the Properties of Polyurethane and Comparison of Vacuum Forming and Solution Casting to Make Artificial Hearts. *Artif. Organs.* 1987 Aug;11(4):323-323.

[148] Hsu YY, Gresser JD, Trantolo DJ, Lyons CM, Gangadharam PRJ, Wise DL. Effect of polymer foam morphology and density on kinetics of in vitro controlled release of isoniazid from compressed foam matrices. *Journal of Biomedical Materials Research.* 1997 Apr;35(1):107-116.

[149] Dubruel P, Unger R, Van Vlierberghe S, Cnudde V, Jacobs PJS, Schacht E, et al. Porous gelatin hydrogels: 2. In vitro cell interaction study. *Biomacromolecules.* 2007 Feb;8(2):338-344.

[150] Van Vlierberghe S, Cnudde V, Dubruel P, Masschaele B, Cosijns A, De Paepe I, et al. Porous gelatin hydrogels: 1. Cryogenic formation and structure analysis. *Biomacromolecules.* 2007 Feb;8(2):331-337.

[151] Tsai SP, Hsieh CY, Hsieh CY, Wang DM, Huang LLH, Lai JY, et al. Preparation and cell compatibility evaluation of chitosan/collagen composite scaffolds using amino acids as crosslinking bridges. *J. Appl. Polym. Sci.* 2007 Aug 15;105(4):1774-1785.

[152] Yannas IV, Burke JF, Gordon PL, Huang C, Rubenstein RH. Design of an Artificial Skin .2. Control of Chemical-Composition. *Journal of Biomedical Materials Research.* 1980;14(2):107-132.

[153] Huang X, Ye M, Zhang W, Wang C, Jiao T. Fabricating auricular prostheses based on rapid prototyping and the FreeForm modelling system. *International Journal of Advanced Manufacturing Technology.* 2004;24(11-12):873-878.

[154] Gibson I, Cheung LK, Chow SP, Cheung WL, Beh SL, Savalani M, et al. The use of rapid prototyping to assist medical applications. *Rapid Prototyping Journal.* 2006;12(1):53-58.

[155] Griffith LG, Wu B, Cima MJ, Powers MJ, Chaignaud B, Vacanti JP. In vitro organogenesis of liver tissue. *Bioartificial Organs.* 1997;831:382-397.

[156] Hopkinson N, Dickens P. Analysis of rapid manufacturing - using layer manufacturing processes for production. *P I Mech. Eng. C-J Mec.* 2003;217(1):31-39.

[157] Sun W, Darling A, Starly B, Nam J. Computer-aided tissue engineering: overview, scope and challenges. *Biotechnology and Applied Biochemistry.* 2004 Feb;39:29-47.

[158] Sun W, Lal P. Recent development on computer aided tissue engineering - a review. *Computer Methods and Programs in Biomedicine.* 2002 Feb;67(2):85-103.

[159] Tsang VL, Bhatia SN. Three-dimensional tissue fabrication. *Advanced Drug Delivery Reviews.* 2004 Sep 22;56(11):1635-1647.

[160] Vacanti CA, Bonassar LJ. An overview of tissue engineered bone. *Clinical Orthopaedics and Related Research.* 1999 Oct(367):S375-S381.

[161] Yang SF, Leong KF, Du ZH, Chua CK. The design of scaffolds for use in tissue engineering. Part II. Rapid prototyping techniques. *Tissue Engineering.* 2002 Feb;8(1):1-11.

[162] Sun W, Starly B, Darling A, Gomez C. Computer-aided tissue engineering: application to biomimetic modelling and design of tissue scaffolds. *Biotechnology and Applied Biochemistry.* 2004 Feb;39:49-58.

[163] Manjubala I, Woesz A, Pilz C, Rumpler M, Fratzl-Zelman N, Roschger P, et al. Biomimetic mineral-organic composite scaffolds with controlled internal architecture. *Journal of Materials Science-Materials in Medicine.* 2005 Dec;16(12):1111-1119.

[164] Chu TMG, Orton DG, Hollister SJ, Feinberg SE, Halloran JW. Mechanical and in vivo performance of hydroxyapatite implants with controlled architectures. *Biomaterials.* 2002 Mar;23(5):1283-1293.

[165] Chua CK, Chou SM, Wong TS. A study of the state-of-the-art rapid prototyping technologies. *International Journal of Advanced Manufacturing Technology.* 1998;14(2):146-152.

[166] Chua CK, Leong KF, Cheah CM, Chua SW. Development of a tissue engineering scaffold structure library for rapid prototyping. Part 1: Investigation and classification. *International Journal of Advanced Manufacturing Technology.* 2003;21(4):291-301.

[167] Chua CK, Leong KF, Cheah CM, Chua SW. Development of a tissue engineering scaffold structure library for rapid prototyping. Part 2: Parametric library and assembly program. *International Journal of Advanced Manufacturing Technology.* 2003;21(4):302-312.

[168] Lin CY, Kikuchi N, Hollister SJ. A novel method for biomaterial scaffold internal architecture design to match bone elastic properties with desired porosity. *Journal of Biomechanics.* 2004 May;37(5):623-636.

[169] Lam CXF, Mo XM, Teoh SH, Hutmacher DW. Scaffold development using 3D printing with a starch-based polymer. *Materials Science and Engineering C-Biomimetic and Supramolecular Systems.* 2002 May 31;20(1-2):49-56.

[170] Charriere E, Lemaitre J, Zysset P. Hydroxyapatite cement scaffolds with controlled macroporosity: fabrication protocol and mechanical properties. *Biomaterials.* 2003 Feb;24(5):809-817.

[171] Khalyfa A, Vogt S, Weisser J, Grimm G, Rechtenbach A, Meyer W, et al. Development of a new calcium phosphate powder-binder system for the 3D printing of patient specific implants. *Journal of Materials Science-Materials in Medicine.* 2007 May;18(5):909-916.

[172] Sachlos E, Reis N, Ainsley C, Derby B, Czernuszka JT. Novel collagen scaffolds with predefined internal morphology made by solid freeform fabrication. *Biomaterials.* 2003 Apr;24(8):1487-1497.

[173] Park A, Wu B, Griffith LG. Integration of surface modification and 3D fabrication techniques to prepare patterned poly(L-lactide) substrates allowing regionally selective cell adhesion. *Journal of Biomaterials Science-Polymer Edition.* 1998;9(2):89-110.

[174] Sherwood JK, Riley SL, Palazzolo R, Brown SC, Monkhouse DC, Coates M, et al. A three-dimensional osteochondral composite scaffold for articular cartilage repair. *Biomaterials.* 2002 Dec;23(24):4739-4751.

[175] Giordano RA, Wu BM, Borland SW, Cima LG, Sachs EM, Cima MJ. Mechanical properties of dense polylactic acid structures fabricated by three dimensional printing. *Journal of Biomaterials Science-Polymer Edition.* 1996;8(1):63-75.

[176] Wu BM, Borland SW, Giordano RA, Cima LG, Sachs EM, Cima MJ. Solid free-form fabrication of drug delivery devices. *Journal of Controlled Release.* 1996 Jun;40(1-2):77-87.

[177] Hutmacher DW, Schantz T, Zein I, Ng KW, Teoh SH, Tan KC. Mechanical properties and cell cultural response of polycaprolactone scaffolds designed and fabricated via fused deposition modeling. *Journal of Biomedical Materials Research.* 2001 May;55(2):203-216.

[178] Woodfield TBF, Malda J, de Wijn JR, Pleters F, Riesleb J, van Blitterswijk CA. Design of porous scaffolds for cartilage tissue engineering using a three-dimensional fiber-deposition technique. *Biomaterials.* 2004;25:4149–4161.

[179] Wang F, Shor L, Darling A, Khalil S, Sun W, Guceri S, et al. Precision extruding deposition and characterization of cellular poly-epsilon-caprolactone tissue scaffolds. *Rapid Prototyping Journal.* 2004;10(1):42-49.

[180] Xiong Z, Yan YN, Zhang RJ, Sun L. Fabrication of porous poly(L-lactic acid) scaffolds for bone tissue engineering via precise extrusion. *Scripta Materialia.* 2001 Oct 10;45(7):773-779.

[181] Pfister A, Landers R, Laib A, Hubner U, Schmelzeisen R, Mulhaupt R. Biofunctional rapid prototyping for tissue-engineering applications: 3D bioplotting versus 3D printing. *Journal of Polymer Science Part a-Polymer Chemistry.* 2004 Feb 1;42(3):624-638.

[182] Landers R, Mulhaupt R. Desktop manufacturing of complex objects, prototypes and biomedical scaffolds by means of computer-assisted design combined with computer-guided 3D plotting of polymers and reactive oligomers. *Macromol. Mater Eng.* 2000 Oct;282(9):17-21.

[183] Landers R, Hubner U, Schmelzeisen R, Mulhaupt R. Rapid prototyping of scaffolds derived from thermoreversible hydrogels and tailored for applications in tissue engineering. *Biomaterials.* 2002 Dec;23(23):4437-4447.

[184] Landers R, Pfister A, Hubner U, John H, Schmelzeisen R, Mulhaupt R. Fabrication of soft tissue engineering scaffolds by means of rapid prototyping techniques. *Journal of Materials Science.* 2002 Aug 1;37(15):3107-3116.

[185] Vozzi G, Flaim C, Ahluwalia A, Bhatia S. Fabrication of PLGA scaffolds using soft lithography and microsyringe deposition. *Biomaterials.* 2003 Jun;24(14):2533-2540.

[186] Vozzi G, Flaim CJ, Bianchi F, Ahluwalia A, Bhatia S. Microfabricated PLGA scaffolds: a comparative study for application to tissue engineering. *Materials Science and Engineering C-Biomimetic and Supramolecular Systems.* 2002 May 31;20(1-2):43-47.

[187] Xiong Z, Yan YN, Wang SG, Zhang RJ, Zhang C. Fabrication of porous scaffolds for bone tissue engineering via low-temperature deposition. *Scripta Materialia.* 2002 Jun 7;46(11):771-776.

[188] Khalil S, Sun W. Biopolyrner deposition for freefonn fabrication of hydrogel tissue constructs. *Materials Science and Engineering C-Biomimetic and Supramolecular Systems.* 2007 Apr;27(3):469-478.

[189] Ang TH, Sultana FSA, Hutmacher DW, Wong YS, Fuh JYH, Mo XM, et al. Fabrication of 3D chitosan-hydroxyapatite scaffolds using a robotic dispensing system. *Materials Science and Engineering C-Biomimetic and Supramolecular Systems*. 2002 May 31;20(1-2):35-42.

[190] Morissette SL, Lewis JA, Cesarano J, Dimos DB, Baer TY. Solid freeform fabrication of aqueous alumina-poly(vinyl alcohol) gelcasting suspensions. *Journal of the American Ceramic Society*. 2000 Oct;83(10):2409-2416.

[191] Dellinger JG, Cesarano J, Jamison RD. Robotic deposition of model hydroxyapatite scaffolds with multiple architectures and multiscale porosity for bone tissue engineering. *Journal of Biomedical Materials Research Part A*. 2007 Aug;82A(2):383-394.

[192] Kodama H. Automatic Method for Fabricating a 3-Dimensional Plastic Model with Photo-Hardening Polymer. *Rev. Sci. Instrum*. 1981;52(11):1770-1773.

[193] Quadrani P, Pasini A, Mattioli-Belmonte M, Zannoni C, Tampieri A, Landi E, et al. High-resolution 3D scaffold model for engineered tissue fabrication using a rapid prototyping technique. *Medical and Biological Engineering and Computing*. 2005 Mar;43(2):196-199.

[194] Kim JY, Lee JW, Lee SJ, Park EK, Kim SY, Cho DW. Development of a bone scaffold using HA nanopowder and micro-stereolithography technology. *Microelectronic Engineering*. 2007 May-Aug;84(5-8):1762-1765.

[195] Ovsianikov A, Chichkov B, Adunka O, Pillsbury H, Doraiswamy A, Narayan RJ. Rapid prototyping of ossicular replacement prostheses. *Appl. Surf. Sci.* 2007 May 30;253(15):6603-6607.

[196] Serbin J, Egbert A, Ostendorf A, Chichkov BN, Houbertz R, Domann G, et al. Femtosecond laser-induced two-photon polymerization of inorganic-organic hybrid materials for applications in photonics. *Opt. Lett.* 2003 Mar 1;28(5):301-303.

[197] Mironov V, Boland T, Trusk T, Forgacs G, Markwald RR. Organ printing: computer-aided jet-based 3D tissue engineering. *Trends in Biotechnology*. 2003 Apr;21(4):157-161.

[198] Mironov V, Reis N, Derby B. Bioprinting: A beginning. *Tissue Engineering*. 2006 Apr;12(4):631-634.

[199] Zein I, Hutmacher DW, Tan KC, Teoh SH. Fused deposition modeling of novel scaffold architectures for tissue engineering applications. *Biomaterials*. 2002 Feb;23(4):1169-1185.

[200] Rai B, Teoh SH, Ho KH, Hutmacher DW, Cao T, Chen F, et al. The effect of rhBMP-2 on canine osteoblasts seeded onto 3D bioactive polycaprolactone scaffolds. *Biomaterials*. 2004 Nov;25(24):5499-5506.

[201] Masood SH, Singh JP, Morsi Y. The design and manufacturing of porous scaffolds for tissue engineering using rapid prototyping. *International Journal of Advanced Manufacturing Technology*. 2005 Dec;27(3-4):415-420.

[202] Darling AL, Sun W. 3D microtomographic characterization of precision extruded poly-epsilon-caprolactone scaffolds. *Journal of Biomedical Materials Research Part B-Applied Biomaterials*. 2004 Aug 15;70B(2):311-317.

[203] Miot S, Woodfield T, Daniels AU, Suetterlin R, Peterschmitt I, Heberer M, et al. Effects of scaffold composition and architecture on human nasal chondrocyte

redifferentiation and cartilaginous matrix deposition. *Biomaterials.* 2005 May;26(15):2479-2489.

[204] Malda J, Woodfield TBF, van der Vloodt F, Kooy FK, Martens DE, Tramper J, et al. The effect of PEGT/PBT scaffold architecture on oxygen gradients in tissue engineered cartilaginous constructs. *Biomaterials*. 2004 Nov;25(26):5773-5780.

[205] Malda J, Woodfield TBF, van der Vloodt F, Wilson C, Martens DE, Tramper J, et al. The effect of PEGT/PBT scaffold architecture on the composition of tissue engineered cartilage. *Biomaterials.* 2005 Jan;26(1):63-72.

[206] Woodfield TBF, Van Blitterswijk CA, De Wijn J, Sims TJ, Hollander AP, Riesle J. Polymer scaffolds fabricated with pore-size gradients as a model for studying the zonal organization within tissue-engineered cartilage constructs. *Tissue Engineering.* 2005 Sep;11(9-10):1297-1311.

[207] Li JP, de Wijn JR, Van Blitterswijk CA, de Groot K. Porous Ti6Al4V scaffold directly fabricating by rapid prototyping: Preparation and in vitro experiment. *Biomaterials.* 2006 Mar;27(8):1223-1235.

[208] Li JP, Habibovic P, van den Doel M, Wilson CE, de Wijn JR, van Blitterswijk CA, et al. Bone ingrowth in porous titanium implants produced by 3D fiber deposition. *Biomaterials*. 2007 Jun;28(18):2810-2820.

[209] Yan YN, Xiong Z, Hu YY, Wang SG, Zhang RJ, Zhang C. Layered manufacturing of tissue engineering scaffolds via multi-nozzle deposition. *Materials Letters.* 2003 May;57(18):2623-2628.

[210] Moroni L, de Wijn JR, van Blitterswijk CA. 3D fiber-deposited scaffolds for tissue engineering: Influence of pores geometry and architecture on dynamic mechanical properties. *Biomaterials.* 2006 Oct;27(7):974-985.

[211] Moroni L, de Wijn JR, van Blitterswijk CA. Three-dimensional fiber-deposited PEOT/PBT copolymer scaffolds for tissue engineering: Influence of porosity, molecular network mesh size, and swelling in aqueous media on dynamic mechanical properties. *Journal of Biomedical Materials Research Part A.* 2005 Dec 15;75A(4):957-965.

[212] Moroni L, Poort G, Van Keulen F, de Wijn JR, van Blitterswijk CA. Dynamic mechanical properties of 3D fiber-deposited PEOT/PBT scaffolds: An experimental and numerical analysis. *Journal of Biomedical Materials Research Part A.* 2006 Sep 1;78A(3):605-614.

[213] Moroni L, Schotel R, Sohier J, de Wijn JR, van Blitterswijk CA. Polymer hollow fiber three-dimensional matrices with controllable cavity and shell thickness. *Biomaterials.* 2006 Dec;27(35):5918-5926.

[214] Yousefi AM, Gauvin C, Sun L, DiRaddo RW, Fernandes J. Design and fabrication of 3D-plotted polymeric scaffolds in functional tissue engineering. *Polymer Engineering and Science.* 2007 May;47(5):608-618.

[215] Adachi T, Osako Y, Tanaka M, Hojo M, Hollister SJ. Framework for optimal design of porous scaffold microstructure by computational simulation of bone regeneration. *Biomaterials.* 2006 Jul;27(21):3964-3972.

[216] Hollister SJ, Lin CY. Computational design of tissue engineering scaffolds. *Computer Methods in Applied Mechanics and Engineering.* 2007;196(31-32):2991-2998.

[217] Ciardelli G, Chiono V, Cristallini C, Barbani N, Ahluwalia A, Vozzi G, et al. Innovative tissue engineering structures through advanced manufacturing technologies. *Journal of Materials Science-Materials in Medicine.* 2004 Apr;15(4):305-310.

[218] Bianchi F, Vassalle C, Simonetti M, Vozzi G, Domenici C, Ahluwalia A. Endothelial cell function on 2D and 3D micro-fabricated polymer scaffolds: applications in cardiovascular tissue engineering. *Journal of Biomaterials Science-Polymer Edition.* 2006;17(1-2):37-51.

[219] Bianchi F, Rosi M, Vozzi G, Emanueli C, Madeddu P, Ahluwalia A. Microfabrication of fractal polymeric structures for capillary morphogenesis: Applications in therapeutic angiogenesis and in the engineering of vascularized tissue. *Journal of Biomedical Materials Research Part B-Applied Biomaterials.* 2007 May;81B(2):462-468.

[220] Vozzi G, Ahluwalia A. Microfabrication for tissue engineering: rethinking the cells-on-a scaffold approach. *Journal of Materials Chemistry.* 2007;17(13):1248-1254.

[221] Cesarano J, Dellinger JG, Saavedra MP, Gill DD, Jamison RD, Grosser BA, et al. Customization of load-bearingy hydroxyapatite lattice scaffolds. *International Journal of Applied Ceramic Technology.* 2005;2(3):212-220.

[222] Xu SL, Li DC, Wang CF, Wang Z, Lu BH. Cell proliferation in CPC scaffold with a central channel. *Bio-Medical Materials and Engineering.* 2007;17(1):1-8.

[223] Taboas JM, Maddox RD, Krebsbach PH, Hollister SJ. Indirect solid free form fabrication of local and global porous, biomimetic and composite 3D polymer-ceramic scaffolds. *Biomaterials.* 2003 Jan;24(1):181-194.

[224] Boland T, Ovsianikov A, Chickov BN, Doraiswamy A, Narayan RJ, Yeong WY, et al. Rapid prototyping of artificial tissues and medical devices. *Advanced Materials and Processes.* 2007 Apr;165(4):51-53.

[225] Wilson WC, Boland T. Cell and organ printing 1: Protein and cell printers. *Anatomical Record Part a-Discoveries in Molecular Cellular and Evolutionary Biology.* 2003 Jun;272A(2):491-496.

[226] Shu XZ, Liu YC, Palumbo FS, Lu Y, Prestwich GD. In situ crosslinkable hyaluronan hydrogels for tissue engineering. *Biomaterials.* 2004 Mar-Apr;25(7-8):1339-1348.

[227] Boland T, Tao X, Damon BJ, Manley B, Kesari P, Jalota S, et al. Drop-on-demand printing of cells and materials for designer tissue constructs. *Materials Science and Engineering C-Biomimetic and Supramolecular Systems.* 2007 Apr;27(3):372-376.

[228] Boland T, Mironov V, Gutowska A, Roth EA, Markwald RR. Cell and organ printing 2: Fusion of cell aggregates in three-dimensional gels. *Anatomical Record Part a-Discoveries in Molecular Cellular and Evolutionary Biology.* 2003 Jun;272A(2):497-502.

[229] Xu T, Jin J, Gregory C, Hickman JJ, Boland T. Inkjet printing of viable mammalian cells. *Biomaterials.* 2005 Jan;26(1):93-99.

[230] Varghese D, Deshpande M, Xu T, Kesari P, Ohri S, Boland T. Advances in tissue engineering: Cell printing. *Journal of Thoracic and Cardiovascular Surgery.* 2005 Feb;129(2):470-472.

[231] Matsusaki M, Yoshida H, Akashi M. The construction of 3D-engineered tissues composed of cells and extracellular matrices by hydrogel template approach. *Biomaterials.* 2007 Jun;28(17):2729-2737.

[232] Calvert P. Printing cells. *Science.* 2007 Oct 12;318(5848):208-209.

[233] Lin ASP, Barrows TH, Cartmell SH, Guldberg RE. Microarchitectural and mechanical characterization of oriented porous polymer scaffolds. *Biomaterials*. 2003 Feb;24(3):481-489.

[234] Ho ST, Hutmacher DW. A comparison of micro CT with other techniques used in the characterization of scaffolds. *Biomaterials*. 2006 Mar;27(8):1362-1376.

[235] Diego RB, Estelles JM, Sanz JA, Garcia-Aznar JM, Sanchez MS. Polymer scaffolds with interconnected spherical pores and controlled architecture for tissue engineering: Fabrication, mechanical properties, and finite element modeling. *Journal of Biomedical Materials Research Part B-Applied Biomaterials*. 2007 May;81B(2):448-455.

[236] Zhang RY, Ma PX. Poly(alpha-hydroxyl acids) hydroxyapatite porous composites for bone-tissue engineering. I. Preparation and morphology. *Journal of Biomedical Materials Research*. 1999 Mar 15;44(4):446-455.

[237] van Lenthe GH, Hagenmuller H, Bohner M, Hollister SJ, Meinel L, Muller R. Nondestructive micro-computed tomography for biological imaging and quantification of scaffold-bone interaction in vivo. *Biomaterials*. 2007 May;28(15):2479-2490.

[238] Jones AC, Arns CH, Sheppard AP, Hutmacher DW, Milthorpe BK, Knackstedt MA. Assessment of bone ingrowth into porous biomaterials using MICRO-CT. *Biomaterials*. 2007 May;28(15):2491-2504.

[239] Zehbe R, Haibel A, Brochhausen C, Gross U, Kirkpatrick CJ, Schubert H. Characterization of oriented protein-ceramic and protein-polymer-composites for cartilage tissue engineering using synchrotron mu-CT. *International Journal of Materials Research*. 2007 Jul;98(7):562-568.

[240] Jones AC, Milthorpe B, Averdunk H, Limaye A, Senden TJ, Sakellariou A, et al. Analysis of 3D bone ingrowth into polymer scaffolds via micro-computed tomography imaging. *Biomaterials*. 2004 Sep;25(20):4947-4954.

[241] Harley BA, Leung JH, Silva ECCM, Gibson LJ. Mechanical characterization of collagen-glycosaminoglycan scaffolds. *Acta Biomaterialia*. 2007 Jul;3(4):463-474.

[242] Kaufman JD, Song J, Klapperich CM. Nanomechanical analysis of bone tissue engineering scaffolds. *Journal of Biomedical Materials Research Part A*. 2007 Jun 1;81A(3):611-623.

[243] Leong DT, Gupta A, Bai HF, Wan GQ, Yoong LF, Too HP, et al. Absolute quantification of gene expression in biomaterials research using real-time PCR. *Biomaterials*. 2007 Jan;28(2):203-210.

[244] Chung CA, Chen CW, Chen CP, Tseng CS. Enhancement of cell growth in tissue-engineering constructs under direct perfusion: Modeling and simulation. *Biotechnology and Bioengineering*. 2007 Aug 15;97(6):1603-1616.

[245] Stephens JS, Cooper JA, Phelan FR, Dunkers JP. Perfusion flow bioreactor for 3D in situ imaging: Investigating cell/biomaterials interactions. *Biotechnology and Bioengineering*. 2007 Jul 1;97(4):952-961.

[246] Cao Y, Liu W, Zhou G, Cui L. Tissue engineering and tissue repair in immunocompetent animals: Tissue construction and repair. *Handchirurgie Mikrochirurgie Plastische Chirurgie*. 2007 Jun;39(3):156-160.

[247] Cai KY, Yao KD, Yang ZM, Qu YL, Li XQ. Histological study of surface modified three dimensional poly(D,L-lactic acid) scaffolds with chitosan in vivo. *Journal of Materials Science-Materials in Medicine*. 2007 Oct;18(10):2017-2024.

[248] Boskey A, Camacho NP. FT-IR imaging of native and tissue-engineered bone and cartilage. *Biomaterials.* 2007 May;28(15):2465-2478.

[249] Paluszkiewicz C, Stoch A. FT-IR microscopic imaging of metal-ceramic joints. *Vibrational Spectroscopy.* 2004 Jun 17;35(1-2):183-187.

[250] Chen YM, Shen KC, Gong JP, Osada Y. Selective cell spreading, proliferation, and orientation on micropatterned gel surfaces. *Journal of Nanoscience and Nanotechnology.* 2007 Mar;7(3):773-779.

[251] Hatakeyama H, Kikuchi A, Yamato M, Okano T. Patterned biofunctional designs of thermoresponsive surfaces for spatiotemporally controlled cell adhesion, growth, and thermally induced detachment. *Biomaterials.* 2007 Sep;28(25):3632-3643.

In: Tissue Engineering: Roles, Materials and Applications ISBN: 978-1-60456-293-4
Editors: S. J. Barnes, L. P. Harris, pp. 191-242

Chapter 8

IN VIVO BONE TISSUE ENGINEERING

Chongyun Bao, Bi Zhang, Yunfeng Li, Lanying Sun, Jinfeng Yao and Xingdong Zhang
State Key Laboratory of Oral Diseases;
West China College of Stomatology; Sichuan University

ABSTRACT

To date, there is no ideal method for treating the segmental bone defect. Autograft bone is considered to be an ideal bone graft material because of its good biological properties and compatible mechanical strength, however, several disadvantages accompany with it. Current bone tissue engineering is using *in vitro* expanded mesenchymal stem cells and synthetic scaffolding materials to produce synthetic bone substitutes comparable to autologous bone. Such an approach is theoretically possible to repair damaged bone tissue with tissue engineered bone in clinics, instead of autologous bone; however, partially due to the nutrition problem it is difficult to produce large living tissue-engineered bone in practice at the moment. Increasing data is showing that osteoinductive synthetic materials can attract mesenchymal stem cells in non-osseous tissues and form ectopic bone. *In vivo* bone tissue engineering, by using osteoinductive synthetic materials to produce bone in soft tissues, would be another approach to produce synthetic "autologous" bone for bone repair. *In vivo* bone tissue engineering means to use the body as bioreactor to construct bone graft with intrinsic osteoinductive biomaterials in the non-osseous or osseous sites in absence of any exogenous growth factors or cells, then transplant the graft to repairing bone defect. Three parts are involved in this chapter. Part I, Osteoinductive materials for *in vivo* bone tissue engineering. The physicochemical properties of materials which are important for osteoinduction will be highlighted. Part II, Biological mechanisms of constructing *in vivo* tissue engineered bone. Histological characteristics of calcium phosphate-induced bone formation will be compared to natural bone regarding biosafety. The material-dependence and animal –dependence of calcium phosphate induced-bone formation will be discussed. Part III, Techniques of constructing and the feasibility of applying in vivo tissue engineered bone. Although limited,current data have already raised the possibility to construct large bone graft with good blood supply and certain mechanical strength with *in vivo* bone tissue engineering technique.

1. INTRODUCTION

1.1. Bone Defect and Its Present Treatment

Bone defect is a common deformity in clinic caused by tumour, injury and congenital malformation. It is necessary to reconstruct bone defect because bone provides the attachment of the muscle and tendons, protects the vital organs in the cranial and internal cavities. In 1998, slightly more than 300,000 bone graft procedures were performed in the United States. Currently, this figure exceeds 500,000 in the US and approximately 2.2 million worldwide, and the numbers are expected to increase as life expectancy increases.The estimated cost of these procedures approach $2.5 billion per year in the US.

Most of the bone grafts proceed in either autograft or allograft tissue. The current golden standard is autograft, which is harvested from the patient, usually from the iliac crest, fibula, rib, and possibly from chin of the mandible. The graft is then transplanted at the injury site. This graft is ideal as a bone graft because it possesses all of the characteristics necessary for new bone growth—namely, osteoconductivity, osteoinductivity, osteogenicity and remolding property, plus satisfactory mechanical properties. Allografts are alternatives to autografts and taken from donors or cadavers, circumventing some of the shortcomings of autografts by eliminating donor-site morbidity and issues of limited supply. Before used, allograft tissue should be treated by tissue freezing, freeze-drying, gamma irradiation, electron beam radiation, ethylene oxide, etc. Plus, xenograft bone represents an unlimited supply of available material. Xenograft bone or xenograft-derived collagen material has been used by some researchers as bone substitute, but the procedure has never gained wider acceptance.

Besides the grafts (autograft, allograft and xenograft), alloplasm is another kind of alternatives for bone substitute. Despite the benefits of autografts and allografts, the limitations of each have necessitated the pursuit of alternatives. Investigators have developed several alternatives, some of which are available for clinical use and others are still in the developmental stage. Many of these alternatives use a variety of materials, including metal, natural and synthetic polymers, ceramics, and composites, whereas others have incorporated growth factor and cell-based strategies that are used either alone or in combination with other materials. Laurencin[1] have suggested a classification scheme of material-based groups. Following are some commercial products: (1) Allograft-based bone graft substitutes involving allograft bone, used alone or in combination with other materials (eg, Allogro [AlloSource, Centennial, Colo], Opteform [Exactech, Inc, Gainesville, Fla], Grafton [BioHorizons, Birmingham, Ala], OrthoBlast [IsoTis OrthoBiologics, Irvine, Calif]); (2) Ceramic-based bone graft substitutes including calcium phosphate, calcium sulfate, and bioglass used alone or in combination (eg, OsteoGraf [DENTSPLY Friadent CeraMed, Lakewood, Colo], Norian SRS [Synthes, Inc, West Chester, Pa], ProOsteon [Interpore Cross International, Irvine, Calif], Osteoset [Wright Medical Technology, Inc, Arlington, Tenn] and BioSS [Geistlich biomaterials Inc, Baden Baden, Germany]). (3) Polymer-based bone graft substitutes, degradable and nondegradable polymers, used alone or in combination with other materials (eg, Cortoss [Orthovita, Inc, Malvern, Pa], open porosity polylactic acid polymer [OPLA], Immix [Osteobiologics, Inc, San Antonio, Tex]).

1.2. Drawbacks of the Present Bone Substitutes

At present, autografts, allografts, xenografts and alloplastic grafts are generally used for the repair of bone defects. Although autograft is the preferred bone substitute, it has several disadvantages. Harvesting the autograft requires an additional surgery at the donor site that could result in its own complications, such as hemorrhage, inflammation, infection, nerve damage, chronic pain and cosmetic disability[2]. Quantities of bone tissue that can be harvested are also limited, thus creating a supply problem. Allografts present risks as well. Although allograft tissue is treated by different methods, the risk of disease transmission from donor to recipient is not completely removed. Some have estimated that the risk of HIV transmission alone with allograft bone is 1 case in 1.6 million population. A case of hepatitis B transmission and 3 cases of hepatitis C transmission have been reported with allograft tissue. More recently, cases of other disease transmission have been reported (by Centers for Disease Control and Prevention). Although many methods can reduce the risk of disease transmission, the treatments used to sterilize the tissue remove proteins and factors, reducing or eliminating the osteoinductivity of the tissue.

Comparing with autografts and allografts, alloplastic grafts including metal, ceramics, polymers and composites, are available in unlimited quantity. A few of allografts are available, however, none of them possess good biological properties and satisfying mechanical strength, there could be used for reconstructing segmental or weight-bearing bone. Since 1980s, tissue engineering[3], one of the major parts of regenerative medicine, combined cell transplantation, materials science and engineering to develop biological substitutes that can restore and maintain normal function. Although bone tissue engineering science has made a great progress in the last 20 years, there are several problems yet, such as: (1) Cell sources. The use of embryonic stem cells encounters an ethical conflict, adult stem cells exhibit some disadvantages such as limited life-cycle and multiplication. (2) Lack of satisfactory scaffolds. The extracellular matrix not only provides support and nutrition, but also adjusts biologic molecules to promote cells migration, attachment, differentiation and proliferation. Scaffolds, which benefit cell culture *in vitro,* may not give bone formation *in vivo*. (3) Limited dimension of grafts because of nutrition. After the cell-scaffold grafts are implanted, most of initial nutrition for cells are just directly diffused from adjacent body fluid before new blood vessels grow into the scaffold. Because the distance of oxygen diffusion is limited in no more than several hundred micron, it is almost impossible to create a graft with enough dimension to fit the actual size of defected bone. (4) Stabilization of the seed cells. It is very important that the seeded cells could differentiate, proliferate and terminate growth as expected. So far, there is no report showing that answered tissue-engineered bone with *in vitro* cultivated cells could be used in clinic.

Until now, no bone grafts except autograft display both the biological and mechanical properties, to treat the massive or segmental bone defect. Researchers never stop seeking for novel bone substitutes and techniques to improve bone repair.

1.3. In Vivo Tissue Engineered Bone As a Hopeful Bone Substitute

As a commonly used bone substitute, calcium phosphate (Ca-P) ceramics have outstanding properties, such as similarity to bone mineral in composition, bioactivity and osteoconductivity. In addition, Ca-P ceramics with appropriate three-dimensional geometry are capable of absorbing and concentrating endogenous bone morphogenetic proteins (BMP) from surrounding circulation and become osteoinductive (capable of osteogenesis). As the theory that Ca-P ceramics were provided with osteoinductivity developed, combining with principles of tissue engineering, in the late 1990s, Zhang and Bao put forword the concept of *in vivo* bone tissue engineering and defined it as "to use the body as bioreactor to construct bone graft with intrinsic osteoinductive biomaterials in the non-osseous or osseous sites, then transplant the graft to repair bone defect without any exogenous growth factors or cells "[4]. This technique relies on the body's own capacity to regenerate itself and it differs from tissue engineering with cells cultured *in vitro*. In 2005, Stevens[5] and his colleagues created a bioreactor between the surface of a long bone (tibia) and the covered periosteum, and the space was filled with calcium-alginate gel. As the researchers reported, a few weeks later, the space was filled with neoosseous tissue. When the formed bone was extracted and transplanted to a bone defect site in the same animal, the new bone integrated seamlessly with the surrounding tissue was observed. Service[6] thinks that this technique is a fresh, new strategy for tissue engineering that relies on the body's own capacity to regenerate itself, and it will have an enormous impact on the field.

In vivo bone tissue engineering, by using osteoinductive synthetic materials to produce bone in soft tissues, would be another approach to produce synthetic "autologous" bone for bone repair. Three parts are involved in this research area: (1) Osteoinductive biomaterials. Biomaterials without any exogenous growth factors or cells can induce bone formation in non-osseous sites. The physical characters and chemical composition of the materials play the most important role and they interact each other. (2) Biological mechanisms of constructing *in vivo* bone tissue engineering. The *in vivo* tissue engineering bone formed with osteoinductive calcium phosphate ceramics had similar histological characteristics to natural bone and this kind of bone graft is biosafety at genetic level. The ratio of osseous or osteoid tissue formation and the amount of bone formed are related to the physicochemical property and animal species. The explorations of biological mechanisms, eg origination of stem cell, are ongoing. (3) Techniques of constructing and applying *in vivo* tissue engineered bone. The studies demonstrate that it is possible to construct large bone graft with good blood supply and certain mechanical strength. *In vivo* tissue engineering bone may provide with a hope of being used to repair segmental bone defect of mandible and to support dental implant in the future.

Overall, osteoinductive material is the key factor for developing *in vivo* tissue engineered bone, which is a hopeful segmental bone substitute.

2. Osteoinductive Scaffolds for In Vivo Bone Tissue Engineering

2.1. Research Devolopment on Osteoinductive Scaffold

Early in 1969, Winter[7] reported the ectopic bone formation induced by a synthetic polymer sponge in the skin of young pigs. Nearly 20 years laters, in 1988, Heughebaert found a bone-like substance formed in porous hydroxyapatite ceramics implanted in soft tissue of hamsters (Heughebaert 1988). Whereafter, several other researchers reported the phenomenon of the material-induced bone formation of a series of biomaterials extraskeletally in a variaty of animal species. Hydroxyapatite (HA)-induced heteroctopic bone formation was also reported in the muscle of baboon (Ripamonti 1991) and dogs (Zhang 1991), in the subcutis of dogs (Yamasaki 1992), and in the muscle of monkey(Vargervik 1992). Since these initial researches, a large series of biomaterials have been shown the capacity of inducing bone formation, including polymer, calcium phosphate ceramics such as HA, biphasic calcium phosphate(BCP), and tricalcium phosphate (TCP), calcium phosphate cement, glass ceramics [8], and bioactivated porous titanium [9] as well, in soft tissue such as in muscles, in the fat, and under subcutis. These suggest that certain materials have the capacity to give rise to bone formation extraskeletally.

The osteoindutive calcium phosphate materials have been documented extensively by more and more researchers. Yuan[10] selected a series of calcium phosphate ceramics in the animal experiments, and presumed that the ceramics with run-through macropores and interconnecting micropores were of good osteoinduction. And rapid osteogenesis in porous biphasic calcium phosphate ceramics implanted in domestic pigs was reported in the literatures. The appearance of a large number of active osteocblasts induced new bone formation and it's confirmed the evidence of the intramembrane osteogenesis in the porous calcium phosphate ceramics implanted in an extra-osseous site in dogs. Gosain AF reported that the rate of osteoinduction was greatest when a porous architecture was maintained, which was best achieved in ceramic rather than cement-paste forms of hydroxyapatite. Porosity and resultant bone formation in cement-paste implants can be improved by combining hydroxyapatite with a rapidly resorbing component, such as tricalcium phosphate in a one year study in adult sheep. The immunolocalization of BMP family members (BMP-3 and OP-1/BMP-7) in cellular material at the hydroxyapatite interface suggests that the sintered ceramic may act as a solid-state matrix for adsorption of endogenously produced BMPs (Ripamonti U). These experiments demonstrate intrinsic osteoinductivity by porous sintered hydroxyapatites implanted in heterotopic sites of adult primates, and that the geometry of the substratum profoundly regulates the expression of the osteogenic phenotype[11]. The incorporation of specific biological activities into biomaterials achieved by manipulating the geometry of the substratum, defined as geometric induction of bone formation, will help engineer morphometric responses for therapeutic osteogenesis in clinical contexts (Ripamonti, 1999). And then, some studies compared the osteoinductive potential of different calcium phosphate in large or small animal models, and fewer compared the potential of the same calcium phosphate between the large and small animal models or the different ceramics in statistics. In 2002, Yuan made the comparison of the osteoinductive potential of hydroxyapatite and biphasic calcium phosphate implanted intramuscularly in goats. And the incidence of material-induced bone formation varies with animal species and is related to the implant material used [10].

From above, it could be concluded that osteoinductivity depends on biomaterials and animal species. In this part, it will be narrated how the biomaterial properties affect the osteoinductivity, and present some methods for fabricating osteoinductive scaffold.

2.2. Chemical Composition of Osteoinductive Scaffolds

2.2.1. Ceramics

Previous studies have indicated that the osteoinductivity depends on biomaterials, and calcium phosphate is the major agent to induce bone formation. A study in our group was further to explore calcium phosphate ceramics with similar physical structure and different chemical constituent affecting osteoinductivity. Morphological observation, 99mTc-MDP absorption evaluation and statistical analysis were employed in the study to optimize and select Ca-P materials with high osteoinductivity and thereby to provide theoretical and experimental evidence for the application of osteoinductive materials.

Calcium phosphate ceramics was prepared by H_2O_2 foaming method. HA was sintered at 1200°C (HA1200), HA/TCP (60/40) at 1200°C (HT1200) and 1100°C (HT1100) respectively for 3 hours with wet-synthetic calcium phosphate powder. The ceramics cylinders with φ5mm×8mm dimensions were prepared and cleaned with de-ionized water by ultrasonic agitation. The chemical composition of the ceramics was analyzed by X-ray diffraction (XRD) method(figure 1). SEM (figure 2) tested the average pore size of the ceramics and the porosity of the ceramics was determined by the Archimedes water displacement methods (table 1).Surgery was performed on 27 dogs under aseptic condition. After anesthesia, four cylinders of each kind of ceramics were implanted in the dorsal muscle of every dog. 9 animals were sacrificed by overdose of pentobarbital sodium solution randomly at the period of 6, 12 and 24 weeks post-implantation. ^{99m}Tc–MDP (1mCi/kg) was injected to all dogs by vein four hours before sacrifice, and then Single Photon Electronic Computer Tomography (SPECT) was employed (figure 3). Samples were harvested for decalcified sections by H.E. staining. Three pathologists determined the judge standard of bone/bone-like tissue. The osteoinductivity and the amount of bone/bone-like tissue formation in different materials were compared by histomorphology and 600CL-Pixcer system under light microscope.

SPECT test showed that the newly formed bone could metabolize in the body early, especially in HT1100 (figure 3). Histological observation of three different ceramics were as follow. In HT1100: bone formation is observed in all the samples of the three periods. The bone trabecula is adjoining to the wall of the pores of the ceramic. As the time prolonged, the new bones become more mature, and osteoclasts imply bone remolding in progress. The cells, which formed the bone trabecula, developed from the osteoblasts, to bone lacuna via osteocytes. The cells in the central tissue of the pores are differentiated from the early star-shaped or rounded mesenchymal stem cells to the mature steariform cells, which have the slender nucleus away from the center; i.e. the marrow structure is formed. In HT1200: in the 6th week, the tissues in the pores are mainly star-shaped cells or slim cells. Bone trabecula can be observed in some samples. And in the 12th week, the unmatured bone trabecula structure appears in most samples, along which there are a belt of osteoblasts. Not until the 24th week, myeloid structure appears along with the maturation of the trabecula and the formation of new vessels. HA1200: In the 6th week, the ceramics are surrounded by the rather thick fibro capsules, and the inside tissues are noumenal tissues, which are composed of fibroblasts and inflammatory cells unorganized. From 12th week to 24th week, the number of samples with more grossus bone trabecula that is not adjoining to the wall of pores is increasing gradually. But the formed tissues still comprise mainly fibroblasts and inflammatory cells(figure 4).

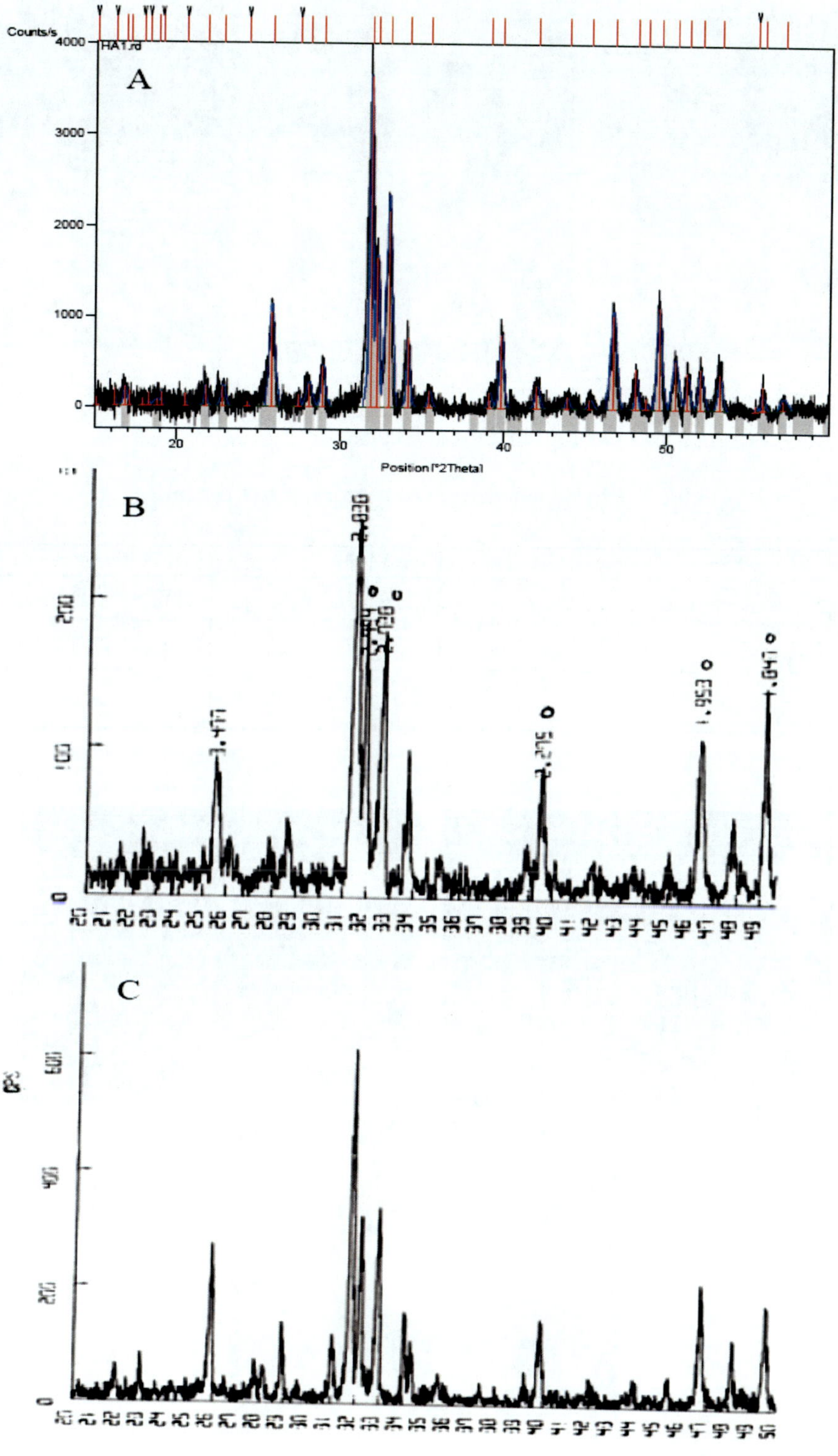

Figure 1. XRD Spectrum of the three ceramics showing the crystal phase of the HA1200(A), HT1200(B) and HT1100(C) respectively.

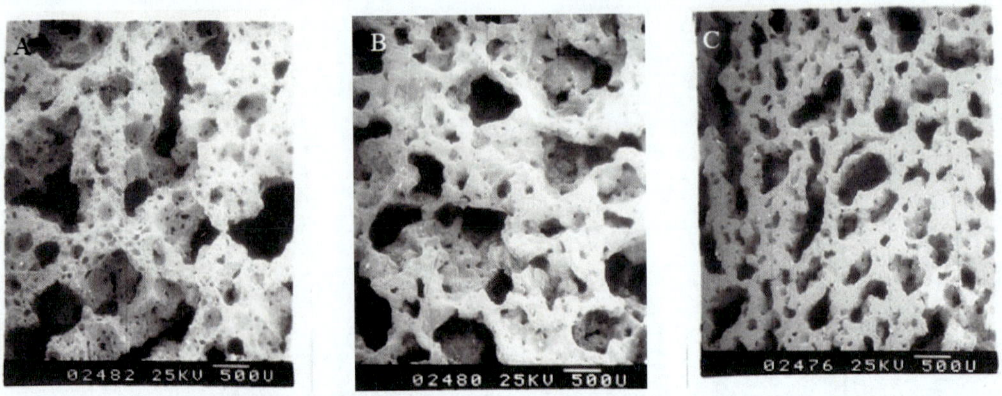

Figure 2. SEM photographs of three ceramics with similar physical property and different chemical constitute. A: HA1200, B: HT1200, C: HT1100; bar=500μm.

Table 1. Physicochemical characters of the ceramics

	HA1200	**HT1200**	**HT1100**
chemistry	HA	60/40 HA/α-TCP	60/40 HA/β-TCP
pore size	450±50μm	450±50μm	400±50μm
Porosity%	68.9	68.2	67.8
Specific weight(g/cm3)	1.20	1.21	1.27

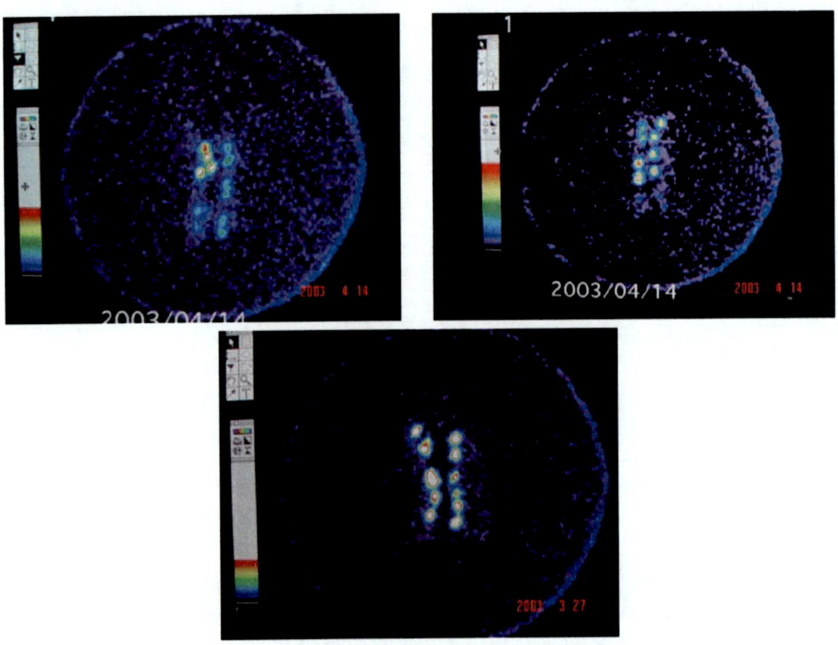

Figure 3. SPECT images of the three ceramic implants in dorsal muscle. left: 6w; right: 12w; lower: 24w. In all the images, the upper four implants of left are HT1100; the upper four implants of right are HT1200; the lower four implants of both sides are HA1200.

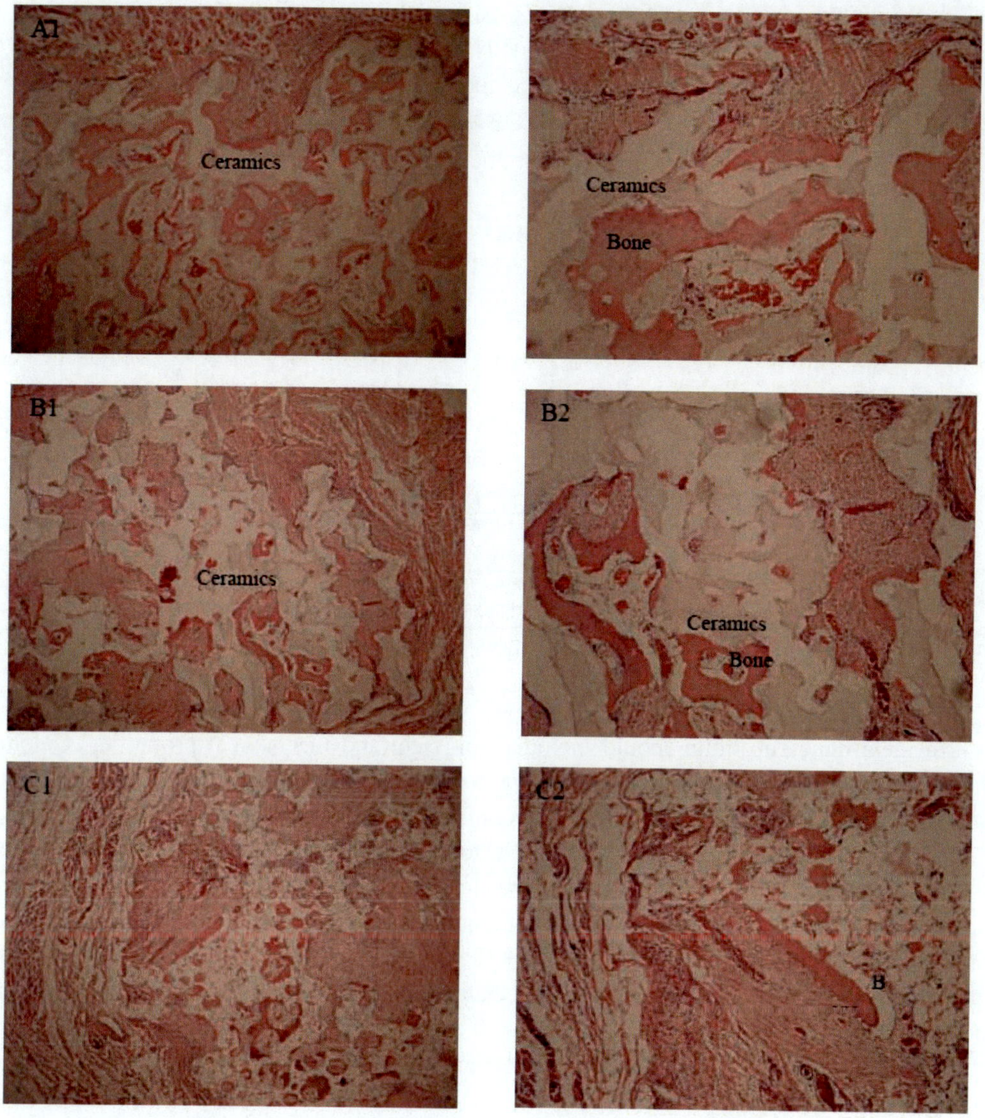

Figure 4. LM of the three ceramics in dog's dorsal muscle of 12 weeks. A: HT1100; B: HT1200; C: HA1200. Decalcified sections, H&E staining. Original magnification of A1, B1and C1 40*, of A2, B2 and C2 100*.

According to the statistics of the relative area of osseous tissue in all osteoinductive samples of dogs and by the analysis of the single-factor covariance, we get the conclusions: (1) in three periods, as for the osseous formation amount, HT1100 has significant difference with the other two groups, whereas in the 12th and 24th week, there were no statistically significant differences in HT1200 or HA 1200. (2). As for the same materials at different periods, both HT1100 and HA1200 performed much better in the 6th week than in the other two periods, which is statistically significant, conversely not statistically significant differences in the 12th and 24th week; however, concerning HT1200, there was statistically significant differences in each period(table 2).

The chemical components of the material have an obvious effect on the osteoinductivity. This study demonstrates that the osteoinductivity of HA/TCP is better than that of HA. Combining the previous studies, it could be indicated that the ranking of osteoinductivity from high to low is: HA/β-TCP>HA/α-TCP>HA>β-TCP>α-TCP. Stable scaffold and some calcium phosphate ingredient release are necessary.

Table 2. stratified statistics of the osseous formation amount of three materials at each time points

VAR	VAR	Mean Difference (1-J)	Std.Error	Sig.	95% Confidence Interval	
					Lower Bound	Upper Bound
HT1100	5	6.7020	1.1067	.000	4.4947	8.9093
	6	5.1811(*)	1.5539	.001	2.0819	8.2803
HT1200	4	-6.7020 (*)	1.1067	.000	-8.9093	-4.4947
	6	-1.5209	1.6092	.348	-4.7305	1.6886
HA1200	4	-5.1811(*)	55393778	.001	-8.2803	-2.0819
	5	1.5209	1.60925527	.348	-1.6886	4.73048

The mean difference is significant at the α=0.05 level.

2.2.2. Composite

An ideal bone substitute should include biological and mechanical properties. Calcium phosphate ceramics as bone substitute have been proved to be good biocompatible and osteoconductive, and in recent years, it was confirmed that calcium phosphate ceramics with special chemical composition and physical character is osteoinductive, which make them have brighter prospect in clinical application. However, unitary calcium phosphate ceramics can not satisfy the requirement of clinical application because of being too brittle and lacking sufficient tensile strength. In order to modify the mechanical property of calcium phosphate ceramics and maintain its bioactivity especially osteoinductivity, optimal design of the composite materials based on calcium phosphate ceramics is required. Previous experiments have shown that the composite materials have better mechanical property than that of unitary calcium phosphate ceramics, but did they have osteoinductivity. The purpose of the following study is to explore the osteoinductivity of composite materials with similar physical character and then to provide experimental evidence for the optimum design of the composite materials as efficient bone substitute.

Two types of materials were prepared. (1) BCP/PLA composite: Two kinds of BCP/PLA composite scaffolds were fabricated through the particulate-leaching technique with BCP (biphasic calcium phosphate) ceramic powder and PDLLA (polylactide) powder in the ratio of 10:90 and 20:80 respectively. Firstly, determined quantity of PDLLA was dissolved in 1, 4-dioxane, then BCP ceramics powder was added and the mixture was stirred until it was uniform. The mixture was then mixed in a ratio of 1:6 with NaCl powder (300-400μm), poured in mold and allowed to dry. Then the solid blocks were removed from the mold and the salt was extracted in de-ionized water. (2) BCP/collagen composite: The calcium phosphate ceramics was prepared using the wet-synthetic calcium phosphate powder through the H_2O_2 foaming method and sintering at 1250 °C for 3 hours. Cleaned with de-ionized water by ultrasonication, the surface of the ceramics was modified with collagen through freeze-drying. 24 composite cylinders with Φ 5×8 mm were prepared in each group respectively. The microstructural properties of the composite materials were tested by SEM. The composition of the BCP/collagen composite was analyzed by infrared radiometer (IR). All the cylinder samples were sterilized by gamma rays before being implanted.

Four dogs were used in this study. After anesthesia with 3% sodium pentobarbital, six cylinder samples of each group were implanted in dorsal muscles of each dog. The specimens were harvested after being implanted for 2, 4, 12 and 24 weeks. Then the specimens were fixed by 10% formalin and thin decalcified sections were prepared to compare the osseous or osteoid tissue formation with different materials chronologically under light microscopy (LM).

The XRD analysis showed that BCP consisted of 60HA/40α-TCP. The porosities of BCP/PLA and BCP/collagen were 80% and 60% respectively. The SEM test showed the average pore-size in both groups was 300-400μm, and the collagen fibers covered the surface of the material in the BCP/collagen group(figure 1). IR spectrum of BCP/collagen illustrated that it was comprised of collagen and calcium phosphate. In these spectra the aminde A, B, I and II bands of collagen appeared at the frequencies of 3336, 3082, 1662 and 1538 cm^{-1} respectively, as well as PO_4^{3-} at the frequency of 1092 $cm^{-1}$and OH^- at 3568 cm^{-1} and 630cm^{-1} (figure 2).

Previous studies have proved that osteoinduction of calcium phosphate ceramics is material-dependent. And the microstructure should affect its capability of osteoinduction, which includes pore size and porosity. Pore size not only offers enough area where cells can cohere and grow, but also benefits the blood vessel formation. It was reported that the ideal pore size should be 200-400μm , while the average pore size of composite scaffolds in this study is 300-400μm. The SEM showed that there were abundant micropores on the inner-wall of the macropore. The previous experiment demonstrated that it is an important factor in the process of bone induction. So the composite materials in this experiment have the similar structure to that of osteoinductive calcium phosphate ceramics. Besides microstructure, the chemistry may be another important factor influencing osteoinductivity of the calcium phosphate ceramics. In this experiment, PDLLA, the widely applied biodegradable and biocompatible polymer, was chosen to make composites with BCP ceramic in order to strengthen tenacity [12]. Collagen, which is a major component of the extra-cell matrix (ECM), was often used as the scaffold for bone tissue engineering .

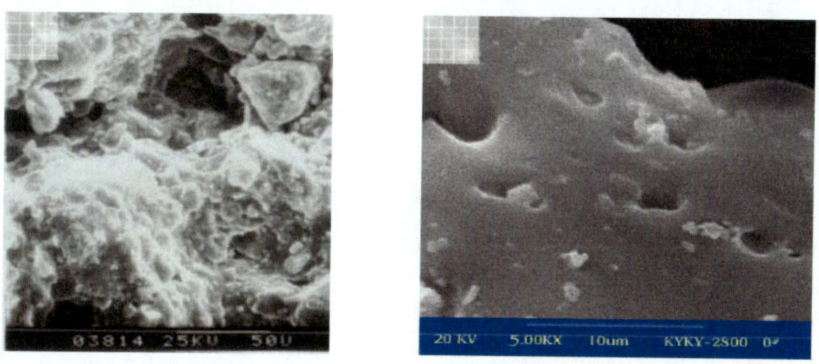

Figure 5. SEM photographs of two types of composite. left: Porous HA/TCP coating with collagen; collagen forming on the surface of BCP; right: HA/PLA composite.

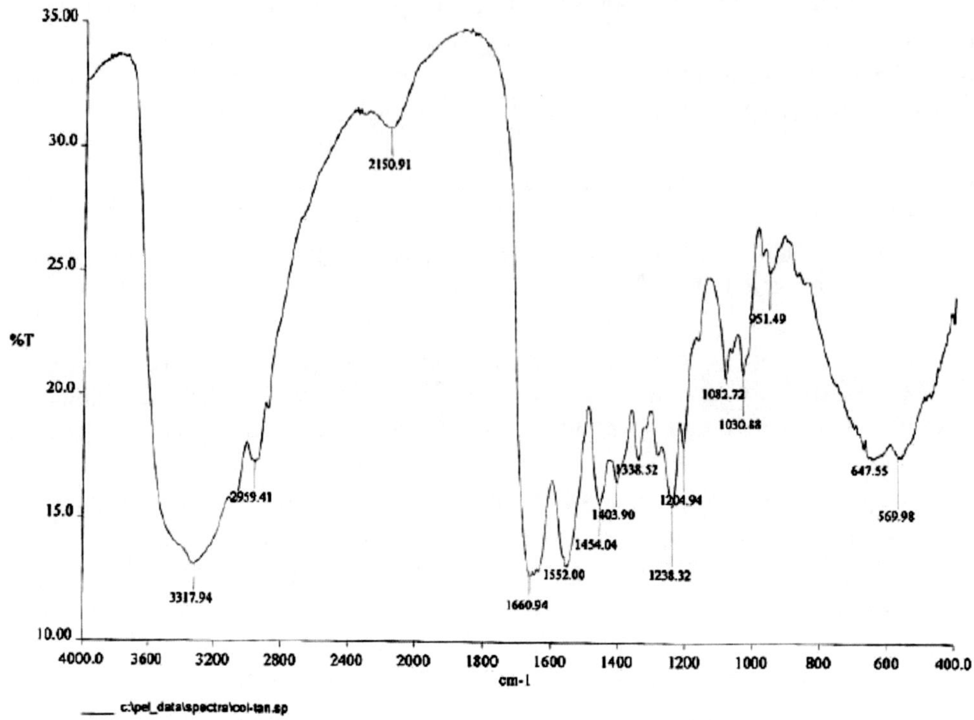

Figure 6. IR spectrum for the composition of BCP/collagen. The aminde A, B, I and II bands of collagen appeared at the frequencies of 3336, 3082, 1662 and 1538 cm^{-1} respectively.

Specimens were harvested from dogs and no infection or inflammation was observed. In BCP/PLA, most of materials were biodegraded so obviously at the 24th weeks post-implantation and their shape changed completely, while those implanted for 12 weeks changed slightly with partial biodegradation. The material integrated with the tissue and the tissue grew into the material. Under LM, all the materials were enveloped by fibrous tissue, in addition to fibrous tissue infiltration into the pore walls. Many inflammatory cells were observed in the materials at 2 weeks post-implantation, and the inflammatory cells decreased obviously along with time, fewest at the 24th week. Otherwise, no bone formation was observed in the tested periods and there was no significant difference between the two samples with different ratio. In BCP/collagen, concerning the cylinders implanted for 4 week, fibrous tissue formed in the pores of cylinders and a few inflammatory cells were observed. At the 12th week, osteoid tissue formed. However, no mature bone was observed in the experimental periods (figure 7).

The results suggest that both types of composite materials in this study are well biocompatible. However, in BCP/PLA, PDLLA takes the high proportion that enables the improvement of the mechanical properties of composite material, but it seems not to improve the osteoinductivity of the materials. However, the degradation speed of PDDLA is so fast that the scaffolds collapse too early to benefit the bone-like apatite layer formation. It is believed that the bone-like apatite layer formation on the surface play crucial role in osteoinduction. Furthermore, BCP and PDLLA are only physical combination and the BCP takes the low proportion. With the degradation of PDLLA, the BCP powder may dissolve in acid environment. Therefore, the composite scaffold could not initiate the osteoinductivity in vivo, though the BCP is osteoinductive. In BCP/collagen, BCP is the major component of composite materials. BCP has good stability *in vivo* and could maintain the porous scaffold for a long time, synchronously with tissue infiltration into pore. Therefore, the shape of materials *in vivo* did not change and the osteoid tissue could be observed. In the study, the results revealed that collagen could enhance the toughness of calcium phosphate ceramics; inversely it could not improve the osteoinductivity of the BCP/collagen composites *in vivo,* although previous studies reported that collagen modification could promote the attachment and proliferation of osteoblast *in vitro.*

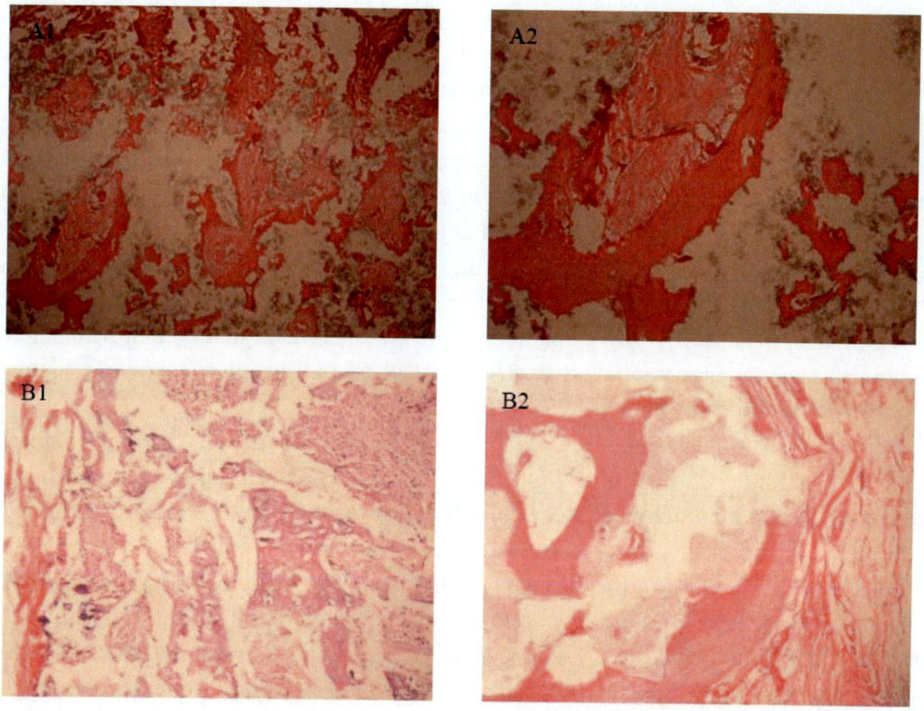

Figure 7. LM photographs of the two kinds of composite impanted in dog's dorsal muscle for 12 weeks. (A) BCP/PLA; (B) BCP/collagen. Original magnification 40* in A1 and B1,100* in A2 and B2, HE staining.

In this experiment, although there was no mature bone formation, the fibro-tissue and osteoid tissue could form in both composite materials[13]. These results implied that they might have the potential of osteoinduction. So it is feasible and prospective to prepare composite materials based on calcium phosphate ceramics in order to improve its mechanical properties and to maintain its bioactivity especially osteoinductivity. The optimizing design of osteoinductive composite materials should be further explored. From the view of preparation technique, it is easy to form porous scaffold and control the chemical and physical properties by the method of preparing BCP/PLA composites.

2.2.3. Polymer and Metal

In 1969, Winter[7] reported that a synthetic sponge made of polyhydroxyethylmethacrylate(polyHEMA: Hydron), when implanted in the skin of young pigs, became filled intially by fibrous connective tissue and later by bone. This is a chemically stable polymer, which was ever used for female breast prostheses in clinic. The authers intented to investigate the tissue reaction of the polymer, in their study bone formation was eventually observed. When they explored the process of osteogenesis induced by the polymer, they discovered calcium / phosphorus mineral was deposited on the sponge before bone formed. Then, One of the research members, Simpson, studied it in rats and other animal species, implantts of polyHEMA sponge which had calcified, but not yet ossified. Some biodegradable polymers, such as polylactine (PLA) and chitin, were implanted in dogs, but no bone formation was observed. The typical appearance was growth of fibroblast and many multinucleate giant cells.

Four types of porous titanium sample treated with different methods were used for osteoinductive research by Shunsuke[9] in 2003. Titanium implants chemically and thermally treated posses micro-porous structure in the macro-pores, and apatite could deposits on these kinds of samples. After implanted in dog's dorsal muscle for 12 months, bone formation was found in the chemically and thermally treated porous samples. On the other hand, the micro-surface of both of the non-treated titanium implants was smooth. No surface morphological changes were observed in the non-treated titanium implants after soaking in the SBF for 7 days, and no bone formation was found in these samples after implantation. So an *in vitro* apatite-forming ability may be a necessary prerequisite for biomaterials to be osteoinductive materials *in vivo*. This study shows that even a non-soluble metal that contains no calcium or phosphorous can be an osteoinductive material when pretreated to form an appropriate macrostructure and microstructure.

Many kinds of biomaterials, such as calcium phosphate ceramics, porous titanium, and composite containing Ca-P composition, with special chemical constituent and physical structure can attract mesenchymal stem cells in non-osseous tissues and form ectopic bone. Porous metal or polymer for osteoinductive scaffold should contain Ca-P or could be deposited with Ca-P easily on the surfaces of their pore. So calcium phosphate constitute is very important to induce bone formation, but materials composition is not the only qualification for osteoinduction. And previous studies have shown that only the calcium phosphate ceramics with special physical structure could induce bone formation, which includes macro-pore size, porosity and micro-structure etc. So the geometry of the biomaterials is one of the most important factors for inducing bone formation.

2.3. Physical Characters of Osteoinductive Scaffolds

The aim of the section is to explore the osteoinductivty of biomatrerials with different physical characters and uniform chemical composition. Three types of bioceramics are used in the study. Starting apatite powder withHA/TCP(60:40) were prepared by a wet precipitation method. The agglomerant used for the pore maker and 30% H_2O_2 were added to the starting apatite powder according to certain ratio, then the mixture were stirred uniformly. The green body were fabricated by pressing the mixture in certain tooting. All the three bioceramics were treated with same process except pore maker and sintered at same temperature 1150℃. According to the different size of the pore maker, the porous ceramic bulks were divided into three groups: A. the size of macropore were 200-500μm; B. the size of macropore were 500-900μm; C. the size of macropore were 1000-1200μm (figure 8).

Three dogs were used in this study. After anaesthesia with 3% sodium pentobarbital, four cylinders with φ5mm×8mm of each kind of ceramics were implanted in the dorsal muscle of one dog. Dogs were sacrificed by overdose of pentobarbital sodium solution randomly at the period of 6, 12 and 24 weeks post-implantation. Samples with the around muscle were harvested for decalcified sections with H.E. staining. The results of histomorphology showed that bone formation was observed in A and B groups, but not at all in C group in the whole experimental stage. In general, the implanted ceramics were covered by a dense connective tissue layer. And in group A and B, bone was formed in the inner pore surfaces of the ceramics and not on the outside surface of the cylinder at the 6th week post-implantation, and became more and more mature along with time, so that bone trabecula with cuboidal layer of osteoblasts was formed in the center of pore; simultaneously vessels crawled into the central of the ceramics along the pore. Inversely, in group C, fibrous tissue infiltrated into the pores of the ceramics chiefly, however, it did not evolved to bone tissue or osteoid tissue. In brief, there was not bone tissue (even osteoid tissue) formation at all in or around the implanted ceramics in the whole experimental stage (figure 9).

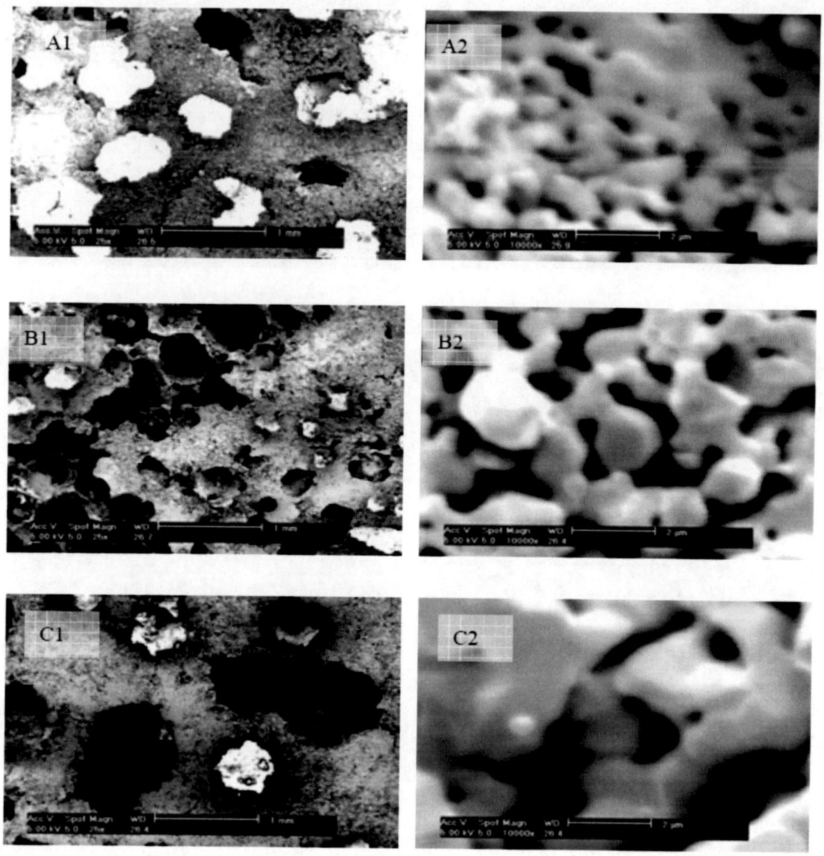

Figure 8. SEM photograph of three ceramics with different physical properties and same chemical constitute.. A1/B1/C1 show macro-pore, different macropore sizes in A: 200-500μm ; B: 500-900μm; C: 1000-1200μm; bar=1mm; A2/B2/C2 show micro-pore, bar = 2μm.

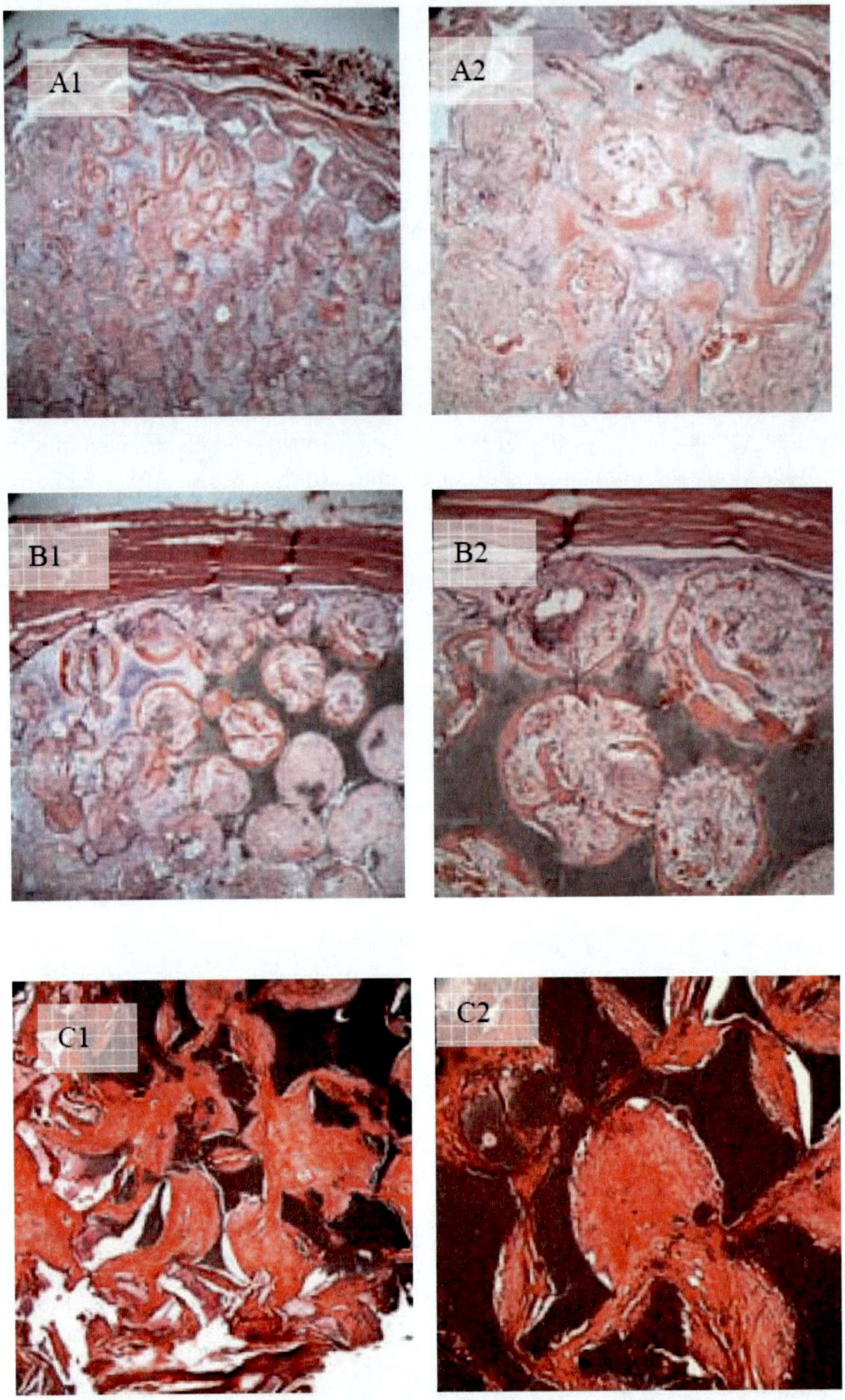

Figure 9. LM photograph of three types of bioceramics impanted in dog's dorsal muscle of 12 weeks. Three bioceramic implants with different macropore sizes in A(200-500μm), in B(500-900μm) and in C(1000-1200μm). The magnification of A1/B1/C1 are 40*, A2/B2/C2 are 100*.

Although the starting apatite powder, sintering temperature and preparing process are the same except pore maker. The SEM photographs show both the macro-pore and microstrucure are different amongst the three bioceramics. The macro-pore maker may affect the microstructure. Two kinds of bioceramics in group A and B respectively possess similar microstructure, namely smaller grain size and more micro-pores(diameter about 1μm).Another bioceramics with 1100μm macro-pore has bigger grain size and less micro-pore. The induced bone can be observed in the two former bioceramics but not in the later. Other researchers[14,15] reported that particle bioceramics could induce bone formation in non-osseous site and macro-pore size did not affect osteoinductivity. So the microstructure might be the most important factor of physical characters for osteoinduction[16]. A proposed explanation for the phenomena may be that micropores enlarge inner surface of the macropores greatly. Great internal surface does favors not only to penetration of body fluid and formation of bone-like apatite (BLA) surface, but also to adsorption of more proteins. The pre-formed BLA and protein adsorbed on the internal surface may facilitate bone formation. Meanwhile, micropores produce rough surface which is favorable to cell adhesion, proliferation and differentiation. These advantages of micropores may make osteoinduction more easily in porous calcium phosphate ceramic. Recently, more and more researches focus on microstructure of biomaterials and osteoinduction. Based on the behavior of rat bone marrow cells growing on surfaces with different pore sizes, Nashimoto reported that the microstructure was a key factor in osteoblast differentiation and proliferation. Takemoto[17] narrated that the peculiar microporous properties of the structure enhanced protein adsorption and cell attachment, and contributes to improve osteoinductivity. The results of Habibovic[18] showed that the microporosity of ceramics affected osteoinductivity.

Both the chemical composition and physical property affect osteoinductivity. Calcium phosphate, which is the inorganical constitute of natural bone, seems to be very important for osteoinductive scaffolds. Osteoinductive scaffold should contain calcium and phosphorus, or cailcium phosphorus could be easily deposited on it *in vivo*. Microstructure, including crystle grain size, the size and distribution of micro-pore, could be the most important factor of the physical charactors for inducing bone formation.

2.4. Some Methods for Fabricating Osteoinductive Scaffold

2.4.1. Preparing Porous Calcium Phosphate Ceramic Scaffold

Porous calcium phosphate ceramics are often used as *in vivo* bone tissue engineering scaffold, and osteoinductive biomaterials seem to be related with calcium and phosphorus composition. Calcium phosphate ceramics include a variety of ceramics such as hydroxyapatite (HA), tricalcium phosphate (TCP), calcium phosphate cement (CPC). Both bioactivity and biomechanical property are required for *in vivo* bone tissue engineering scaffold[19]. Present studies show that porous HA/TCP with ratio of 60/40 could be a good osteoinductive ceramics. In this section, some methods for fabricating porous HA/TCP with high porosity, high pore interconnectivity, and controlled pore size will be narrated.

Wet-synthetic method can be used to produce various calcium phosphate powders depending on the ratio of calcium and phosphorus. Some chemical agents add the calcium phosphate powder to form porous calcium phosphate ceramics. These chemical agents include: ① organic agents, such as polyvinybutyl, polymethylmethacrylate, naphthalene, etc. ② foaming agents including H_2O_2 and acid liquid to obtain CO_2. While, ceramics manufactured by adding pore maker only have uniformed pores and a dense wall without any micropores. The two kinds of agents can be combined together to be used. Polymer foams, such as polyurethane, can be burnt off to produce interconnective porous calcium phosphate ceramics too. After ceramic green body is prepared, the sintering temperature could range from 1050℃ to 1250℃. The compressive strength is between 3 to 10 Mpa.

Porous calcium phosphate ceramics with high porosity, high pore interconnectivity, and controlled macro-pore size could be prepared with above methods. But high osteoinductivity and acurrately controlled microstructure could not be realized presently. Based on the traditonal methods for fabricating *in vivo* bone tissue engineering scaffold, new methods should be devoloped.

2.4.2. Ceramic Scaffold Treated in Varoius SBF (Simulated Body Fluid)

Coating bone-like apatite on bioceramics through biomimetic process is considered as a useful method to improve their bioactivity, since bone-like apatite, particularly calcium-deficient and carbonated apatite, has been proven to be highly beneficial for bonding to bone compared with the current bioactive ceramics[20,21,22]. In addition, it was recently demonstrated that the bone-like apatite layer formed on the surface played crucial role in osteoinduction of bioceramics. SBF (synthetic/simulated body fluid) solutions were found to be able to induce apatite-like calcium phosphate formation on numerous metallic, ceramic or polymeric surfaces (only after proper surface pretreatments) immersed in them[23]. The *in vivo* process of apatite formation is usually simulated *in vitro* by precipitation of calcium phosphates from SBF solutions, which try to mimic the ionic concentrations of human blood plasma, therefore, is supersaturated with respect to the precipitation of carbonated, non-stoichiometric, poorly-crystallized or nanocrystalline apatite-like calcium phosphate. Dynamic method, introduced by Duan[24] recently, is closer to physiological condition than static immersion assay. However, in addition, the simulation *in vitro* is different from physiological condition *in vivo*. SBF would release CO2 gas from the fluid, causing a decrease in $HCO3^-$ concentration, and an increase in pH value, in which calcium phosphates precipitate to form non-stoichiometric hydroxyapatite, when the storage period was long. Furthermore, some authors stated that SBF would not be suitable for long-term use in the biomimetic coating processes owing to its instability[25]. To simulate the behavior of calcium phosphate ceramics veritably, a modified CO_2 supplying device was designed to adjust the pH value and to enhance buffer ability of solution, and porous biphasic calcium phosphate ceramics were immersed in a revised synthetic/simulated body fluid (R-SBF) introduced by Kim et al.

Biphasic porous calcium-phosphate ceramics (BCP) (HA/α-TCP=40/60) was sintered at 1250°C for 3 hours with wet-synthetic calcium phosphate powder by H2O2 foaming method. The ceramics cylinders with φ5mm×8mm dimensions were prepared and cleaned with de-ionized water by ultrasonic agitation.

Table 3. comparison of Ion concentrations (mM) of the blood plasma and SBF

Ion concentration	Na^+	K^+	Ca^{2+}	Mg^{2+}	HCO_3^{2-}	Cl^-	HPO_4^{2-}	SO_4^{2-}
blood plasma	142	5	2.5	1.5	27	103	1	0.5
rSBF	142	5	2.5	1.5	27	103	1	0.5
iSBF	142	5	3.8	1.5	27	103	1.5	0.5

Compared with the conventional RSBF, the iSBF has Ca^{2+} and HPO_4^- concentrations 1.5 times to those of RSBF (table 3.) Experiments on the formation of bone-like apatite in revised simulated body fluid (i-SBF) were performed at 37°C and a stable pH (7.35-7.45), which was controlled by filling CO2 gas into R-SBF storage tank at a suitable rate. i-SBF flowed at a rate of 2 ml/min by a peristaltic pump (figure 10). HA/a-TCP was immersed for 2 days in R-SBF. After that, samples were removed, washed with distilled water for several times, then, dried at 50°C. The surface morphologies of the samples were observed by SEM. Figure 11 shows SEM photographs of cross section of scaffolds after incubation. It was found that a dense layer of needle-shaped apatite formed on the surface of the pores. Moreover, in order to examine cell responses to the formed bone like apatite, the composite of modified scaffold and osteoblast cell, derived from the marrow mesenchymal stem cells of dogs, was co-cultured. The SEM showed that cells on the surface showed a high degree of spreading, and it was considered that the enhanced cell attachment might be related to the morphology of bone like apatite layer on the ceramics.

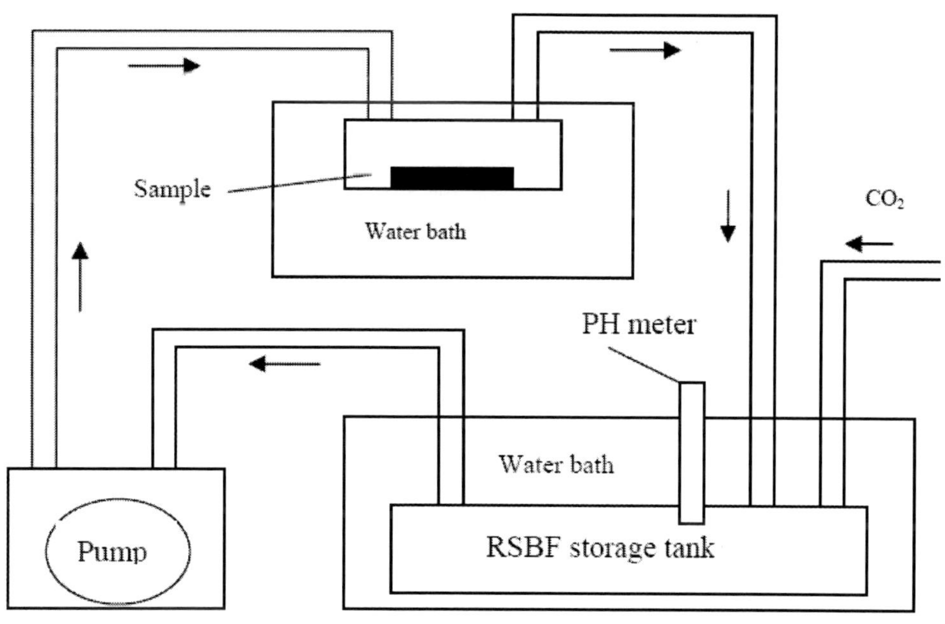

Figure 10. Schematic drawing of dynamic immersion system.

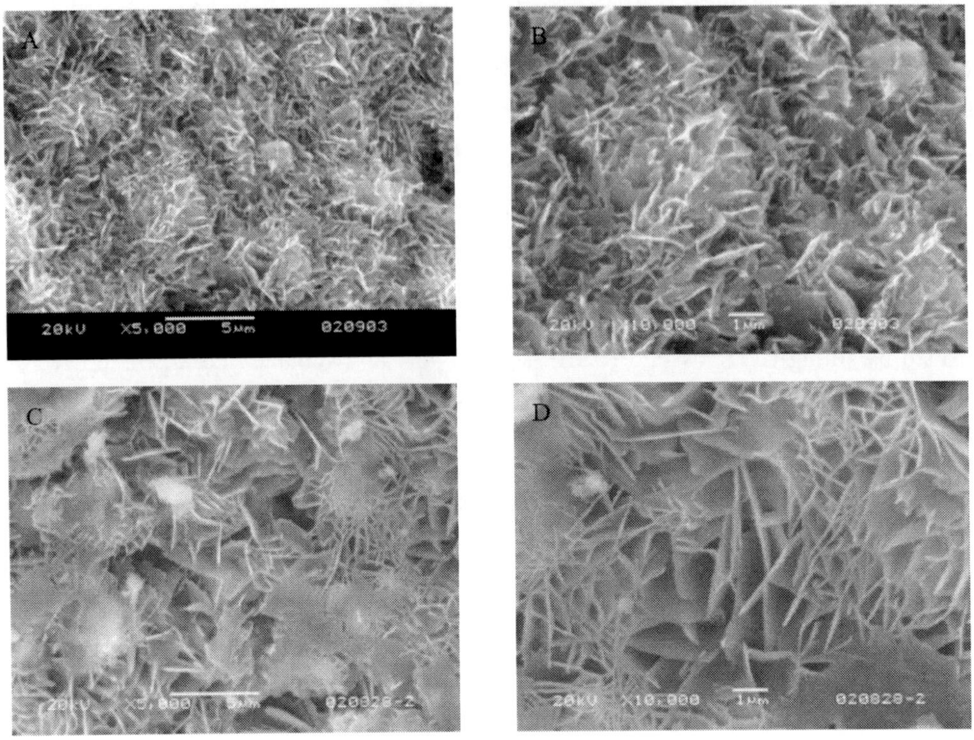

Figure 11. SEM micrographs of the modified BCP in the dynamic iSBF. (A) modified surface, bar=5μm;(B) modified, bar=1μm;(C) modified, bar=5μm;(D) modified, bar=1μm.

2.4.3. Fabricating Scaffold with Controlled Microstructure Accurately

Until now, the mechanisms of microstructure for osteoinduction remains unclear, and the further investigation of the crucial micro-size and micro-geometry for bone induction are required. Of course, it is necessary to explore the accurate fabricating technique and to establish the biological evaluating system for osteoinductive materials.

Techniques for fabricating microstructure originate from percutaneous devices because these devices arc often involved in microorganism attachment and infection. Surface properties of biomaterials significantly govern the first steps of bacterial adhesion processes. Roughness and surface composition can be modified through appropriate micro/nanofabrication techniques to study the influence of these properties on bacterial adhesion. This technique was used to study the bacterial adhesion on topography first. Then some studies about cell behaviors on micropattern were reported, and various techniques were developed to fabricate microstructure. Bernke[26] used silicon wafer micropattern mold to cast polymeer on it, then separated and released the polymer film to fabricate porous micropatterned scaffolds. Phase separation micromolding is an easy method to design micropatterned porous 2-D scaffold sheets in one step.The micropatterned sheets are very promising as basis for building a 3-D scaffold with respect to porosity, interconnectivity and control of tissue orientation. In general, the substrates used in microstructure are fabricated from silicon, glasses, metals, ceramics or polymers. Of these materials, polymer offers the lowest cost and ease of ues for fabrication of substrates with complex patterns over multiple length scales. At present, several techniques are available for micro- and nanopatterning.

(1)lithography with electron beam, ion beam, laser, DeepUV or X-rays; (2) physical vapour deposition (PVD) of metals and chemical vapour deposition (CVD); (3) mechanical micromachining (milling, drilling, turning etc.) normally in metals; (4) laser ablation and cutting in polymers, silicon, glasses, metals and ceramics; (5) rapid prototyping (RPT) techniques such as micro stereolithography or building up moulds from metals, silicon, glasses and ceramics; (6) micro injection moulding compression, hot embossing casting and knife coating microthermoforming nanoimprint lithography (NIL) techniques. Traditional microfabrication process readily creates microscale features, but is expensive and time consuming,especially when if a big sample is required. Molding and other replication technologies can be used to fabricate either microscale or nanoscale features in polymer substrates with high throughput and low cost compared to electron beam lithography, optical lithography, and other traditional fabrication techniques. Marcus[27] adopted that the multiple hot embossing steps formed the cell substrate surface topography of microscale and nanoscale surface features. The master templates used to form the micron-scale features were fabricated from silicon using standard photolithography and etching techniques. The templates had microgroove patterns either 2 μm grooves on 4 μm pitch, or 10 μm grooves on 20 μm pitches. This work presents a technique to fabricate cell substrates with increased pattern complexity to further understand the interactions between cell response and patterning.

Based on the present studies, it could be concluded that both chemical composition and physical characters affect osteoinductivity, and the microstructure may be the most important factor. The further study will aim to screen the critical micro-pore size and the shape of micro-structure for osteoinductive scaffold. Two parts will be included. One is to explore the fabricating technique for microstructure mold, the other is to establish the biological system to evaluate how the microstructure affect osteoinductivity *in vitro*.

3. Biological Mechanism of Constructing in Vivo Tissue Engineering Bone

3.1. Hypothesis of Osteoinductive Phenomenon

Increasing data is showing that osteoinductive synthetic materials can attract mesenchymal stem cells in non-osseous tissues and form ectopic bone. How does the phenomenon of osteoinductivity take place? According to Urist's theory, three conditions are essential for osteoinduction: existence of inducing agent; existence of precursor cells, namely target cells of inducing agent; eligible circumstance for bone formation. The following phenomenons can be seen in a lot of researches: □formation of thrombus in the implantation environment; □ ingrowth of blood vessels and formation of granulation tissue; □ concentration of mesenchymal cells; □ differentiation of mesenchymal cells into osteoblast and form bone tissue. Biomaterial may play important role in osteoinduction: they induce mesenchymal cells differentiate into bone precursor cells by adsorbing bone growth factors and elevate their concentration; they provide scaffold and place for bone formation; their degradation provides calcium and phosphate ions for bone maturity. After being implanted into body, biomaterials discriminate each other with different agents in body fluid. Serial physicochemical and biochemical reaction occur on the internal surface of the material while biomolecule such as water, ion and protein et al were adsorbed. A special biological modification surface and local microenvironment were formed finally, which mediate the interaction between material and cells by redistribution and recombination of biomolecule, such as extracellular matrix, types of cytokines, ions and so on. Based on a great quantity of experimental results [28], Xingdong Zhang[29] put forword the hypothesis for the progress and biological mechanism of osteoinduction of calcium phosphate (Ca-P) ceramics: after being implanted in the non-osseous site of body, Ca-P ceramics adsorb endogenous bone growth factors, such as bone morphogenetic proteins (BMPs), which interact with homologous receptors of the adjacent mesenchymal cells. Cascade effects take place via cell signal transduction system, which induce expression of certain gene. Mesenchymal cells differentiate into osteoblast close behind and excrete specific adhesive molecule among osseous cells. Osteoblasts identify Ca-P ceramic with special construction as natural bone by mistake, immigrate and adsorb on the internal surface moreover. Certain differentiation, proliferation, maturation and bone formation occur on osteoblast themselves by excreting growth factors like BMPs.

3.1.1. BMP Adsorption May Play an Important Role in Osteoinduction by Ca-P Ceramic

In order to study how mineral matter deposits on collagen during natural bone development, Urist[30] observed new bone formation after implanting demineralized rabbit bone in musle. So he presumed there was some special molecule that played roles and named it BMP. In 1976[31,32], BMP was successfully extracted, and galactic progress was also got in its purification, preparation,, application, mechanism of action etc. Genetic engineering technique was also used to get high osteoinductive human recombinant bone morphogenetic protein by cloning in eukaryocyte, such as rhBMP-2[33] and rhBMP-7[34]. Ripamonti U commented that the immunolocalization of BMP family members (BMP-3 and OP-1/BMP-7) in cellular material at the hydroxyapatite interface suggests that the sintered ceramic may act as a solid-state matrix for adsorption of endogenously produced BMPs. In normal conditions, lots of organs like bone, kideny, lung have the function of excreting BMPs.The concentration of BMPs in blood is at the level of several nanograms per millilitre, which usually can not function on mesenchymal cells. Only if the density of BMPs at cetain place reaches certain level can induce differentiation of mesenchymal cells into osteoblast. ZouPing assured that HA/TCP could forcefully adsorb BMPs by in vitro experiment. Qu[35] discovered that there were BMP-2 and BMP-4 in cell matrix adjacent to Ca-P ceramics through immunohistochemical method in animal experiment, which indicated that BMPs played important role in osteoinduction by Ca-P ceramic. Autocrine is supposed to facilitate new bone formation after osteoinduction, but lack of dependable evidence. Good carrying agents, keeping active in preparation and controlled release in application play important roles in BMP's mechanism of action.Urist thinks human cellulose[36], decalcified bone matrix and HA/collagen[37] may be good supporter for BMP. Niedhart[38] deems TCP is good carrying agent for BMP. Bone defect was repaired in rat tibia with complex of rhBMP and TCP, new bone formation was observed three weeks after operation,but controlled group not.In Seeherman's[39] overview, inorganic material, synthetic polymer, natural polymer and the three above-mentioned compounds are four kinds of controlled release material system for BMP. These materials should be matrix for inducing osteoblast and provide space for bone formation. Identification of BMP's biological activity in compound materials should be made both *in vitro* and *in vivo*.BMP has high homology in different animal species. HYPO, muscle, interval muscle, fat tissue and fossa orbitalis periphery are all feasible site for implantation of BMP, Okubo[40] deems interval muscle can educe the osteoinductive property of BMP most quickly. JJ Du studied osteogenesis induced by rhBMP-2 while enhanced by TGF-β *in vivo*. Mice receiving rhBMP-2/TGF-β/carrier served as the subject and those receiving respectively rhBMP-2/carrier and carrier served as the controls. In order to investigate the process of osteogenesis, histological analysis was performed at 1,2,3,5,7,8,9,11,14,17,21 days after implantation; Meanwhile alkaline phosphatase activity and calcium content were measured at 7,14,21 days after implantation. Proliferation and differentiation of mesenehvmal cells, chondrogenesis and osteogenesis induced by rhBMP-2/TGF-β/carrier were prior to those induced by rhBMP-2/carrier and carrier. The extent of osteogenesis was greater than those in the groups of rhBMP-2/carrier and carrier. Peak of ALP activity in the experimental group was prior to those in the controls. Calcium content was higher than those in the controls obviously. According to results above, Du thinks rhBMP-2 and TGF-β exert a synergistic action in regulating osteoinduction and osteogenesis *in vivo*.

3.1.2. Original of Stem Cells Remaining an Open Question

If Ca-P ceramic without any exogenous growth factors and cells was implanted in non-osseous site in animals, osteoid tissue can be seen in materials one week after implantation, bone trabecula formed one month after implantation, matured bone formed three months after implantation. According to light microscope and transmission electron microscope observation, the internal surface of micropores were surrounded by blood capillaries and mesenchymal cells.This result indicated that mesenchymal cells inducing bone formation may be pericyte of blood capillary that ingrowed in the pore. However, the source of mesenchymal cells may not be single, and bone marrow stem cells may be one of them according to some research uing immunohistochemical tecnique. The source of stem cells for osteoinduction is still not well identified up to now.

3.2. Osteoinduction Depending on Animal Species

Considerable studies on osteoinduction of Ca-P ceramic concentrate on whether materials with different physicochemical characteristics have osteoinductive property or not in different animal species. However, there is no statistical evidence of the efficiency of osteoinduction, namely quantity of bone formation. In order to explore the osteoinductve property of the same biomaterial in different animals, the bioceramics was implanted in the muscles of rat, rabbit and dog. It is affirmed that osteoinductive property of Ca-P ceramic is influenced by animal species after statistical analysis. That means ratio of osseous or osteoid tissue formation and the amount of bone formed are related to animal species.

In the study[41], 27 dogs, 42 New Zealand rabbits and 54 SD rats were used. Ceramics were implanted in muscles of dogs, rabbits and rats according to statistical principle. All of these animals were sacrificed by overdose of pentobarbital sodium solution at the 6th, 12th and 24th weeks respectively. The osteoinductivity and the amount of bone/bone-like tissue formation in different animals were compared by morphological observation, 99mTc-MDP absorption evaluation and statistical analysis. The judgement of bone/bone-like tissue was determined by three pathologists. The quantity of newly formed bone was determined with an

Image-Pro-Plus (IPP, Media cybernetics,USA) image analyser, and the data were analysed by SPSS11.5. Samples of Ca-P ceramics implanted in different animal species of different periods were decided by statistics.

Calcium phosphate ceramics was prepared by H_2O_2 foaming method. HA/TCP(60/40) sintered at 1100°C(HT1100) for 3 hours with wet-synthetic calcium phosphate powder. The ceramic cylinders of dimensions with φ5mm×8mm, φ4mm×6mm and φ2mm×3mm were prepared respectively and cleaned with de-ionized water by ultrasonic agitation. The chemical composition of the ceramics was analyzed by XRD. The average pore size of the ceramics was tested by SEM and the porosity of the ceramics was determined by the Archimedes water displacement methods.

After anaesthesia with 3% sodium pentobarbital, four cylinders with φ5mm×8mm of the ceramics were implanted in the dorsal muscle of one dog, three cylinders with φ4mm×6mm of a ceramics were implanted in the dorsal muscle of one rabbit, and two cylinders with φ2mm×3mm of one ceramics were implanted in the leg muscle of one rat. All of these animals were sacrificed by overdose of pentobarbital sodium solution at the 6th, 12th and 24th weeks respectively.

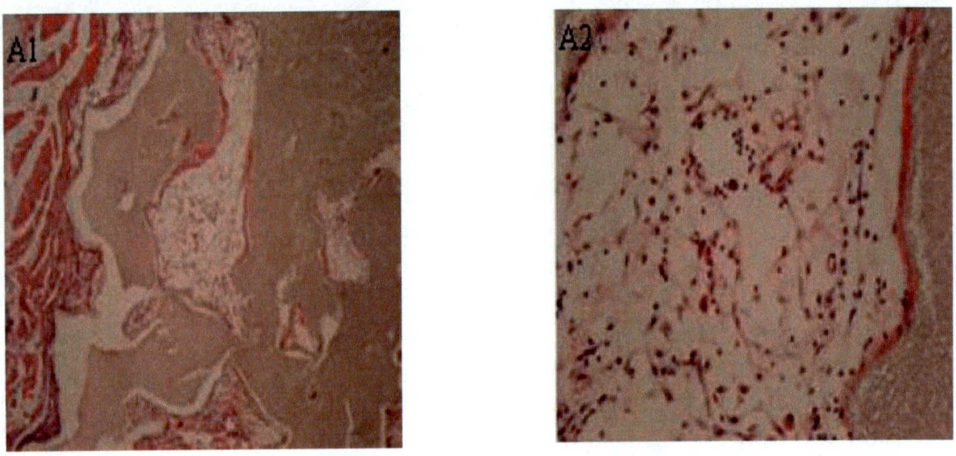

Figure 12. Light micrographs of decalcified sections in rats after being implanted for 12 weeks. H&E staining, A1, 100*;A2, 400*.

Rat: There are thin layers of fibrous capsulos around the ceramics, but no osseous tissues. Until 12 weeks after the implantation, osteoid tissues are observed only in some samples. Exactly, there is a tissue band of compacted cell adjoining to wall of pores, which is osteoblast-like, and there are sparse cells which are fibroblast-like or fibrocyte-like (figure 2).

Rabbit: In the 6th week, the tissue inside the pores of HT1100 has developed into compact cell belt, with the cell nucleuses stained deeply and a lot of cytoplasm, moreover some cells appear to be osteogenic cells; the central cells are loose relatively, polygon in shape, looking like being the fibroblasts or osteoblasts(figure 13). The phenomenon in the 24th week is almost the same as that of the 12th week, more osteoid tissues are observed in more samples; however, no bone trabecula is formed in different periods, except in the 12th week, bone trabecula is observed in the three samples in only one rabbit(figure 14).

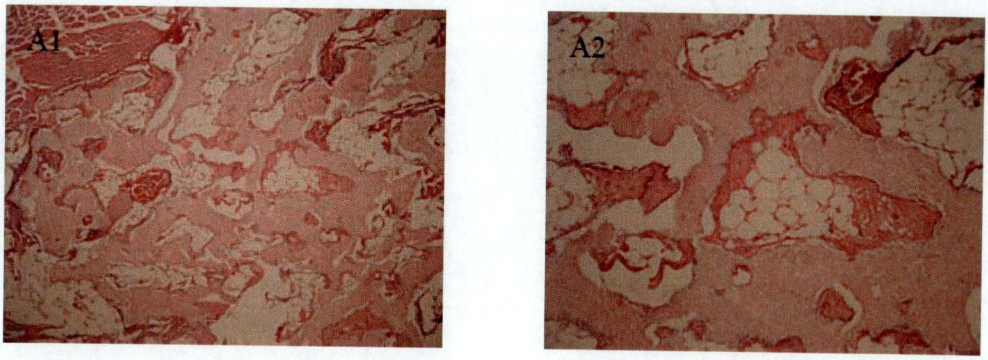

Figure 13. Light micrograph of decalcified sections of HT1100 after being implanted for 12 weeks in rabbits, H& E staining. A1, 40*;A2,100*.

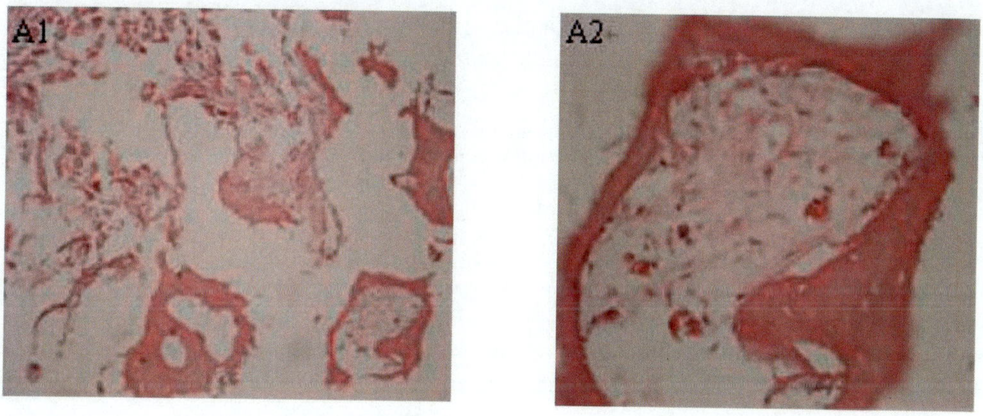

Figure 14. LM of decalcified sections of HT1100 at the 12th week postimplantation in rabbits. H&E staining. A1, 100*, A2, 400*.

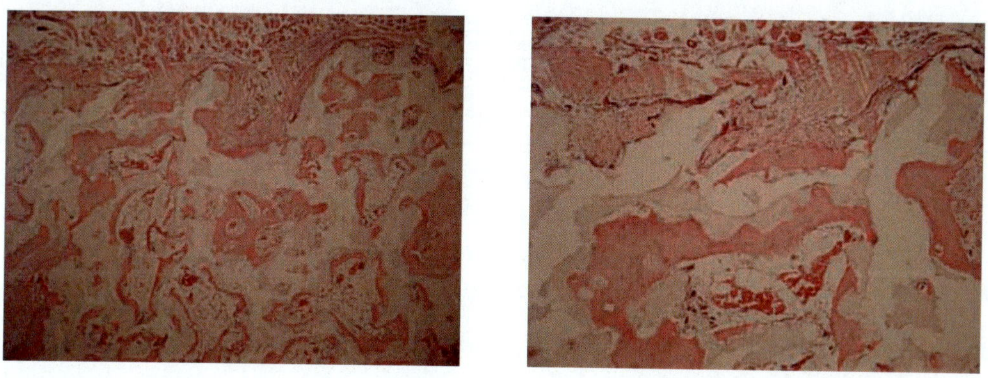

Figure 15. LM of decalcified sections of HT1100 in dogs at the 12th week after implantation. H&E staining. Left, 40*; right, 100*.

Dog: Bone formation is observed in all the samples of the three peroids. In each period, the osteogenic or osteoid tissues that are formed in ceramics are more mature than that of rabbit and rat. As for the process and form of the osteoinductivity, the inflammation reaction is slight in each period for HT1100, the lymphocytes are rarely soaked, and the tissues in the material differentiate directly from the mesenchymal stem cells to osteoblasts, forming the bone tissue adjoining to the pores of the ceramic. In this material, the bone formation is the typical intramembranous ossification(figure 15).

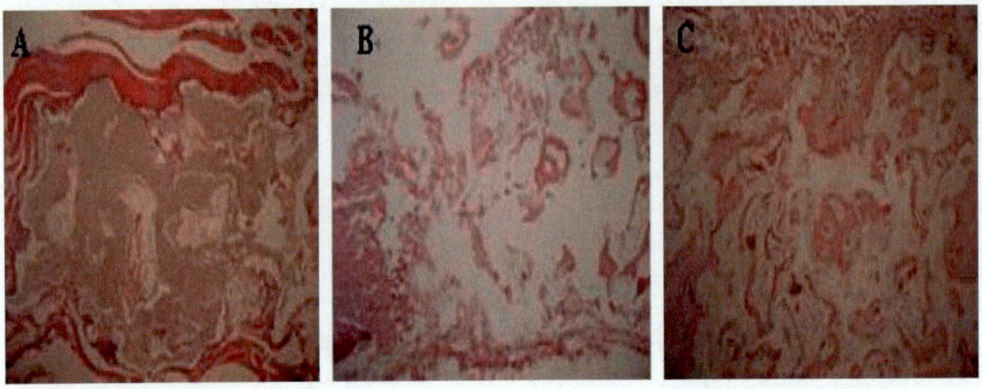

Figure 16. LM of decalcified sections of HT1100 at the 12th week postoperation. H&E staining. Magnification 40*.The amount of bone formation was most in dogs and least in rats. A:Rat B:Rabbit C: Dog.

Through LM observation of the decalcified slices of the samples of the same materials in the same period, but on different animals, we get the statistical data of the samples of the osteoinductivity for osseous or osteoid tissue formation(table 4). It could be seen that at the same period and in different animals, as for the osseous or osteoid tissue formation ratio, the sequence is as follows: dog, rabbit and rat . The analysis is carried out in three groups on the same material, in the same period, but on different animals(table 5).

Table 4. osseous formation amount statistics of HT1100 in three periods in each sample groups

	N	Mean	Std.Deviation	Std.Error	95% Confidence Interval for Mean	
					Lower Bound	Upper Bound
6wks		8.67970682	4.52992332	.76569679	7.12362372	10.23578992
12wks	36	13.6684301	4.31235551	.71872592	12.20933989	15.12752135
24wks	36	14.7341421	4.24209074	.70701512	13.29882513	16.16945915
Total	107	12.3951620	5.06269440	.48942914	11.42482120	13.36550295

Table 5. osseous formation amount statistics of HT1100 in three periods in each sample groups

VAR	VAR	Mean Difference (1-J)	Std. Error	Sig.	95% Confidence Interval	
					Lower Bound	Upper Bound
HT1100	4	-4.98872335(*)	1.03534256	.000	-7.04184649	-2.93560020
	7	-6.05443532(*)	1.03534256	.000	-8.10755846	-4.0013127

The judgement of the final effect of the implanted material is influenced by several factors, such as the differences of animal model, the operation conduction, and the judging standards, etc. In order to minimize the system error and increase the credibility, thus avoiding the unnecessary waste, this research ensures that the checkout efficiency is no less than 0.75, and determines the sample amount according to the statistical principle. The material implanting is conducted by the same operator. Under several methods like SPECT and Tissue Morphology and the two morphological standards -- bone tissues and bone like tissue formation, we make the standards for judgement with the professional pathologist. As for the statistic of the relative area of bone formation amount, it is calculated according to the area of bone trabecula; the LM image of the sample is gotten by 600CL-Pixcer digital image system, and is analyzed by IPP. SPSS11.5 statistic analysis software is used for the analysis of osteoinductive ratio of osseous and osteoid tissue formation, and the bone mass. The data are processed by X^2 inspection and covariance analysis.

The incidence of material-induced bone formation varies with animal species, in other words, the osteoinductivity of biomaterials is animal species dependent. Material-induced bone formation reported in canine, bovid, suid, and primate species does not often occur in lagomorph or rodent models. Early in 1969, Winter reported the ectopic bone formation induced by a synthetic polymer sponge in the skin of young pigs. Nearly 20 years laters, in 1988, Heughebaert found that a bone-like substance formed in porous hydroxyapatite ceramics implanted in soft tissue of hamsters. Studies compared the osteoinductive potential of different calcium phosphate in large alonely or small animal models, but few compared the potential of the same calcium phosphate between the large and small animal models or the different ceramics in statistics. In 2002, Yuan made the comparison of the osteoinductive potential of hydroxyapatite and biphasic calcium phosphate implanted intramuscularly in goats. In Yuan's study, Material-induced bone formation were also tested in subcutaneous pockets of mice and intramuscular pockets in rats, rabbits, and dogs. All scaffolds are of similar size, and all animals were sacrificed at 90 days post-implantation. In dogs (N = 8), all implants showed bone formation in contrast to rabbits and mice. Material-induced bone formation in 6 of 8 scaffolds implanted in 4 rabbits and 3 of 16 scaffolds implanted in 16 mice were observed, whereas it did not induce bone formation in any of the 8 rats. The results presented herein suggested that the incidence of material-induced bone formation varied with animal species. Same Ca-P ceramic has different osteoinductive property in different species of animals. Some researchers think it is much easier for osteoinduction in Primate, Susscrota domestica and canine tooth animal. Different osteoinductive phenomenon may be related to different metabolism and degradation-adsorbtion, or to different observation time and evaluation criterion. For example, bone formation was observed in gnawer which was considered to be non-osteoinductive if observation time was prolonged.

One of the Winter's research members, Simpson, has studied the sponge in rats, implants of polyHEMA sponge which have calcified, but not yet ossified after six months. Implants have been put into a number of other species, but until the results of these studies are available, the specificity of this reaction remains an open question.

From results above all, it is demonstrated that the same material is highly osteoinductive in dogs while low in rats. Only certain material in certain period can induce bone formation in rats in this study. So rat may be an ideal animal model for osteoinduction study of different biomaterials, for example whether they have osteoinductive property and the intensity.

3.3. Multiple Profiles to Explore *in Vivo* Tissue Engineered Bone

Despite the fact that considerable researches have been directed toward exploring how biomaterials regulate cellular functions, the mechanism by which osteoinductive materials induced *in vivo* ectopic bone formation has not been definitely elucidated. In order to do a comparison between bone graft constructed with osteoinductive calcium phosphate ceramics *in vivo* and naturally healing bone, furthermore to explore the application prospect of *in vivo* tissue engineered bone graft. Porous calcium phosphate ceramics without any addition of living cells or growth factors were implanted into the dorsal muscles of six dogs in our research. As control, one molar tooth was pulled out in each dog to create bone defect for naturally healing bone at the same time (figure 17). The dogs were sacrificed at 1, 2, 4, 8, 12 and 24 weeks post-operation respectively. Samples were harvested for SPECT, histological observation and detection of osteogenesis-associated genes of *in vivo* tissue engineered bone graft and naturally healing bone.

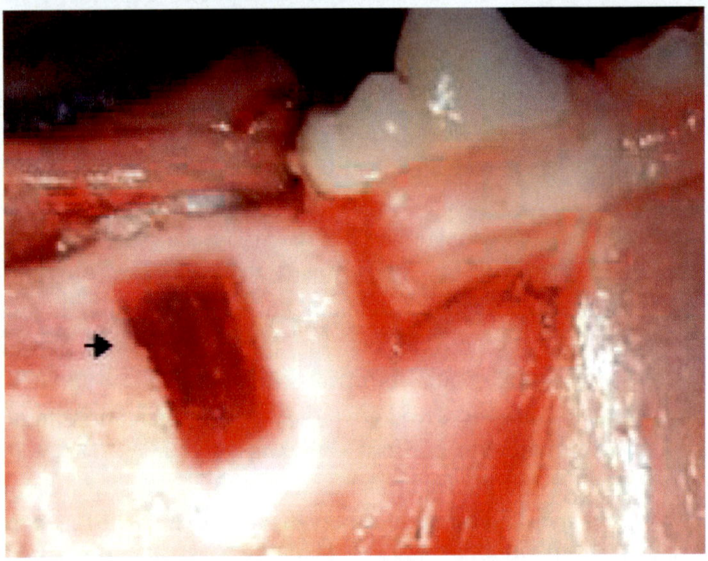

Figure 17. Bone defect (arrow) created for naturally healing bone. Its size was 5×5×5 mm.

3.3.1. Histological Observation

Specimens for histological observation were fixed by 10% formalin. Four specimens at each time point were dehydrated and embedded in methyl methacrylate and 15 μm undecalcified sections were made. The other four specimens were decalcified and 4 μm thin sections were prepared. The undecalcified sections were stained with HE and decalcified sections with Masson for light microscope (LM) analysis to observe and compare the osseous or osteoid tissue formation in different periods.

Histological sections showed that the muscular tissue around implants grew normally. Fibrous connective tissue, tube-like tissue which indicated the beginning of angiogenesis and neutrophils close to the surface of the pores of implants were observed at 1 week after implantation. The above structures distributed wider in the internal pores regions of implants at 2 weeks post-operation. Besides, polymorphic cells with a massive volume and deeply stained by hematoxylin aggregated mainly around the capillary. The histological morphology features (figure 18), which cubical osteoblast-like cells were lined adjacent to some homogeneous matrices excreted directly on inner pore surface of implants at 4 weeks post-operation, indicated the beginning of osteogenesis. Osteocytes and bone marrow cavity presented on the surface of the ceramic pores after implantation for 6 weeks. Since then, the newly formed bone became more and more mature. At 12 weeks post-operation, Harversian system had began developing in the pore and complete Harversian system was formed after implantation for 24 weeks. The trace of bone remodeling was observed. No cartilage and chondrocyte were found during this osteogenesis process.

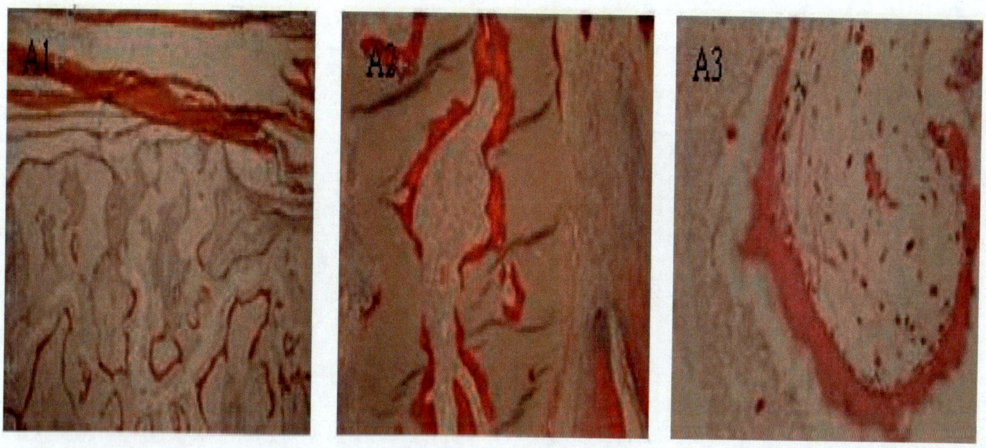

Figure 18. Light micrograph of HT1100 after being implanted for 6 weeks, new bone with cubical osteoblast-like cells rimming 6th weeks.A1: decalcified slice, magnification 40*, Masson staining ; A2: sclerous tissues slice, 100*, Masson staining; A3: decalcified slice, 400*, HE staining.

And in the earlier experiment, osteoid tissue could be found in HT1100 after the implantation of the 1th weeks. The bone trabecula was adjoining to the wall of the pores of the ceramic. As the implanted times prolonged, the new bones became more mature, and reconstruction could be observed on the bone trabecula; the cells that formed the bone trabecula developed from the osteoblasts to bone lacuna, via osteocytes. The cells in the central tissue of the pores were differentiated from the early star-shaped or rounded mesenchymal stem cells to the mature lipocyte-like cells which had the slim and long nucleus and departed from the center, then the marrow structure was formed, and the osteoclast appeaed.

Lots of studies manifest that there are different forms of bone formation induced by osteoinductive ceramic. Most reports about patterm of osteoinduction by Ca-P ceramic is membrane bone formation. Bao observed that the composition and morphous of ingrowth tissue were related to osteoinductive property of biomaterial; and cells tend to form bone directly in good osteoinductive environment; on the contrary, in low osteoinductive conditons, fibrous tissue formed firstly with inflammatory cells infiltrating, then bone formed by membrane bone formation or fibrous degeneration. Patterm of bone formation is os cartilaginea if ectogenic BMP or osteoblast is added in[42, 43]. Other researchs presumes that bone formed directly if blood supply was sufficient, and cartilage was formed if insufficient. New bone formation in long-term intramuscular implant of Ca-P biomaterial was investigated in XF Lu's experiment. After implanted into dog dorsal muscle for 15 months, a thin fibrous membrane that wrapped HA/TCP implant was still observed obviously. Three types of tissues, i.e. mesenchymal tissue, bone and bone marrow, regularly distributed in different pores of implant. Nearly all the pores of implants were occupied by bone. Bone in the pores located in the central region of implant was mature lamellar bone characterized by obvious lacuna and rich bone marrow. However, bone in the peripheral pores was immature woven bone without bone marrow formation. Furthermore, mesenchymal tissues only existed in the peripheral pores and usually were connected with immature woven bone. Well-organized distribution of mesenchymal tissue, bone and bone marrow in the pores suggest that bone morphogenesis in the implant should obey a specific space-time program.

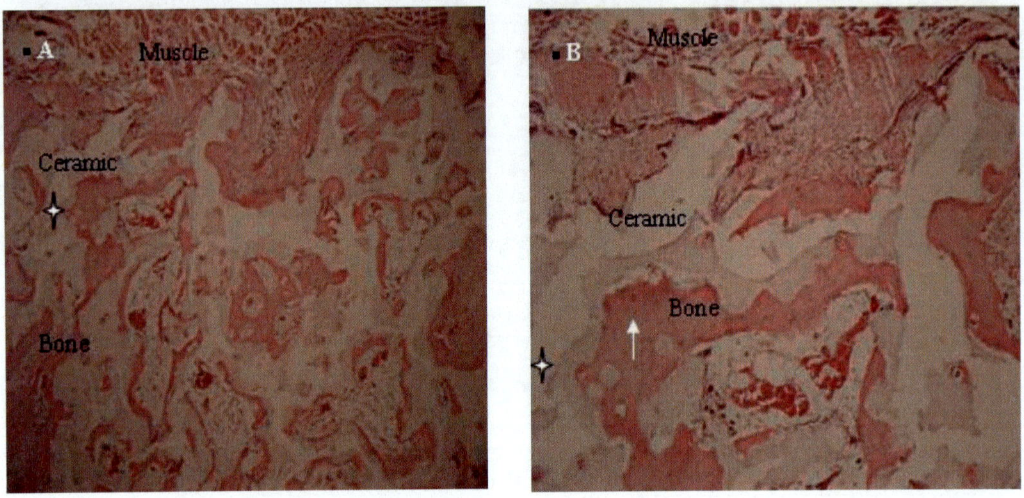

Figure 19. Light micrograph of decalcified sections after being implanted for 24 weeks, H&E staining, showing mature woven bone occurred in direct contact with the HA/TCP ceramics. Osteocytes (arrows) embedded randomly in the bone matrix. Original magnification: A 40*; B 100*. Stars mean the same site.

Histological morphological evidence showed that the microstructure of bone graft with mature Harversian system and bone matrix was similar to that of the native bone. As is previously reported, the newly formed bone was confined within the ceramics and no bone was found on the outer-surface and in the surrounding muscles. The phenomenon that no cartilage or chondrocyte were found during this osteogenesis process indicated that it should be an obvious intramembranous ossification process about osteoinduction of HA/β-TCP ceramics.

Previous studies had showed that the *in vivo* tissue-engineered bone formed with osteoinductive calcium phosphate ceramics had similar histological characteristics to natural bone[44]. The newly formed bone did not disappear or over-grow in 2.5 years. The newly formed bone was observed in specimens of 24 weeks (the longest period in this experiment). The results were in agreement with the previous reports that the newly formed bone was mature bone with osteocytes embedded randomly. Because of carrying living cells and having good blood supply, the newly formed bone participated in the bone metabolism of host as a bone graft for reconstructing segmental or large bone defect[4]. Compared with the *in vitro* tissue engineered bone graft involved cells culturing, the *in vivo* tissue engineered bone should have broad prospect for clinic application. So it is necessary to narrate the biosafety of *in vivo* tissue engineered bone at gene level.

3.3.2. Analysis of Metabolism of Bone Graft Constructed with Osteoinductive Calcium Phosphate Ceramics

The osteoinductivity and the amount of bone/bone-like tissue formation in different materials and in different animal species were compared by morphological observation and 99mTc-MDP absorption evaluation under 600CL-Pixcer. 99mTc–MDP (1mCi/kg) was injected to all dogs in vein four hours before sacrifice. Samples were fixed in 4% buffered formaldehyde (PH=7.4) for calcified sections. 99mTc-MDP was always detected in the implants throughout the tested period. The photo of SPECT indicated that incorporation of 99mTc-MDP into bone graft was increased over time (figure 20). The peak in 99mTc-MDP uptake was observed at the 24th week.

99mTc-MDP has been used for diagnostic imaging of new bone formation because of their special predilection for foci of bone remodeling. In the bone, 99mTc-MDP dissociates into its technetium and methylenediaphosphonate moieties, which are then incorporated into the organic and inorganic phases, respectively. So the amount of 99mTc-MDP uptake reflects the quantity of bone formation. In the present study, the photo of SPECT, which 99mTc-MDP in bone graft was increased over time, demonstrated that more and more new bone was formed in the HA/β-TCP ceramics and the newly formed bone participated in the metabolism of host. The photo of SPECT also showed that no bone was formed in the surrounding muscles.

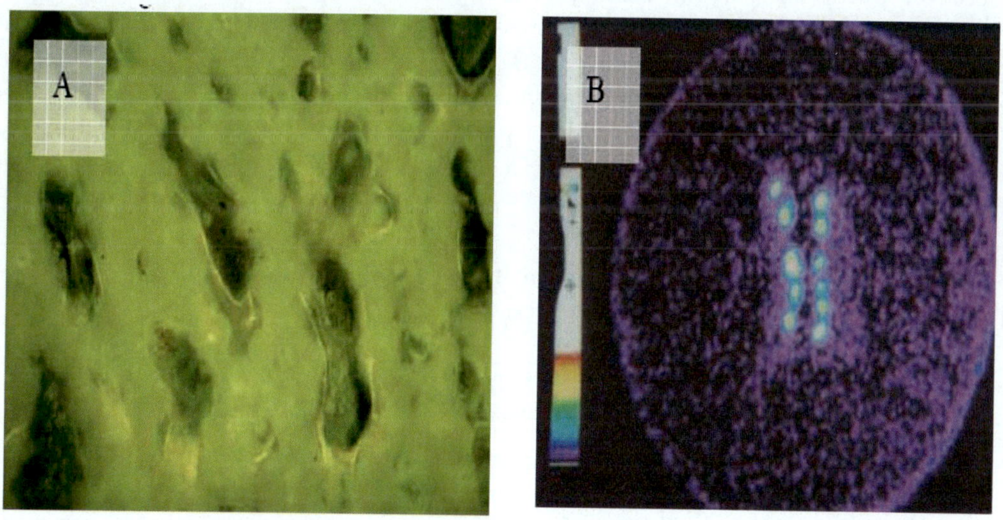

Figure 20. Tetrecycline labling (A) and SPECT images (B) of the HA/TCP implants at 24 weeks post-operation.

3.3.3. Detecting mRNA Expression of Bone-Specific Gene

Total RNA extracted from dog's frozen naturally healing bone and specimens with TRIzol® Reagent (Invitrogen, Carlsbad, CA, USA) was primed with oligo $(dT)_{18}$ primers and reverse transcribed with ReverAid™ First Strand cDNA Synthesis Kit (Fermentas, Lithuania) into single stranded according to the manufacturer's instructions. The synthetical first strand cDNA was stored at –20°C. A set of bone-specific genes, COL-I, ALP, BMP-4, and house keeping gene GAPDH mRNA were measured using real-time quantitative polymerase chain reaction (PCR) analysis based on SYBR Green I methodology according to standard protocols. Characteristics of gene specific primers for real-time PCR were summarized in table 2. The real-time PCR reactions were set up according to manufacturer's instructions with SYBR® Premix Ex Taq™ kit (TaKaRa, Japan) and run on a ABI Prism® 7300 Real-Time PCR System instrument (Applied Biosystems, USA). PCR reaction mixtures consisted of 12.5 μl of 2× SYBR® Premix Ex Taq™ (TaKaRa, Japan), 0.5 μl of forward primer (10 μM), 0.5 μl of reverse primer (10 μM), 0.5 μl of ROX Reference Dye (50×), 2.0 μl of cDNA template and 9 μl of dH_2O in a reaction volume of 25 ml. For negative control, 2.0 μl dH_2O rather than cDNA was used. Thermal cycling conditions comprised of initial denaturation at 95°C for 10 s followed by 40 cycles of denaturation at 95°C for 5 s, annealing and extension at 60°C for 31 s when data was collected. Dissociation curve analysis which was carried out at 95°C for 15 s, followed by 60°C for 30 s and 95°C for 15 s was performed at the end of each extension to measure dsDNA specific fluorescence. The relative copy numbers of target genes were adjusted based on the expression of GAPDH as an internal control for each plate. Experiments were performed in triplicate.

Osteogenesis-associated genes, COL-I, ALP and BMP-4, which were detected in naturally healing bone were also positive in the bone graft constructed with osteoinductive HA/TCP ceramics during the tested period. The relative amounts of COL-I were significantly different at different time point whereby its gene expression increased steadily from week 1 to a peak at week 12 followed by a sharp decrease at week 24 [Graph.2 (a); $p<0.001$, respectively]. ALP mRNA, with a similar gene expression pattern to COL-I mRNA, reached a marked peak at week 8 when the gene expression level was significantly higher than that at any other time point [Graph.2 (b); $p<0.001$]. BMP-4 mRNA, with a different gene expression pattern from COL-I and ALP mRNA, had a peak at week 1 and its gene expression level was significantly different at different time point [Graph.2 (c); $p<0.001$]. However, in naturally healing bone all the target genes had similar gene expression pattern with a marked peak at week 1 (Graph.4,5 ; $p<0.001$, respectively).

Table 6. Primer sequences, theoretical melt temperature (Tm), length, product size and GenBank accession number used in quantitative PCR

Gene	Accession No.	Primer Sequence	Tm (°C)	Len (bp)	Product Size (bp)
GAPDH	AB038240	5'-ggtgctgagtatgttgtggagtc-3'	60.2	23	174
		5'-gctgacaatcttgagggagttg-3'	58.1	22	
COL-I	NM_001003090	5'-aagaaccccaaggagaagagac-3'	58.1	22	168
		5'-cttgcagtggtaggtgatgttc-3'	58.1	22	
ALP	XM_535374	5'-gagatggacaagttcccttacg-3'	58.1	22	177
		5'-ctcgtttccctgagtcgtgtt-3'	57.8	21	
BMP-4	XM_859035	5'- gggcttccaccgaataaaca -3'	55.4	20	142
		5'-cgcagggctcacatcaaa-3'	54.3	18	

For each primer sequences, the first sequence is forward primer, and the second is reverse primer.

When a comparison was done, the amount of gene expression of COL-I, ALP and BMP-4 of the bone graft constructed with osteoinductive HA/TCP ceramics was significantly different from that of the naturally healing bone. In detail, the gene expression level of COL-I of bone graft was obviously higher than that of naturally healing bone at week 2, 4, 8 and 12, and obviously lower than the latter at week 1 and 24 [Graph 1(a); $P<0.001$]. Similar to that of COL-I, the relative amount of ALP of bone graft was significantly larger than that of naturally healing bone at week 4 and 8 and smaller than the latter at any other time point [Graph 1(b); $P<0.001$]. The gene expression level of BMP-4 of bone graft, however, was always significantly higher than that of naturally healing bone throughout the experimental period [Graph 1 (c); $P<0.0001$]. Nevertheless, mRNA expression tendency of COL-I, ALP and BMP-4 of the bone graft and the naturally healing bone shared a similar unimodality curve with an expression peak at specified time (Graph 2,3 and 4).

Bone formation is an intricate and ordered cascade of matrix proteins and calcium phosphate in a continuously renewed biological environment and regulated by growth factors[44-48]. In osteogenesis, bone–related genes are strictly regulated to ensure correct chronological order and each gene has a unique expression profile. Therefore, an ideal bone graft induced by HA/TCP ceramics should have similar chronological order of gene expression to naturally healing bone. COL-I is a major organic component of bone matrix. During osteogenesis, it might play a key role in triggering cell differentiation. ALP, as the only membrane-bound protein, is an important osteoblastic differentiation marker. It acts as an initiator and regulator in the calcification of bone matrix[47-52]. Both COL-I and ALP are highly expressed in the early stage of bone formation[53]. After reaching an expression peak at specified time, the mRNA expression of COL-I and ALP of bone graft were sharply down-regulated, which was the same as that of naturally healing bone. This demonstrated that COL-I and ALP of bone graft were not over-expressed. However, in bone graft, the gene expression of COL-I and ALP were firstly up-regulated with a peak at week 12 for COL-I mRNA and at week 6 for ALP mRNA and then down-regulated. The appearance of peak time of ALP mRNA was in accordance with the deposition of bone matrix. While in naturally healing bone, both COL-I and ALP expression peaked at week 1 and their expression was down-regulated throughout the tested period, which meant that the formation of naturally healing bone was earlier than that of bone graft.

Studies have shown that *in vivo* BMP-4 is one of the main contributing factors in new bone formation during the early phase of fracture healing and can improve bone regeneration[36-43]. The events stimulated by BMP-4 begin with stem cells chemotaxis and proliferation, followed by their subsequent differentiation into bone-forming cells. After that, BMP-4 increases the osteoinductivity of these differentiated cells[54-57]. In present study, the gene expression pattern of BMP-4 in the bone graft was the same as that in the naturally healing bone, while the amount of BMP-4 mRNA was much larger in the former than in the latter, which may contribute to ectopic bone formation.

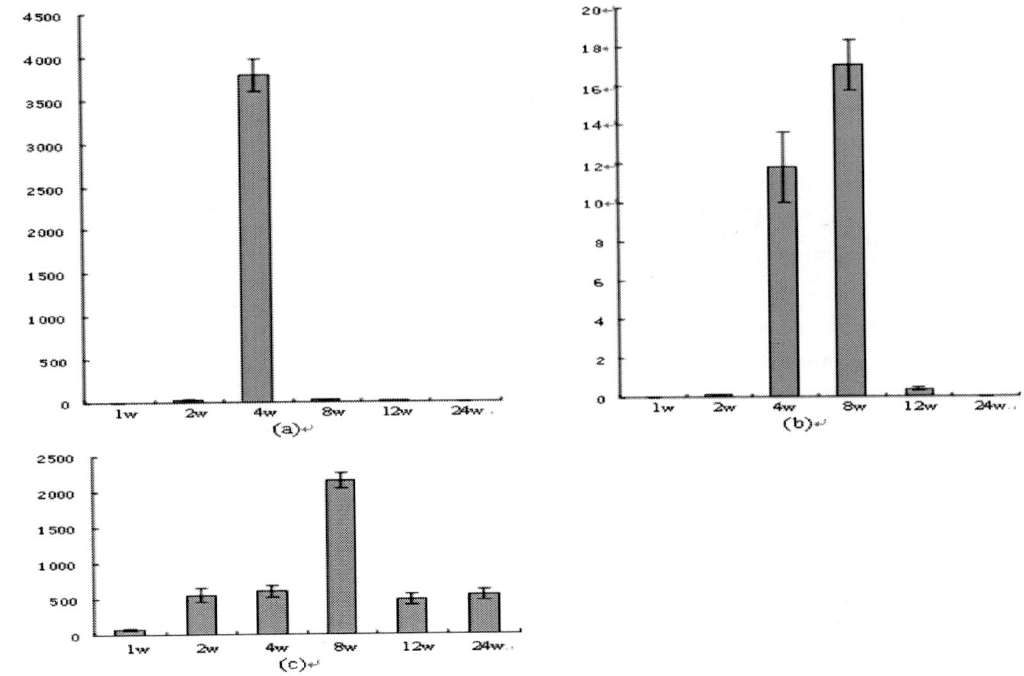

Graph 1. Relative expression profiles for target genes of bone graft. The horizontal axes represent the time of post-surgery and the vertical axes represent gene expression normalized to GAPDH and relative to that of naturally healing bone. Bars =means ± standard deviation. (a) COL-I, (b) ALP, and (c) BMP-4(w stands for week).

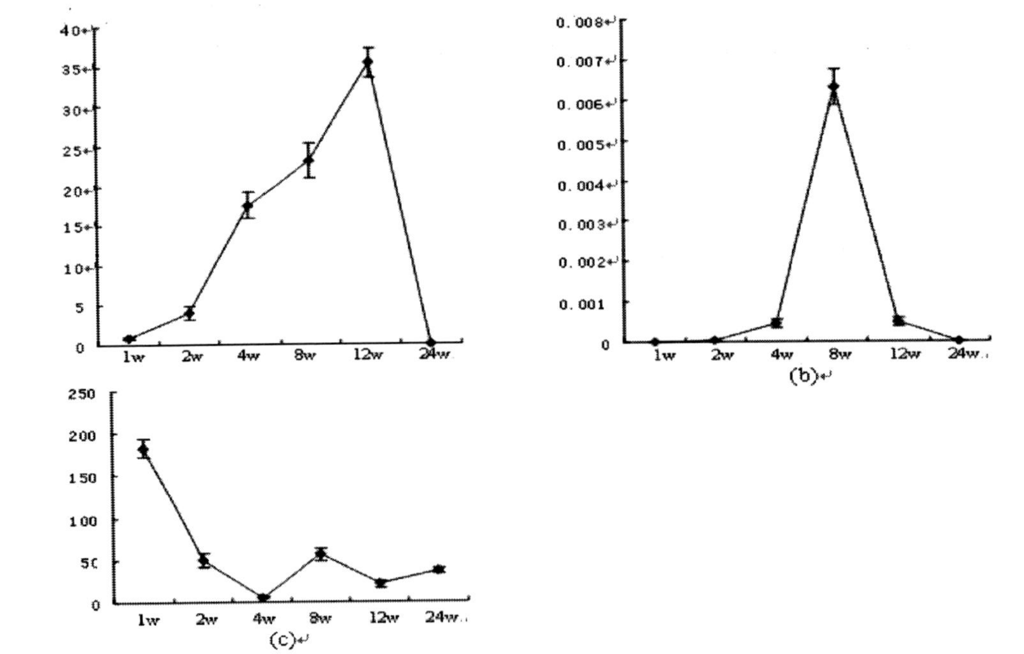

Graph 2. Expression profiles for target genes of naturally healing bone. The horizontal axes represent post-implantation period and the vertical axes represent the relative gene expression level normalized to GAPDH. All values are mean ± standard deviation of triplicates. (a) COL-I, (b) ALP, and (c) BMP-4. w stands for week.

Until now, considerable experiments have been performed in order to elucidate the mechanism by which HA/TCP ceramics induced ectopic bone formation *in vivo*[59-61]. These findings suggested that HA/TCP ceramics scaffolds stimulated chemotaxis of endogenous BMPs in protein form and served as the substratum for precipitation of BMPs and other biological apatite. Then under the induction of BMPs, stem cells proliferated and differentiated into osteoblastic cells producing bone tissue. In current study, however, a large amount of BMP-4 mRNA rather than protein were detected in the earlier phase of ectopic bone forming, which means at least most of the BMP-4 acting in the earlier phase of ectopic bone formation was not from endogenous body fluid. Therefore, the mechanism by which ectopic bone was formed *in vivo* was challenged. Immunohistochemistry have demonstrated that it is the stem cells which could differentiate into bone forming cells rather than osteoblast that synthesized BMP-4. Accordingly, the mechanism by which HA/TCP ceramics induced ectopic bone formation *in vivo* may be formulated as follows. After implantation in a muscular site, the HA/TCP ceramics might be considered as the normal inorganic components of bone and trigger chemotaxis of mesenchymal stem cells, muscle-derived stem cells, and other stem cells and their expression of BMPs mRNA, followed by its increased transcription and translation. The local concentration of BMPs induced the corresponding stem cells to proliferate and differentiate into bone-forming cells which precipitated on the wall of HA/TCP ceramics pores, synthesized and secreted COL-I and ALP to form a mature bone. Therefore, the bone graft constructed with osteoinductive HA/TCP ceramics became more and more tenacious from week 1 to week 24. However, the study did little help in understanding why the peak amplitude of COL-I and ALP in the bone graft appeared later than that in the naturally healing bone.

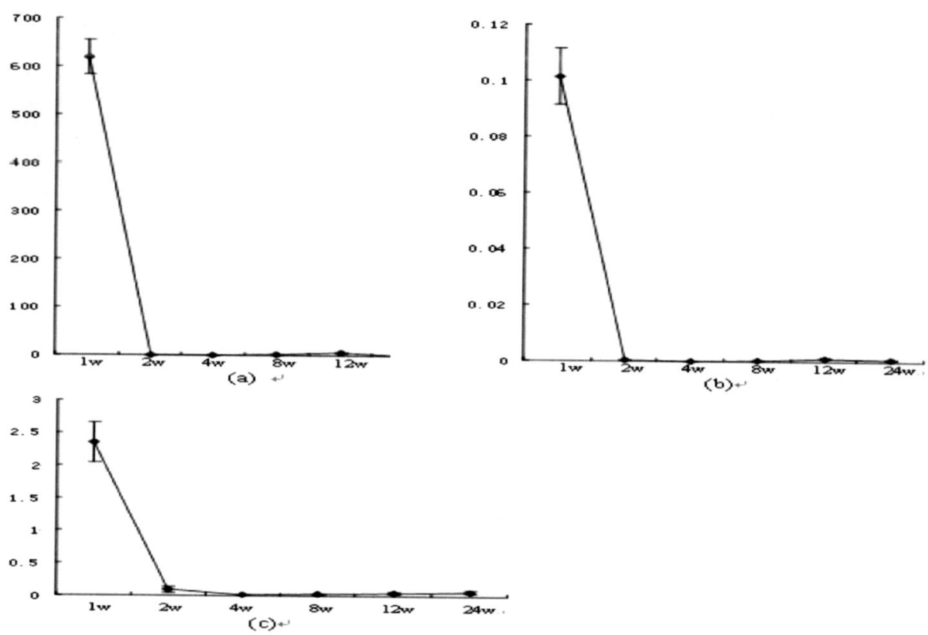

Graph 3. Expression profiles for target genes of naturally healing bone. The horizontal axes represent post-implantation period and the vertical axes represent the relative gene expression level normalized to GAPDH. All values are mean ± standard deviation of triplicates. (a) COL-I, (b) ALP, and (c) BMP-4. w stands for week.

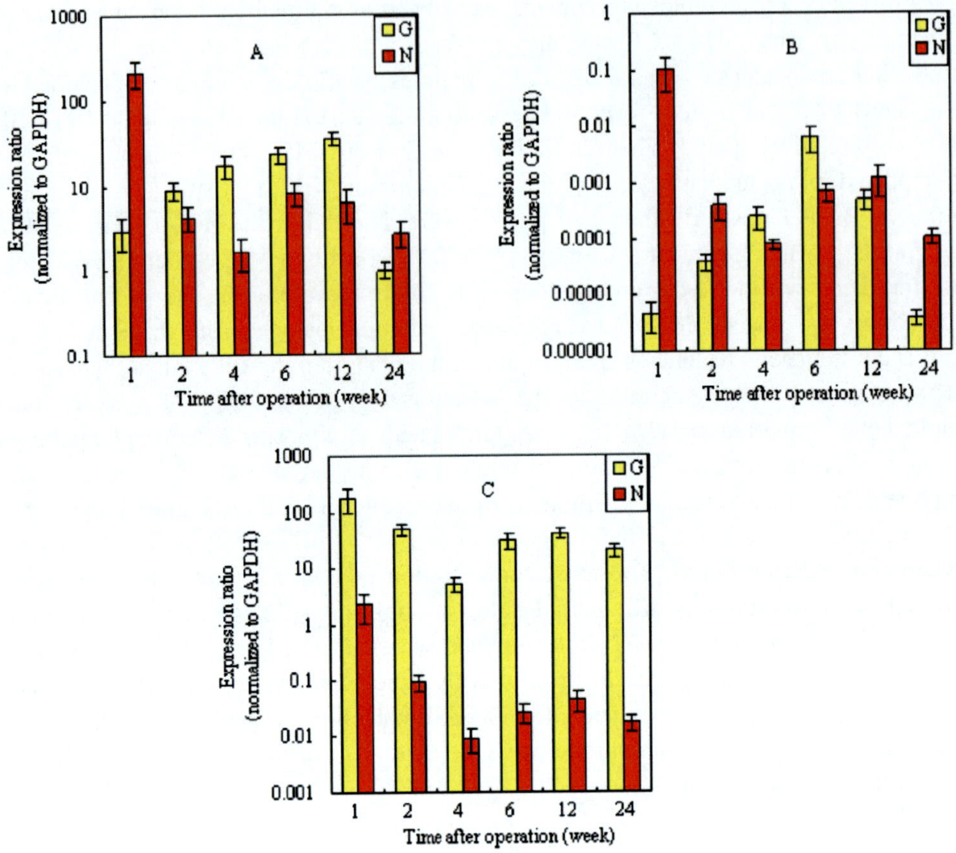

Graph 4. Relative quantitative real-time PCR expression in bone graft (G) and naturally healing bone (N) of type I collagen (a), ALP (b), and BMP-4 (c). Data are expressed normalized to GAPDH. Bars represent were arithmetic mean value ± standard deviation. In order to highlight possible differences in different ranges of expression, a lg scale is used.

The results of SPECT, histological observation and real-time PCR were in concordance. The bone formation in HA/TCP ceramics involved a complex process beginning with the angiogenesis and chemotaxis of BMP4, second the polymorphic cells aggregation and differentiation towards osteoblast, third the expression and secretion of COL-I and ALP and finally mineralization of matrix. The bone graft became more and more mature and similar to naturally healing bone from the aspect of morphology, metabolism and gene expression after 24 weeks post-operation as time passed since HA/TCP ceramics were implanted in the dorsal muscles. The bone graft constructed with HA/TCP ceramics had similar morphological structure and mRNA expression pattern of BMP4, COL-I and ALP to naturally healing bone and participate the metabolism of the body. Therefore, it should have great prospect in clinical application.

Based on this part, it could be concluded that osteoinduction is relevant to animal species. Inducing bone could be easier found in Primate, Susscrota domestica and canine tooth animals than rodents. So the rodents could be used for optimizing high osteoinductive biomaterials. Initiately, BMP adsorption of osteoinductive biomaterials may play an important role, then trigger stem cell to differenciate into osteoblast. But what kind of stem cells differentiate into osteoblast and then form bone remains unclear. The bone graft constructed with HA/TCP ceramics had similar morphological structure and mRNA expression pattern of BMP4, COL-I and ALP to naturally healing bone and participate the metabolism in the body. Therefore, it should have great prospect in clinical application.

4. Techniques of Constructing and Applying in Vivo Tissue Engineered Bone

4.1 Research on Application of Osteoinductive Biomaterials

Multiple methods have been studied for the restoration of bone defects in the field of reconstructive surgery and orthopedic surgery. Fresh autogenous bone graft has been used to carry this out historically, but it is associated with substantial morbidity, including infection, malformation, pain, and malfunction [61,62]. The complications resulting from graft harvest have inspired the development of alternative strategies for the repair of clinically significant bone defects. A previous approach to this problem focused on the development of various artificial materials that might be used, to replace the autogenous bone grafts. Most of these artificial materials are not osteogenic or osteoinductive; therefore, it is considered that their potential for bone regeneration is determinate. Another alternative approach is the implantation of living cells that are directly osteogenic. It is thought possible to regenerate various tissues using living cells and the appropriate scaffolds according to tissue engineering concepts. However classical tissue engineering met with its dilemma in fields of seed cells, scaffold materials and constructing technique. Additionally, partially due to the nutritional problem, it is difficult to produce large living tissue-engineered bone in practice at the moment. Meanwhile current bone tissue engineering technique is expensive, therefore far from the clinics. So another new strategy is expected to come to being because of the demand.

As early as 1911, Wells reported that calcium salts facilitate bone formation in the book Achives of Internal Medical. In 1988, Heughebeart reported that bone-like deposites were found on the surface of Ca-P ceramics which did not carry with any growth factors or living cells when implanted in non-osseous site of animals. Recently, increasing data shows that many kinds of biomaterials without any exogenous growth factors or cells could induce bone formation in non-osseous sites or osseous sites. And more and more focus on the Ca-P ceramics because of their intrinsic quality. In 1991, Xingdong Zhang[63] and Ripamonti[64] respectively reported that new heterotopic bone tissue formed after porous hydroxyapatite was placed in dogs' and baboons' non-osseous sites.

The invasion by host tissue is mostly facilitated by porous materials which have the advantage of allowing circulation of body fluids and of increasing the potential for firm attachment of body tissue[65]. In general, biomaterials with interconnective pores are considered to be superior to biomaterials containing dead-end pores, because a spatially continuous connection of the pore system has a decisive meaning for the ingrowth of new bone[66], especially in long-term tissue interface maintenance[68]. So a series of calcium phosphate ceramics in the animal experiments involved were porous, and it was presumed that the ceramics with run-through macropores and interconnecting micropores were of good osteoinduction[68,69]. And beta-TCP has osteoinductivity after implantation in dog dorsal muscles without use of bone marrow cells or osteoinductive cytokines through the chronological histology. The appearance of a large number of active osteoclasts induced new bone formation[69]. It's confirmed the evidence of the intramembrane osteogenesis in the porous calcium phosphate ceramics implanted in an extra-osseous site in dogs [70,71].

The present study demonstrates that it is possible to construct large bone graft with good blood supply and certain mechanical strength. The *in vivo* tissue engineered bone has been provided with a hope of being used to repair segmental bone defect of mandible and to support dental implant in the future. And our previous works had showed that vascularized bone grafts could be constructed with osteoinductive calcium phosphate ceramics *in vivo* near a branch of femoral blood vessel in the muscle of dogs[4], whereafter the bone graft had been applied to repair the periodontal bone defect and the box-like bone defect of mandible in dogs, not or for supporting the dental implants, resulting in the pretty restoration [72,73]. Furthermore, Stevens reported another wonderful way of engineering bone in the osseous site, and when the new bone transplanted into the contralateral tibia defects in the same animal, the defect healed and were indistinguishable from the original[5]. And it was considered that it's a fresh, new strategy for tissue engineering that relies on the body's own capacity to regenerate itself [6].

It is confirmative now that porous Ca-P ceramics is osteoinductive. Attempt to explore the possible applications of Ca-P ceramics in segmental load-bearing bone reparation in our group, HA/TCP ceramics sintered at 1250°C were fabricated to cylinder of Φ15mm×30mm with a central canal of Φ4mm. Compounded with the fixation of net-cage-structured TC4, the HA/TCP cylinders were used to repair 30mm long segmental bone defect made in the middle of femur(figure 21). Stress was analysis by ANSYS to optimize the thickness of TC4 cage.

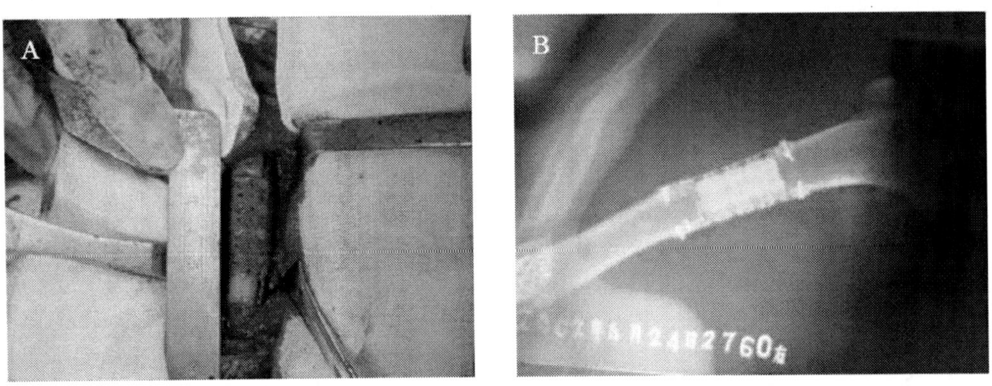

Figure 21. Pictures of the repairing the artificial femur defect of 30mm length with HA/TCP aided with TC4 net cage (A). X ray film of the reparation(B).

As a result, stress analysis showed that the thickness of TC4 net cage was 0.3mm. All dogs shifted their weight from the treated leg to the contra-lateral pelvic limb after surgery. Weight bearing forces were redistributed equally between the pelvic limbs at 4 months. The contour recovery of the bone defect was good and function was restored after 2 months gradually. The results indicated that new bone formed in the osteoinductive HA/TCP, and the composite was a useful device for repairing segmental load-bearing bone defect directly[74].

Unfortunately, the major drawbacks to the use of Ca-P ceramics are their adverse mechanical properties. They are brittle, have low impact and fracture resistance, and low tensile strength; consequently, they fail when directly used in a structural capacity. The mechanical properties of both TCP and HA depend on their physical structure, primarily their porosity. As porosity increases, strength decreases [75]. In fact, compressive strength is inversely proportional to the volume percentage of porosity [76] Consequently, it is only with external support, until adequate bony ingrowth occurs, that they can be used as a structural material. Also, the unnatural pathways that are characteristic of intact ceramic matrices do not favor the normal process of bone ingrowth and remodeling that occur after bone transplants[77].

4.2. Constructing *In Vivo* Tissue Engineered Bone

The *in vivo* bone tissue engineering is another exciting and novel concept of the tissue engineering, which is the generation of bone tissue induction by implantation of smart biomaterials that on their own can induce de novo bone formation from the host tissues without the addition of exogenously osteogenic factors, powerful inducer of the tissue morphogenesis including angiogenesis and osteogenesis. Meanwhile unlike the procedure of conventional bone tissue engineering, the procedure of the *in vivo* bone tissue engineering seems extremely simple, just involving the implantation of the biomaterials and transplantation for the reparation; there are no harvest/expansion of the cell, seeding, use of growth factors, and a variety of the costly equipments, and so on.

In the literatures, the *in vivo* tissue engineered bone could be constructed in the osseous and the non-osseous site such as the muscle of leg, dorsal muscles, abdominis muscle, subcutis, and the artificial space between the tibia and periosteum in animal models. In general, the number of mesenchymal cells depended on the sites of the transplantation; it was high in the muscles and bone [78]. And the massive volume polymorphic cells aggregation in the pore innner surface of the biomaterials may come from the surrounding tissue such as muscles, bone, subcutis, adipose tissue. So the constructing location mostly involved muscles. In heterotopic sites, after anaesthesia, the operative area should be shaved and the skin was cleansed, the longitudinal incision was made and the muscles were exposed, subsequently longitudinal muscles incision were made, and the muscular pouches were created pivotally by blunt separation, into which ceramics cylinder were placed, the incision was closed in layers eventually using suture. And when the biomaterials were to place in the subcutis, the surgical procedure was even simpler and easier. When happened in the osseous sites, it seemed a little distinct. Specifically, in the context of bone engineering, it was not hypothesis that the pluripotent cells population in the periosteum and biomolecular signals necessary for the formation of bone could be locally derived. And Stevens[5] created a wonderful ''space'' as an ''*in vivo* bioreactor," where the biocompatible calcium-alginate gel was injected in, between the surface of a long bone and periosteum within the New Zealand White rabbit. Postoperatively, Using the ''*in vivo* bioreactor'', the engineered bone was easily harvested that is biomechanically identical to native bone. And when the osteoinductive calcium phosphate ceramics cylinder were implanted near a branch of femoral blood vessel in the leg muscles, the vascularized bone grafts encapsulated with rete vasculosum could be constructed in our study[4](figure 22).

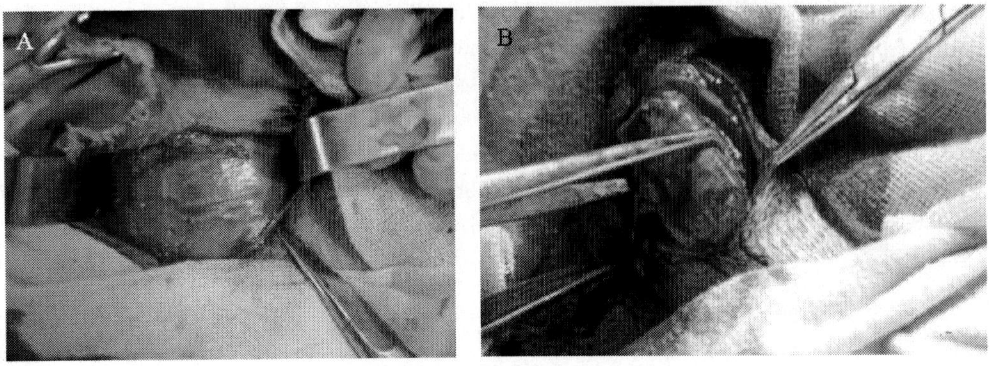

Figure 22. The procedure of constructing the vascularized *in vivo* tissue engineering bone grafts. Isolation and exposure of one muscular branch of the femoral artery and vein(A, published in[89]), and muscle pouch was made in the circumscription of the blood supply and the ceramics was inserted in it. The bone graft encapsulated rete vascukosum were formed at the 6th weeks (B).

The specimens were harvested with the surrounding tissues postoperatively at the certain time points. When some special detection was to be employed, the specified operations should be carried out at the certain time before the harvest. For instance, ^{99m}Tc-MDP deposited specifically in bone tissue, indicating the metabolism state and neogenesis of bone qualitatively and quantitatively[79]; ^{99m}Tc-MDP (1mCi/kg) was often injected to all dogs in the vein respectively, three to four hours before the animals were sacrificed in our group, when ^{99m}Tc-MDP absorption evaluation under the SPECT were to be employed. Affinis procedure was involved in tetracycline fluorescence testing.

4.2.2. Mechanical Properties of In Vivo Tissue Engineered Bone

In spite of the similarity to the mineral phase of bone, absence of antigenicity, excellent biocompatibility, osseointegration with the host bone tissue, the porous calcium phosphate ceramics were of some shortcomings for the stress bearing structure, such as the difficulty of molding, degranulation, strength, friability, mal-ductility, and so on. The ceramics with run-through macropores and interconnecting micropores were of good osteoinductivity. And osseous tissue and blood vessel were easy to grow into the porous ceramics and the mechanical properties could be improved after implantation.

In order to test the mechanical improvement, Bao did some beneficial work. In his study, osteoinductive calcium phosphate cylinders (consisted of 60% HA and 40% β-TCP)was prepared with 10mm in diameter and 20mm, 30mm in length respectively, Φ2mm canal in the center and the average pore size was about 395μm in average and the porosity rate was 60%(figure 23） BLA layer was formed on the wall of ceramic pores in revised simulate body fluid (RSBF). The prefabricated biomaterials, compounded with fresh marrow and cancellous bone or not, were implanted in leg muscles. As a result, bone formation was found in the ceramics, at 6 and 12 weeks respectively after operation. Vascular pigmentation with 3% methylthioninium chloride showed that these bone grafts turned blue 6 weeks after operation. Histological observation and tetracycline labeling of cylinders showed obvious new bone formation. SPECT showed that ^{99m}Tc–MDP concentrated at the sites of ceramics cylinders implanted. Compared with not-implanted ceramics, two kinds of implanted materials during different periods were provided with higher and higher compressing strength and bending strength with time [4].

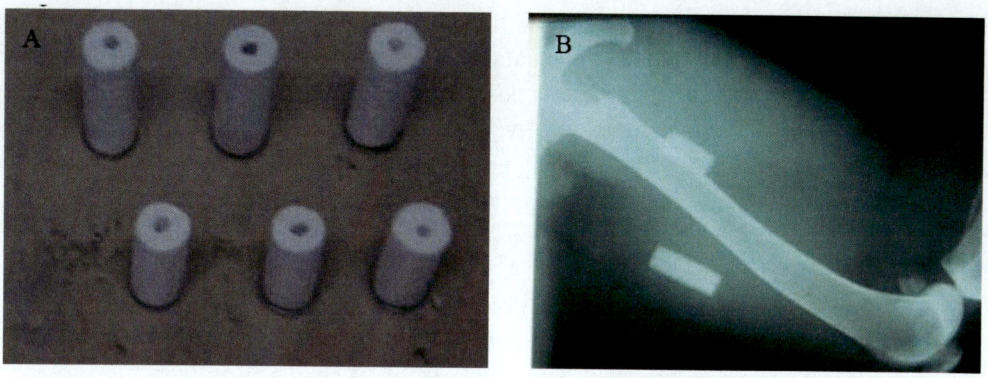

Figure 23. Two etalon of the HA/TCP cylinder sintering at 1250℃ with Φ10mm×30mm and Φ10mm×20mm and a Φ2mm run-through canal for constructing *in vivo* tissue engineering bone grafts(A). The X ray film of the constructed *in vivo* tissue engineering bone grafts in the thigh of dog at the 12th week post-implantation(B).

Similar to the cancellous bone in structure, compressive stress of the engineered bone (ceramics cylinders) varied from 4.15 MPa to 8.12 MPa after 6 weeks and to 13.81 MPa after 12 weeks post-implantation; bending stresses varied from the original 4.13 MPa to 7.93 MPa after 6 weeks, and to 12.54 MPa after 12 weeks. The compressive strength and rupture strength of calcium phosphate ceramics were within the range which were typical of the cancellous bone (50MPa and 3.5MPa of human) respectively. And it was believed that the biomechanics could be improved to same extent as time went on. Although the strength had not met the clinical diversified requirements (weight-bearing or not), but the engineered bone has overcome the shortcoming of the ceramics, and could be sculpted for the application [80]. Bone grafts did not crack under mechanic property test just like natural bone. Between two kinds, HA/TCP+BLA with marrow and cancellous bone or not, the morphology was just a little better in the former but their biochemical property were comparable.

Stenvens[5] determined the mechanical properties of the engineered neo-bone and found that its compressive strength [Young's modulus of 17.4±3.9 GPa] was within the range that is typical of compact bone (15–29 GPa); and the mean values of 115.2± 17.1 MPa for the ultimate strength of the neo-bone specimens also fall within the range for compact bone (100–300 MPa, human: 193 MPa), which was comparable to native bone. It was revolutionary, it seemed that it could have overcome the tough problem; however the shape and volume were confined by the bioreactor.

4.2.2. Observation of In Vivo Tissue Engineered Bone

Unlike bone formation induced by bone morphogenetic proteins, material-induced bone formation is not often observed in small animal models, such as rabbits, rats, and mice; however, it was affirmative in the primates, pigs and canine animals. And literatures showed that the osteoindutive bone formation seemed most easily in the baboon. Maybe, a receivable explanation was that it was correlated to difference velocity of the substance interchange and metabolism of the materials different in diversified animal models, besides the materials factors.

Usually, in the whole, there was no or little inflammatory reaction locally post-implantation, which showed the excellent biocompatibility. However, macrophage and multinucleated giant cell containing the material particle often could be observed near the implanted materials or in the local lymphoid node, which might be correlated to the style of the degradation of the materials. The population of phagocytosing cells consisted mainly of multinuclear cells and osteoclasts. There is no consensus on which cells play the key role in the resorption process. Lane has stated that the cell that is involved in the partial resorption of hydroxyapatite seems to be the foreign body giant cell but not the osteoclast[81]. But strong evidence exists that the osteoclast is responsible for the resorption process in calcium phosphate ceramics. Osteoclastic resorption has been demonstrated in biphasic calcium phosphates with different HA/TCP ratios[82], Others have shown that macrophages are involved in the phagocytosis of calcium phosphate ceramics as well[83]. In a study by Frankenburg et al[84], a combination of macrophages and osteoclasts has been demonstrated to be involved in the phagocytosis of an injectable carbonate apatite (Dahllite).

There were thin layers of fiber bursa (a dense connective tissue layer) around the porous materials with less osteoinduction; on the contrary almost hardly any fiber bursa around those with the more osteoinduction, and the around tissue which was steariform or chondriform were interlaced with the ceramics.

Radionuclide bone imaging is a sensitive modality for actively growing area of bone, and it is suitable for effect observation. For instance, ^{99m}Tc-MDP deposited specifically in bone tissue, and absorbed dose was correlated to blood flow and metabolism state of the bone [4]. In our group, ^{99m}Tc-MDP (1mCi/kg) was injected to all dogs in the vein respectively, three to four hours before the animals were sacrificed, when ^{99m}Tc-MDP absorption evaluation under the SPECT were to be employed. The color distribution in the implanted bioceramics and in the adjacent host bone was coincident, which showed that the de novo bone formation could metabolize in the body, as well as the host bone(figure 23).

As for HA/β-TCP (60/40) sintering in 1100℃ in our experiment, osteoid tissue could be found after the implantation earlier in the 1st weeks in dogs. The bone trabecula was adjoining to the wall of the inner pores of the ceramic. As the implanted times prolonged, the new bones became more mature, and reorganization could be observed on the bone trabecula; the cells that form the bone trabecula changed from the osteoblasts into bone lacuna, via osteocytes. In addition, HA/β-TCP (60/40) cylinders sintering in 1250℃ on the surface of which bone like apatite formed in RSBF, loading or not with the autogenic fresh marrow and cancellous bone, were implanted in the dorsal muscle of dogs. However the results showed that the autogenic fresh marrow and cancellous bone loaded on the cylinder did not improve the osteoinductivity of the HA/β-TCP (60/40) cylinders significantly by histomorphology, SEM, tetracycline fluorescence microscope, SPECT (figure 24).

Collagen, a highly heterogeneous protein, could play a key role in the modulation of the cell surface and triggering of the differentiation process [85]. IHC was employed to confirm the expression of collagen type I in the newly formed tissue in Ca/P ceramics after implantation of non-osseous sites, IHC staining has shown that osteoblast secrete collagen I in the newly formed bone tissue, which demonstrates that this kind of the osteogenesis process was similar to bony osteogenesis, and it is the evidence of the intramembrane osteogenesis in the porous calcium phosphate ceramics implanted in an extra-osseous site in dogs[86].

Figure 24. ^{99m}Tc-MDP SPECT of the HA/TCP+BLA implanted in the leg muscles: negative results in the implanted areas immediately post-implantation (A); at the 6th weeks post-implantation, the color distribution in the ceramics was coincident to that in the femoral diaphysis of the host(B).

The modus of the osteoinductive bone formation was diverse. The typical intramembrane osteogenesis(ossification) was observed in materials with the excellent osteoinductivity, the osteoblast in the inner pore of the material might be differentiated directly from the recruited MSCs, and the bone was formed near the ceramic pore walls(figure 24). However, fibroblasts or like-fibroblast cells, the inflammatory cells in the early period, and the osseous and osteoid tissues in the later period on the wall of pores or in the center of the pore tissues were appearanced sequentially in the materials with the less osteoindutivity. The form of osteoinduction was not typical, and it supposed that there may be cartilaginous osteogenesis or fibrosis osteogenesis as well as intramembranous osteogenesis.(figure 25)

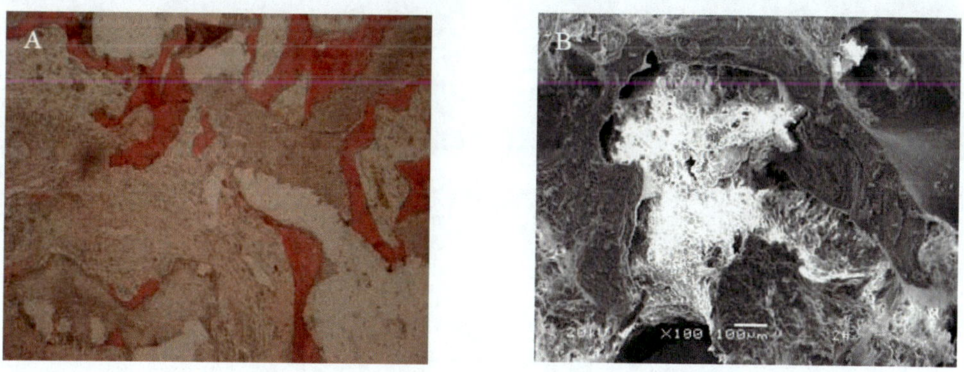

Figure 25. Light micrograph (A) and SEM micrograph (B) of HA/TCP+BLA at the 12th weeks post-implantation. New bone trabecula(A) and osseous matrix(B) filled in pore of ceramics.

The study by Stenven showed that the bioreactor space was filled with spindle-shaped periosteal cells from inner cambium layer of the periosteum and capillaries in the 3rd day, subsequently woven bone appeared at 2 weeks, and eventually evolved into compact bone tissue at a later time point, and the stage of neo-bone formation appeared to follow a path akin to intramembraneous bone formation in the presence of a good blood supply. However, the presence of hypertrophic chondrocytes implied that the endochondral pathway could also occur in the process of the neo-bone formation, although the cartilage template would be remolded to yield bone rapidly in the 10th day[26].

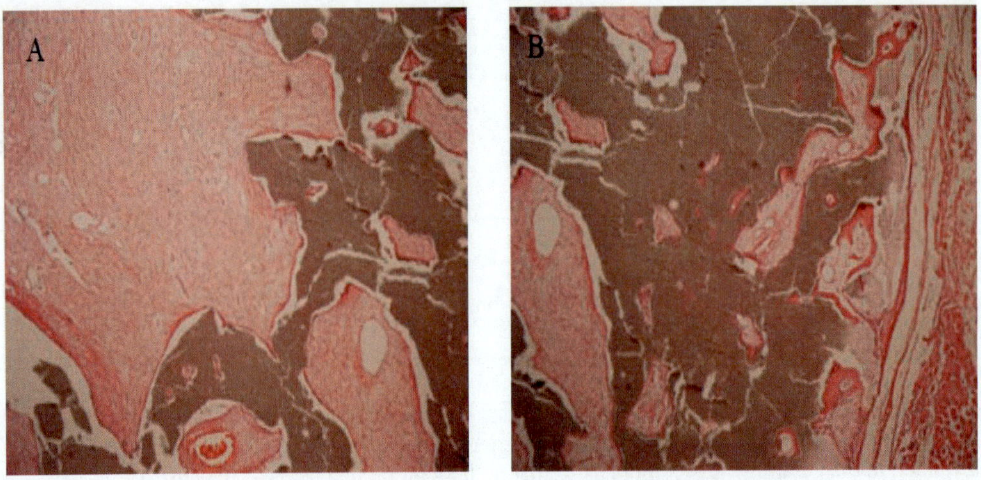

Figure 26. LM observation of the response to the HA/TCP+BLA incorporated with autogenic fresh marrow and cancellous bone in muscle of dogs at the 6th weeks post-implantation in decalcified HE staining section. Compact fibrous tissue in the central pores, with blood vessel ingrowth(A); and bone trabecula in the periphery pores and on the surface of ceramics adjoining the muscle(B).

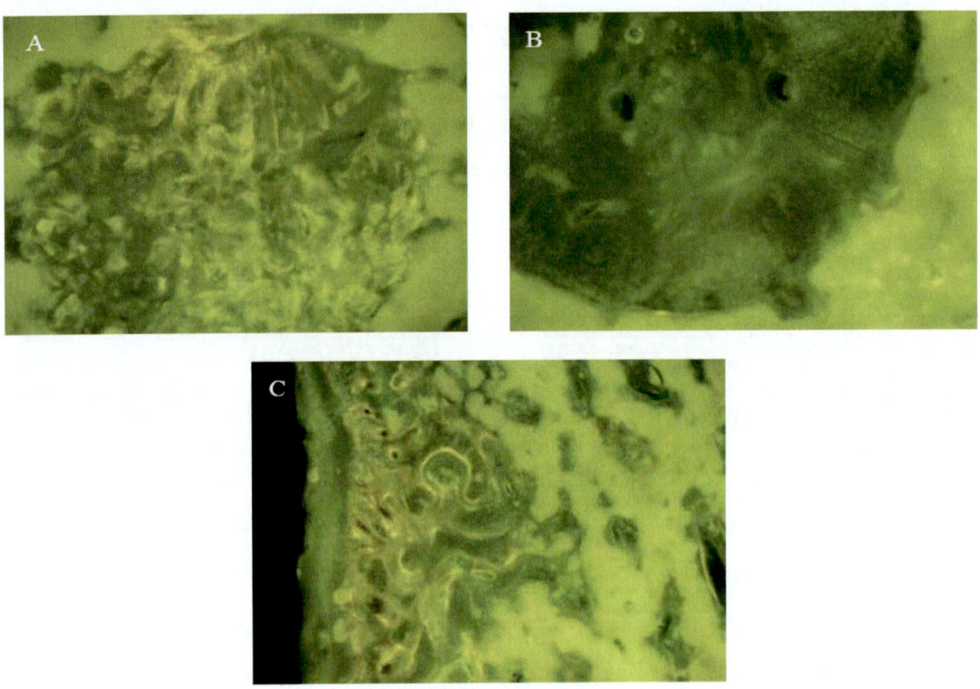

Figure 27. LM of Tetracycline fluorescence labeling post-operation: strong dispersed fluorescence in the central of canal and weak in the surface of pore in HA/TCP+BLA with run-through Φ4mm canal in the central at 2nd month(A); inversely weak fluorescence in the central of canal(B) and strong in the surface of pores(C) in ceramics at 8th month. Magnification 40*.

4.3. Repairing Box-Like Bone Defect of Mandible and Supporting Dental Implant with In Vivo Tissue Engineering Bone

Stenvens et al had showed that a $7\times 3\times 1.5$ mm (length width depth) cortical bone defect created in the proximal diaphysis of the anteromedial aspect of the left tibia was repaired by the neoosseous tissue harvested from the right tibia of four rabbits 6 weeks after the wonderful bioreactor creation in the same rabbit. After another period of 6 weeks the engineered bone had remodeled and integrated with the surrounding bone in the defect[5].

Until now, in our group, it was believed that the bone graft could be constructed *in vivo* with the osteoinductive phosphate calcium ceramics in some animal models and the engineered bone grafts could metabolize in the body; moreover, the size of the *in vivo* engineered bone graft could be more than 20mm×10mm×10mm in our group, so that the bone defect with the size beyond the critical size defect (CSD) could be repaired. And the *in vivo* tissue engineered bone has been applied in the reparation of the periodontal bone defect and the mandible box-like defect[87, 89], not or for supporting the implant in dogs in our group. And the selected porous calcium phosphate ceramics sintered in 1100℃ was comprised of 60%HA and 40%TCP, with the average pore size at 350±50um, porosity at 70.8%, which had demonstrated the more osteoinductivity in our previous studies. The bone graft could metabolize in the bone defects in the early. In the reparation of the periodontal bone defects with the bone grafts in size of 8mm×6mm in dogs, 8 weeks post-operation, the in vivo tissue engineering bone could improve the restoration of configuration of periodontal bone defects as early as possible and accelerate bone regeneration[89]. In the restoration of mandible box-like defect with bigger size (20mm×10mm×10mm) in our group, the result showed that the *in vivo* tissue engineered bone grafts could participate in the bone metabolism of host by SPECT, eventually fused with the host bone and restored the configuration of the mandible excellently (figure 27).

Furthermore, the engineered bone graft in the restoration of the box-like defects could immediately support the prefabricated dental implant in the mandible of dogs in another study, with the titanium internal fixation system. In this research, 16 osteoinductive Ca-P ceramics samples with a size of 20mm×10mm×8mm were soaked in RSBF for 72 hours before implanted. One ceramics was implanted in each femoral muscle of each dog, to constructing *in vivo* tissue engineering bone grafts.

Six weeks after implantation, the bone grafts was constructed, enveloped with the vascular rete. The box-like bone defects of 20mm in length and 10mm in depth were formed in mandible bilaterally of the dogs along the alveolar bone, where the teeth had been extracted in advance. Meanwhile constructed bone grafts were sculptured and transplanted immediately to reconstruct the box-like bone defects (figure.28A). Simultaneously one prefabricated dental implant was implanted in each graft as well as in alveolar bone, with internal titanium fixation system.

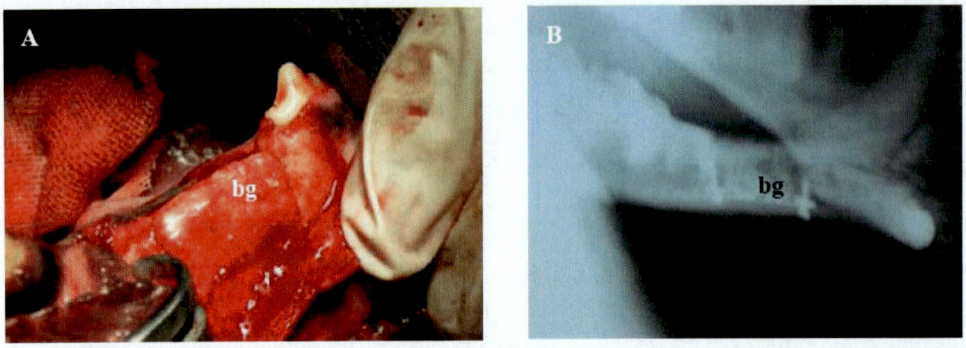

Figure 28. The procedure (A) of transplanted bone graft, and X ray film (B) of the bone graft (bg) osseointegrated with the mandible.

The bone graft fused with the host bone and restored the shape of mandible. Roentgenograms showed that the density of bone graft was similar to the host bone and the boundaries between host bone and transplant became vague (figure 28B). Of the SPECT results, the color distribution in implanted ceramics and in host bone was coincident, which revealed that bone grafts could metabolize in the transplanted bone area(figure 29 B). Histological observation showed that the active bone regeneration was found in the boundaries between the host bone and transplant and new bone grew into the pores of ceramics(figure 29A). Osteoblast rimming, even overlapping around the developing bone grafts could be detected although thin connective tissue was also seen. Inside of transplant, mineralized bone matrix with osteocytes filled the pores of ceramics and the macrophage gathered into ball appeared in the ceramics. Haversian system was clearly observed and the matured trabeculae bone connected with each other.

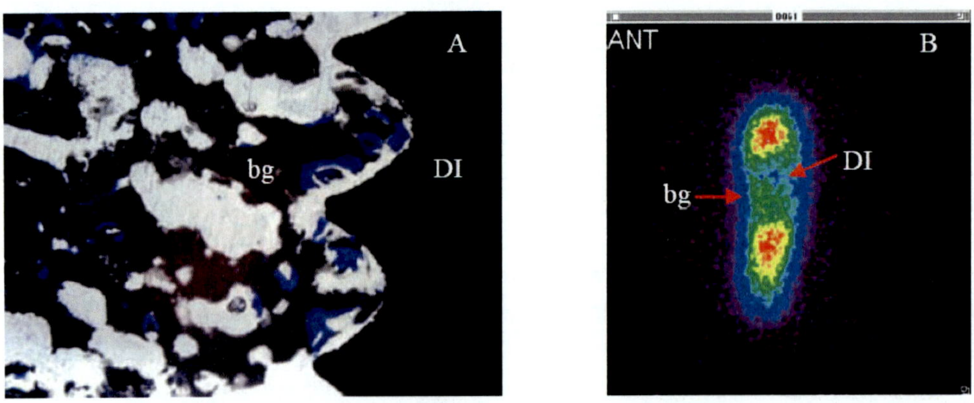

Figure 29. LM obvservation (A) and ^{99m}Tc-MDP SPECT (B) of the bone graft supporting the dental implant(DI) in mandible of dog at the 12th week post-reparation. A: Toluidin blue staining, magnification=40*; B: the transplanted bone graft (bg) metabolizing in the host bone.

In these studies, bone graft was constructed directly by the technique of *in vivo* bone tissue engineering. Compared with the traditional methods, the superiority of *in vivo* bone tissue engineering is significant. First, the complicated procedures *in vitro*[87]are omitted and replaced by relatively simple methods *in vivo*. Second, it is well known that Ca-P ceramics are frequently used as bone substitute because of their excellent properties such as biocompatibility, osteoconductivity and osteoinductivity; as the bone grows into the pores of ceramics, as a result, the tissue engineering bone *in vivo* modified the poor biomechanics property of Ca-P ceramics, such as high elastic modulus and low flexibility. Importantly, the *in vivo* tissue engineering bone with living auto-cells and good blood supply, of which biomechanics properties and bioactivity were approximate to the autografts', could participate in the bone metabolism of host as early as possible. Third, because it is difficult to treat the massive or segmental bone defect in clinic nowadays, the *in vivo* bone tissue engineering might be another strategy to treat the massive or segmental bone defect.

5. Conclusion

Osteoinductive material is the footstone for devoloping *in vivo* bone tissue engineering, and osteoinduction depends on both the chemical composition and physical characters of materials. The microstructure may be the most important factor for osteinduction, but the critical size of grain and micro-pore and the shape of mocrostructure for osteoinductinve materials remains unclear. Compared to naturally healing bone, *in vivo* tissue engineered bone has similar histological characteristics and bone-specific mRNA expression pattern. The biological mechanism of constructing *in vivo* tissue engineered bone, such as the original of stem cell, should be further explored. The limited data have already shown the possibility to construct large bone graft with good blood supply and certain mechanical strength with *in vivo* bone tissue engineering technique. Because it is difficult to treat the massive or segmental bone defect in clinic nowadays, the *in vivo* bone tissue engineering might be another strategy to treat the massive or segmental bone defect.

6. Acknowledgements

These researches were surpported by Natural Science Fund of China (NSFC:30672337) and New Century Excellent Talent of Chinese Educational Ministry (NCET-06-0782). The authors thank Dr. HP Yuan, the Senior Scientist of Progentix, the Neterlands, for his suggestion on editing this chapter; Dr. Roman Truckenmueller and Prof. Clemens van Balitterswijk, Department of Tissue Regeneration, University of Twente, The Netherlands, for their help and support.

7. References

[1] Laurencin C, Khan Y, El-Amin SF. Bone graft substitutes. *Expert Rev. Med. Devices.* 2006; 3(1): 49-57

[2] von Arx T, Cochran DL, Schenk RK, Buser D. Evaluation of a prototype trilayer membrane (PTLM) for lateral ridge augmentation: an experimental study in the canine mandible. *Int. J. Oral Maxillofac. Surg.* 2002;31:190–9

[3] George FM, Chizu N, Linda GG. Engineering principles of clinical cell-based tissuee engine-ering. *J. Bone and Joint Surg.* 2004; 86:1541-58

[4] CY Bao, H Fan, X Zhang, et al. Experimental study on construction of vascularized bone gra-ft with osteoinductive calcium phosphate *in vivo*. *Key Eng. Mat.* 2004;255: 801-4

[5] Stevens MM, Marini RP, Schaefer D, et al. In vivo engineering of organs: the bone bioreact-or. *Proc. Natl. Acdi. Sci.* 2005;102(32):11450-58

[6] Robert F. Service. Tissue engineering: technique uses body as 'bioreactor' to grow new bone. *Science.* 2005; 309 :683

[7] Winter GD. Heterotopic bone formation in a synthetic sponge. *Plastic Surgery.* 1970; 63: 11-5

[8] Yuan H, Yang Z, de Bruijn JD, et al. Material-dependent bone induction by calcium phosphate ceramics: a 2.5-year study in dog. *Biomaterials.* 2001;22:2617–23

[9] Shunsuke Fujibayashi, Masashi Neo, Hyun-Min Kim, et al. Osteoinduction of porous bioactive titannium matal. *Biomateials.* 2004; 25: 443-50

[10] Yuan HP, Van Blitterswijk CA, De Groot K, et al. Cross-species comparison of ectopic bone formation in biphasic calcium phosphate (BCP) and hydroxyapatite (HA) scaffolds. *Tissue Engineering.* 2006;12 (6): 1607-15

[11] Ripamonti U. Soluble osteogenic molecular signals and the induction of bone formation. *Biomaterials.* 2006;27(6):807-17

[12] Brekke JH, Toth JM. Principles of tissue engineering applied to osteogenesis. *J. Biomed. Mater Res.* 1998; 43: 380-94

[13] Yao JF, Bao CY, Sun R, et al. Exploring the Osteoinductivity of Calcium Phosphate Ceramics/PLA or Collagen Composites. *Key Eng. Mat.* 2007; 336-338: 1638-41

[14] Le Nihouannen D, Daculsi G, Saffarzadeh A, et al. Ectopic bone formation by microporous calcium phosphate ceramic particles in sheep muscles. *Bone.* 2005; 36(6): 1086-93

[15] Ripamonti U, Crooks J, Kirkbride AN. Sintered porous hydroxyapatites with intrinsic osteoinductive activity: geometric induction of bone formation. *South African Science,* 1999,95(8):335-8

[16] Nashimoto M, Matsuzaka K, Yoshinari M, et al. The effect of surface pore size on the differentiation of rat bone marrow cells: morphological observation and experession of bone related protein mRNA. *Bull. Tokyo Dent. Coll.* 2004; 45:201-11

[17] Kondo N, Ogose A, Tomaki H, et al. Osteoinduction with highly purified beta-tricalium phosphate in dog muscles and the proliferation of osteoclast before heterotopic bone formation. *Biomaterials.* 2006; 27(25): 4419-27

[18] Habibovic P, Yuan HP, Van den Doel M, et al. Relevance of osteoinductive biomaterials in critical-size orthotopic defect. *J. of Orthopaedic Research.* 2006; 24(5): 867-76

[19] Miao X, Hu Y, Liu J,Wong AP. Porous calcium phosphate ceramics prepared by coating polyurethane foams with calcium phosphate cements. *Materials Letters.* 2004;58:397-402

[20] Holand W, Vogel W, Naumann K. Interface reaction between machinable bioactive glass-ceramics and bone. *J. Biomed. Mater Res.* 1985;19:303-12

[21] Ohura K, Nakamura T, Yamamuro T. Bone-bonding ability of P2O5-free CaO· SiO2 glasses. *J. Biomed. Mater Res.* 1991;25:357-65

[22] Neo M, Kotani S, Nakamura T. A comparative study of ultrastructure of the interfaces between four kinds of surface-active ceramic and bone. *J. Biomed. Mater Res.* 1992;26:1419-32

[23] Tadashi Kokubo Apatite formation on surfaces of ceramics, metals and polymers in body environment. *Acta Materialia.* 1998; 46(7):2519-27

[24] Duan YR, Zhang ZR, Wang CY, et al. Apatite formation on HA/TCP ceramics in dynamic sumilated body fluid. *Key Eng. Mat.* 2004;254-256: 131-4

[25] Oyane A, Onuma K, Ito A, Kim HM, Kokubo T, Nakamura T. Formation and growth of clusters in conventional and new kinds of simulated body fluids. *J. Biomed Mat. Res. Part A.* 2003; 64A (2): 339-48

[26] Bernke JP, Laura V, Lydia AM, et al. One –step fabrication of porous micropatterned scaffolds to control cell behavior. *Biomaterials.* 2007; 28(1): 1998-2009

[27] Marcus TE, Joseph LC. Nanoimprint fabrication of polymer cell substrates with combined microscale and nanoscale topography. *J. Vacu. Sci. and Tech. B.* 2007; 25(4): L31-4

[28] H Yuan,Y Li, JD de Bruijin, et a1. Tissue responses of calcium phosphate cement: a study in dogs. *Biomaterials,* 2000;21: 1283-89

[29] Xingdong Zhang, Huipin Yuan, K.e Groot. Calcium Phosphate Biomaterials with intrinsic Osteoinductivity. Notebook: Wrokshop 1#, Biomaterials With Inteinsic Osteoinductivity, The 6th World Biomaterials Congress, May,2000,Hawaii,USA

[30] Urist MR. Bone formation by autoinduction. *Science.* 1965; 150(4):893

[31] Urist MR, Mikulski A, Lietze A, et al. Solubilized and insolubilized bone morphogenetic protein. *Proc. Natl. Acad. Sci.(USA).* 1979;76(3): 1828-32

[32] Urist MR, Delange RJ, Finerman GA. Bone cell differentiation and growth factors. *Science.* 1983; 222: 680-6

[33] Niederwanger M, Urist MR. Demineralized bone matrix supplied by bone banks for a carrier of recombinant human bone morphorgenetic protein (rhBMP-2): a substitute for autogenetic bone grafts. *J. Oral Implantol.* 1996; 22(3-4):210-5

[34] Cook SD, Baffes GC, Wolfe MW, et al. Recombinant human bone morphogenetic protein-7 induces healing in a canine long-bone segmental defect model. *Clin. Orthop.* 1994 Apr; (301):302-12

[35] SX Qu, Y Leng, X Guo, et al. Histological and ultrastructral analysis of heterotopic osteogenesis in porous calcium phosphate ceramics. *J. Mater Sci. L.* 2002; 21:153-5

[36] Kawamura M, Urist MR. Human fibrin is a physiologic delivery system for bone morphogenetic protein. *Clin. Orthop.* 1988; 235:302-10

[37] Itoh S, Kikuchi M, Takakuda K, et al. The biocompatibility and osteoconductive activity of a novel hydroxyapatite/collagen composite biomaterial, and its function as a carrier of rhBMP-2. *J. Biomed. Mater Res.* 2001 5; 543: 445-53

[38] Niedhart C, Maus U, Redmann E, et al. Stimulation of bone formation with an in situ setting tricalium phosphate/rhBMP-2 composite in rats. *J. Biomed. Mater Res.* 2003; 65A(1):17-23

[39] Seeherman H, Wozney J. Li R. Bone morphogenetic protein delivery system. *Spine.* 2002;27(16 suppl 1):S16-25

[40] Okubo Y, Bessho K, Fujimura K. osteoinduction by recombinant human bone morphogenetic protein-2 at inntramuscular, intermuscular,subcutaneous and intrafatty sites. *Int. J. Oral Maxillofac. Surg.* 2000,29(1): 62-6

[41] CY Bao, HL Wang, HS Fan, et a1. Statistical study on osteoinductivity of the calcium phosphate with different physicochemical characters in different animal species. The 7th World Biomaterials Congress, Sydney, 2004; 516

[42] Yuan,H., Kurashina, K., de Brujin,J.D., Li, Y., de Groot,K., and Zhang,X. A preliminary study on osteoinduction of two kinds of calcium phosphate ceramics. *Biomaterials.* 1999; 20: 1799-1806

[43] De Brujin, J.D, Yuan,H., Dekker, R., and van Blitterswijk,C.A. Osteoinduction by biomimetic calcium phosphate coating and their potential use as tissue engineering scaffolds. In Davis J.E.,eds. *Bone engineering.* Toronto:emsquared incorporated,2000, pp421-31

[44] Fisher LW, Termine JD. Noncollagenous proteins influencing the local mechanisms of calcification. *Clin. Orthop.* 1985;200:362–85.

[45] Giraud Guille MM, Mosser G, Helary C, Eglin D. Bone matrix like assemblies of collagen: From liquid crystals to gels and biomimetic materials. *Micron.* 2005;36:602–8.

[46] Mundlos S, Otto F, Mundlos C, Mulliken JB, Aylsworth AS, Albright S, Lindhout D, Cole WG, Henn W, Knoll JHM, Owen MJ, Mertelsmann R, Zabel BU, Olsen BR. Mutations involving the transcription factor CBFA1 cause cleidocranial dysplasia. *Cell.* 1997;89:773–9.

[47] Gerstenfeld LC, Gotoh Y, McKee MD, Nanci A, Landis WJ, Glimcher MJ. Expression and ultrastructural immunolocalization of a major 66 kDa phosphoprotein synthesized by chicken osteoblasts during mineralization *in vitro*. *Anat. Rec.* 1990;228:93–103.

[48] Sodek J, Zhang Q, Goldberg HA, Domenicucci C, Kasugai S, Wrana JL, Shapiro H, Chen J. Non-collagenous bone proteins and their role in substrate-induced bioactivity. In: Davies JE, editor. *The bone–biomaterial Interface*. Toronto: University of Toronto Press; 1991. p 97–110.

[49] Strauss PG, Closs EI, Schmidt J, Erfle V. Gene expression during osteogenic differentiation in mandibular condyles in vitro. *J. Cell Biol.* 1990;110:1369–78.

[50] Pockwinse SM, Wilming LG, Conlon DM, Stein GS, Lian JB. Expression of cell growth and bone specific genes at single cell resolution during development of bone tissue-like organization in primary osteoblast cultures. *J. Cell Biochem.* 1992;49:310–23.

[51] Sodek J, Cheifitz S. Molecular regulation of osteogenesis. In: Davies JE, editor. Bone engineering. Toronto, Canada: Em. Squared Inc; 2000. p 31–43.

[52] Wang C, Duan Y, Markovic B, Barbara J, Howlett CR, Zhang X, Zreiqat H. Phenotypic expression of bone-related genes in osteoblasts grown on calcium phosphate ceramics with different phase compositions. *Biomaterials.* 2004;25:2507–14.

[53] Luk KDK, Chen Y, Cheung KMC, Kung HF, Lu WW, Leong JCY. Adeno-associated virus-mediated bone morphogenetic protein-4 gene therapy for *in vivo* bone formation. *Biochem. Biophys. Res. Commun.* 2003;308:636–45.

[54] Yamaguchi A, Ishizuya T, Kintou N, Wada Y, Katagiri T, Wozney JM, Rosen V, Yoshiki S. Effects of BMP-2, BMP-4, and BMP-6 on osteoblastic differentiation of bone marrow-derived stromal cell lines, ST2 and MC3T3-G2/PA6. *Biochem. Biophys. Res. Commun.* 1996;220:366–71.

[55] Suzawa M, Takeuchi Y, Fukumoto S, Kato S, Ueno N, Miyazono K, Matsumoto T, Fujita T. Extracellular matrix-associated bone morphogenetic proteins are essential for differentiation of murine osteoblastic cells in vitro. *Endocrinology.* 1999;140:2125–33.

[56] Reddi AH. Bone morphogenetic proteins, bone marrow stromal cells, and mesenchymal stem cells. Maureen Owen revisited. *Clin. Orthop.* 1995;313:115–9.

[57] Jaiswal N, Haynesworth SE, Caplan AI, Bruder SP. Osteogenic differentiation of purified, culture-expanded human mesenchymal stem cells *in vitro. J. Cell Biochem.* 1997;64:295–312.

[58] Hing KA.Bioceramic bone graft substitutes: Influence of porosity and chemistry. *Int. J. Appl. Ceramic Tec.* 2005;2:184–99.

[59] Hing KA. Bone repair in the twenty-first century: biology, chemistry or engineering? *Philos. Trans. R Soc. Lond. A.* 2004;362:2821–50.

[60] De Groot J. Carriers that concentrate native bone morphogenetic protein *in vivo. Tissue Eng.* 1998;4:337–41.

[61] Younger EM, Chapman MW. Morbidity at bone graft donor sites. *J. Orthop. Trauma.* 1989;3:187–91.

[62] Laurie SWS, Kaban LB, Mulliken JB, et al. Donor-site morbidityafter harvesting rib and iliac bone. *Plast. Reconstr. Surg.* 1984;73:933–38.

[63] Zhang XD, Zou P, Wu C, et al. A study of porous block HA ceramics and its osteogenesis. In: Bioceramics and the Human Body. ED by Ravaglioli and Krajewski. Elsevier Applied Science. P408-415,1991

[64] Ripamonti U. The morphogenesis of bone in replicas of porous hydroxapatite obtained from conversion of calcium carbonate exoskeletons of coral. *J. Bone Joint Surg.* 1991; 73:692-703

[65] Roy DM, Linnehan SK. Hydroxyapatite formed from coral skeletal carbonate by hydrothermal exchange. *Nature.* 1974;247:220 –2

[66] Osborn JF, Newesely H. The material science of calcium phosphate ceramics. *Biomaterials.* 1980;1:108 –11.

[67] Lemons JE. Ceramics: past, present, and future. *Bone.* 1996;19:121S–128S.

[68] Yuan,H., Li, Y., de Brujin, J.D., de Groot, K., and Zhang, X. Tissue reponses of calcium phosphate bone cement: A study in dogs. *Biomaterial.* 2000;21:1283-90,

[69] Yuan,H., de Brujin, J.D., Li,Y., Feng, J., Yang,Z., de Groot,K.,and Zhang, X. bone formation induced by calcium phosphate ceramics in soft tissue of dogs: A comparative study between α-TCP and β-TCP. *J. Mater. Sci. Mater. Med.* 2001;12:7-13,

[70] Yuan,H., de Brujin, J.D., Zhang,X., van Blitterswijk,C.A., and de Grooot, K. Bone induction by porous glass ceramics made from Bioglass(45S5). *J. Appl. Biomat.* 2001;58:270-6,

[71] Kurashina,K., Kurita,H., Wu, Q., Ohtsuka, A.,and Kobayashi, H. Ectopic osteogenesis with biphasic ceramics of hydroxyapatite and tricalcium phosphate in rabbits. *Biomaterials.* 2002;23:407-12,

[72] JF Yao, YZ Zhang, CY Bao, LY Sun, XM Hao, HS Fan, XD Zhang. Reconstructing box-like bone defect of mandible with *in vivo* tissue engineering bone. *Key Eng, Mat.* 2007, *vol.*330-332: 1165-8,

[73] B Zhang, XJ Zhang , CY Bao , Q Wang , JF Yao , HS Fan , CD Xiong , XD Zhang Repairing Periodontal Bone Defect With in vivo Tissue Engineering. *Key eng. Mat.* 2007;Vol.330-332:1121-24,

[74] CY Bao, YZ Zhang, HL Wang, JM Luo, YF Tan, HS Fan, XD Zhang. Stress analysis and optimizing of osteoinductive Ca-P ceramics and Net-Cage-Structure titanium alloy in dog segmental femoral defect reparation. *Key Eng. Mat.* 2006; vol.309-311: 231-234

[75] de Groot K. Effect of porosity and physicochemical properties on the stability, resorption, and strength of calcium phosphate ceramics. *Ann. NY Acad. Sci.* 1988;523:227–33.

[76] Williams DF: The biocompatibility and clinical uses of calcium phosphate ceramics, in Williams DF (ed): *Biocompatibility of Tissue Analogs,* Boca Raton, FL, CRC Press, 1985, vol II: 43–69

[77] Gazdag AR, Lane JM, Glaser D, et al.: Alternatives to autogenous bone graft: Efficacy and indications. *J. Am. Acad. Orthop. Surg.* 1995, 3:1–8

[78] Urist MR, Hay PH,Dubee F, Buring K. Osteogenesis competence. *Clin. Orthop.* 1969;64:194-201,

[79] Kobayashi W, Kobayashi M, Nakayama K, et al.Free omental transfer for osteoradionecrosis of the mandible. *Int. J. Oral and maxillofac.. Surg..* 2000; 29 (3): 201-6

[80] Encyclopedic Handbook of Biomaterials and Bioengineering. New York: Marcel Dekker, Inc; 1995

[81] Lane JM, Bostrom MP. Bone grafting and new composite biosynthetic graft materials. *Instr. Course Lect.* 1998; 47:525–34

[82] Dennis JE, Haynesworth SE, Young RG, Caplan AI. Osteogenesis in marrow-derived mesenchymal cell porous ceramic composites transplanted subcutaneously: effect of fibronectin and laminin on cell retention and rate of osteogenic expression. *Cell Transplant.* 1992;1:23–32

[83] Gross UM, Muller-Mai CM, Voigt C. The interface of calcium-phosphate and glass-ceramic in bone, a structural analysis. *Biomaterials.* 1990;11:83– 5.

[84] Frankenburg EP, Goldstein SA, Bauer TW, Harris SA, Poser, RD. Biomechanical and histological evaluation of a calcium phosphate cement. *J. Bone Joint Surg. Am.* 1998; 80:1112–24.

[85] Wlodarski, KH. Bone formation in the soft tissue. In :Hall BK, editor.Fracture repair and regeneration.Boca Raton,FL:CRC press Inc;1991.p313-37

[86] Qu SX, Guo X, Weng J, et al. Evaluation of the expression of collagen type I in porous calcium phosphate ceramics implanted in an extra-osseous site. *Biomaterilas.* 2004; 25 (4): 659-67

[87] Hutmacher DW, Garcia AJ. Scaffold-based bone engineering by using genetically modified cells. *Gene.* 2005; 347(1):1-10

In: Tissue Engineering: Roles, Materials and Applications ISBN: 978-1-60456-293-4
Editors: S. J. Barnes, L. P. Harris, pp. 243-257

Chapter 9

GELLAN/GELATIN-BASED BIOMATERIALS FOR TISSUE ENGINEERING

Valeria Chiono**[*,1,2]**, Gianluca Ciardelli**[1,2]**,
Giovanni Vozzi**[2]**, Franca Bertoni**[1,2]**,
***Niccoletta Barbani**[2]** and Paolo Giusti**[2,3]

[1] Department of Mechanics, Politecnico in Turin, Corso Duca degli Abruzzi 24, 10129 Turin, Italy. Fax: +39-011-5646969
[2] Department of Chemical Engineering, Industrial Chemistry and Materials Science, University of Pisa, Via Diotisalvi 2, 56126 Pisa, Italy. Fax +39/050/511266
[3] C.N.R. Institute for Composite and Biomedical Materials, Via Diotisalvi 2, 56126 Pisa, Italy. Fax +39/050/511266

ABSTRACT

Blends between gellan (GL) and gelatin (GE) with various compositions (GL/GE 0/100, 25/75, 50/50, 75/25, 100/0 wt/wt) were produced, as candidate materials for biomedical applications. GL/GE films obtained by casting were subjected to a chemical treatment (CT) in an acidic coagulation bath at a pH promoting ionic interactions between GE and GL. A subsequent dehydrothermal treatment (DHT) was performed, allowing the formation of amide and/or ester bonds between the macromolecules. The effects of blend composition and the type of treatment on the physicochemical properties of the samples were evaluated by scanning electron microscopy, infrared analysis and release tests. Interactions between GE and GL by H-bonding were found to increase with decreasing the GL amount and, for each composition, by applying CT+DHT. Infrared spectra of samples after CT+DHT displayed a new adsorption band which was indicative of ester bonds formation. Release tests evidenced that resistance of GL/GE blends to water dissolution greatly increased after CT+DHT. Mechanical properties of GL/GE samples, determined through stress-strain tests, displayed significant variations as a function of composition. GL/GE two-dimensional scaffolds were fabricated by micro-molding (μML), a technique based on the use of a poly-(dimethylsiloxane) (PDMS) mould to create patterns with micro-scale resolution on cast films. The influence of blend composition on the adhesion of primary endothelial cells was studied: GL/GE blends

* Valeria Chiono: e-mail: valeria.chiono@polito.it

with a high GE content (50 and 75 wt. %) were found to support cells adhesion and proliferation which made them promising candidates for uses in the field of tissue engineering. Coating of GL/GE samples with gelatine enhanced endothelial cell response only for blends containing a high (75 wt%) GE content. Cells tests using endothelial cells were preliminary to future investigations on the potentiality of GL/GE blends as constituent materials for vascular grafts.

Keywords: gelatin; gellan; micro-moulding; soft lithography; tissue engineering.

1. Background

The object of tissue engineering is to reconstruct natural tissue from living cells to replace defective or lost tissues and organs. Scaffolds provide the necessary support for cells to proliferate and maintain their differentiated function [1,2]. Scaffolds should have a three dimensional structure with high porosity and an interconnected network for cell growth and flow transport of nutrients and metabolic waste. These features have to be associated with (i) biocompatibility and bioresorbability with a controllable degradation and resorption rate; (ii) suitable surface chemistry for cell attachment, proliferation and differentiation and (iii) similar mechanical properties to those of the tissues at the site of implantation. Another fundamental aspect in tissue engineering is the control of the polymer scaffold architecture at the micro-scale which should mimic the microscale structure of the tissue to be repaired. Soft lithography is a well known microfabrication technique pioneered by Whitesides et al.[3-5] to create micropatterned substrates for spatially positioning and controlling the shape and function of mammalian cells. This technique is optimal for the control of surface chemistry at a molecular level [6,7].

Natural polymers, such as proteins and polysaccharides, being component of, or structurally similar to, the glycosaminoglycans (GAGs) in the extra-cellular matrix (ECM) are promising materials for tissue engineering applications [8,9].

In this work we produced blends between gellan (GL), an anionic exocellular polysaccharide, and gelatine (GE), a protein derived from collagen, with the aim to obtain a material with suitable mechanical and physico-chemical properties for scaffold production. Our aim was that to verify the possibility to exploit the synergism of these biopolymers for the realization of innovative systems for tissue engineering.

Gellan is produced by fermentation of *Pseudomonas elodea* and consists of tetrasaccharide repeating units of β-D-glucose, β-D-glucuronic acid and α-L-rhamnose in molar ratios of 2:1:1. Gellan has numerous industrial applications as a high viscosity biogum and as a suspending or gelling agent, whereas gellan derivatives (gellan benzyl esters) have been found to provide a convenient means for the controlled release of encapsulated drugs [10, 11].

Gelatin is a high molecular weight polypeptide, derived from collagen, the primary protein component of animal connective tissues. Industrial preparation of gelatin involves the controlled hydrolysis of the organised structure of collagen to obtain soluble gelatin. As compared to collagen, gelatin exhibits no antigenicity and it is less expensive. Gelatin can be dissolved in hot water forming a gel on cooling which has been widely used in food applications [12]. Moreover, gelatine is one of the key structural proteins found in the extracellular matrices of many tissues, thus having the ability to affect cellular attachment and growth

One of gelatin and gellan drawbacks for tissue engineering applications is their solubility in aqueous media, therefore gelatin- and gellan-containing structures for long-term biomedical applications need to be crosslinked. The main limitation in the use of chemical crosslinkers arises from the presence of traces of unreacted crosslinker inside the scaffolds and from the risk of formation of toxic products by reaction between gelatin and/or gellan with the crosslinking agent during *in vivo* biodegradation. Physical crosslinking methods may overcome these drawbacks; they include microwave energy [13], dehydrothermal treatment (DHT) [14-18] and UV-irradiation [19]. Dehydrothermal crosslinking is a temperature and time dependent technique for protein crosslinking [14-16], which promotes a condensation reaction between carboxyl and amino or hydroxyl groups by the application of vacuum at high temperature [14]. The main limitation of dehydrothermal method is that a high crosslinking density can generally be achieved only through a long time treatment (even more than one day) [15]. The efficacy of DHT may be increased by blending gelatin with a polymer containing a relatively high amount of reactive groups (e.g. gellan), thus favouring the kinetics and the yield of the crosslinking reaction.

In the present work, we produced a series of gellan/gelatin blends (GL/GE) with various compositions, which were treated in an acidic coagulation bath to promote ionic interactions between GL and GE carboxylic groups and GE amino groups and then crosslinked by a thermal treatment under vacuum (DHT).

Physico-chemical properties of crosslinked blends were investigated by scanning electron microscopy (SEM), release tests by ultra-violet analysis (UV) and Fourier transform infrared total reflectance spectroscopy analysis (FTIR-ATR). We also studied (GL/GE) mechanical properties through stress-strain tests.

Finally, GL/GE blends were used for the production of two-dimensional scaffolds with a micrometric resolution by micro-molding (μML), which is a microfabrication technique derived from soft lithography and making use of a patterned poly-dimethylsiloxane (PDMS) mould [20-21]. Primary endothelial cells from human umbilical cords were seeded on crosslinked substrates obtained by μML to investigate their biocompatibility.

2. Experimental Part

2.1. Materials

Gelrite gellan gum (GL) was purchased from Sigma – Aldrich Chemical (St. Louis, USA).

Type A gelatin from porcine skin (GE; approx. 300 Bloom gel strength) was supplied from Aldrich (St. Louis, USA). All commercially available solvents and reagents were of analytical grade and used without further purification.

2.1.1. Preparations of Blend Films

GL/GE blends with 0/100, 25/75, 50/50, 75/25, 100/0 (wt/wt) ratios were prepared from aqueous solutions of the polymers (in the weight ratio of the blend) with an overall concentration of 1% (wt/v). Each system was kept under magnetic agitation at 70°C till complete dissolution. Blend solutions were cast on Petri dishes and then left drying at 37°C for 48 h in a vented oven.

2.1.2. Preparation of Two-Dimensional Microstructures by Micro-Molding (μML)

Micro-molding technique (μML) is a variant of soft lithography, based on the utilisation of a poly(dimethylsiloxane) (PDMS) mould. The first step of the technique consists of the obtainment of a silicone master for the realisation of a PDMS mould with a defined geometry.

Silicone wafers were spin-coated with EPON-SU8 photoresist (Microresist Technology GmbH, Berlin, Germany). They were then baked and exposed to ultraviolet light in an Ultraviolet Crosslinker (CL-100 Ultraviolet Crosslinker) through a high-resolution transparency mask, drawn with Coreldraw 9.0 and printed on a plastic slide using a commercial Linotronic-Hercules 3300 dpi high-resolution line printer. Exposed photoresist was then developed (SU8 developer, Microresist Technology GmbH, Berlin, Germany) and subsequently the wafers were baked.

PDMS prepolymer was prepared by mixing the commercially available prepolymer and catalyst (Sylgard 184 kit, Dow Corning) in a 10:1 wt/wt ratio. The mixture was degassed under vacuum to eliminate bubbles created during mixing. The prepolymer solution was cast on the master and placed under vacuum once again to remove bubbles created by pouring and then baked in an oven for 2 hours at 65° C.

After solvent evaporation, the wafer was removed by lifting an edge of the master and pulling off the elastomer. The mould was then washed with 70% ethanol (Aldrich) and sonicated for 5 min prior to use. A PDMS mould composed of square-shaped grooves with 150 μm side and 200 μm distance between sides of adiacent grooves was prepared by this technique.

The production of the two-dimensional scaffolds by micro-moulding started with the deposition of a little amount of blend solution in water (with 1% (w/v) overall concentration) on the PDMS mould. The system (polymer solution and PDMS mould) was placed under vacuum for 2 min to allow the filling up of mould microchannels by the polymer solution. Then, the system was put in a vented oven at 37 °C for 48 h to evaporate the solvent. After drying, the polymer pattern was gently removed from the mould with a pair of tweezers.

Then, the micropatterned films were subjected to a *chemical treatment* (CT), based on their immersion in a coagulation bath containing acetic acid (Aldrich) /water at pH=3 for 4 h, followed by several water baths of 1 h till neutral pH of the medium and, finally, by freeze drying at -20°C for 24 h (Christ alpha 1-4; B. Braun).

After CT, a *dehydro-thermal treatment* (DHT) was applied to the micropatterned films, based on keeping samples in a vacuum oven at 90°C for 90 min, followed by a 6 h treatment at 130°C (CT + DHT).

2.2. Methods

2.2.1. Scanning Electron Microscopy (SEM)

Scanning electron microscopic analysis (SEM) was performed by Jeol JSM-5600 LV apparatus on film specimens fractured under liquid nitrogen and coated with gold by the Edwards Sputter Coater B150S equipment.

2.2.2. Fourier Transform Infrared Attenuated Total Reflectance Spectroscopy Analysis (FTIR-ATR)

The Fourier transform infrared-attenuated total reflectance spectroscopy (FTIR-ATR) spectra were recorded at room temperature in a Perkin Elmer Spectrum One Spectrometer in the 4000-600 cm^{-1} range.

2.2.3. Release Test

The amounts of gelatine and gellan released from GL/GE blends after 24 h immersion in a water medium were evaluated by spectrophotometric methods. Precisely, the amount of released gellan was measured using the well-known carbazole method [22-23], while the quantity of released gelatin was derived from UV absorbance values at 280 nm [24]. Absorbance measurements were carried out on a UV-2100 recording spectrophotometer (Shimadzu). All measurements were carried out using standard 1 cm thick quartz cells, at room temperature.

2.2.4. Mechanical Properties

The mechanical properties were measured using an isotonic force transducer (model 7006, UGO Basile Biological Research Apparatus, Italy), where the applied force has a resolution of 1 mN Test specimens were dry cast films with 2 cm length, 0.5 cm width and a height of around 100 μm. Images of test specimens were recorded through an optical microscope (Olympus AX 70) equipped with a digital camera (ccD Camera HIRES, DTA; Italy), to obtain data on initial dimensions and geometry. The traction force was applied along the length of the samples. Three specimens for each composition were tested. A stress-strain curve was plotted based on the apparent stress σ (MPa) and strain (ε)% values, determined dividing the load values by the initial cross-sectional area of each test specimen and the deformation values by the initial specimen height, respectively. The Young's modulus of each sample was calculated from the slope of the initial linear portion of the stress (σ)-strain (ε) curve.

2.2.5. Cell Adhesion and Proliferation Tests

Cell adhesion and proliferation tests were performed in order to verify the cell compatibility of studied materials, in the perspective of their application as scaffolds for tissue reconstruction. Primary endothelial cells from human umbilical cords were used for cell adhesion tests. Cells were cultured in F199 medium (Sigma, Italy) containing 10% of foetal bovine serum, 1% of glutamine, 1% of antibiotics (Eurobio, Italy). In addition, the medium also contained 1% of endothelial growth factor and 0.5 % of hydrocortisone (Sigma,Italy).

Culture was mantained in an incubator equilibrated with 5% CO_2 at 37°C. Once the cells has grown to confluence, the culture medium was removed under a sterile hood and culture flasks were washed twice with phosphate buffered saline (PBS, Sigma). Cells detachment was performed by adding 0.01 % trypsin (Cambrex, Italy)/0.01% EDTA (Cambrex, Italy) solution (3 ml) to each culture flask and incubating at 37°C for 3 min. Trypsin was inactivated by adding a volume of culture medium three times higher than that of trypsin/EDTA solution. Cells pellets were recovered from cell dispersion by centrifugation (Hereus, Switzerland) at 1000 rpm for 10 min and then dispersed again in a new culture medium at a concentration of 100,000 cell·ml^{-1}.

The polymeric structures were sterilised according to the protocol at ref [25] and then seeded with cell suspension in a 24-well tissue culture plastic plates. At various time intervals after seeding (2, 8, 24 h), culture medium was removed and substrates with attached cells were rinsed with PBS. Attached cells were fixed by addition of 4% (v/v) formaldehyde (Sigma, Italy) solution in PBS for 10 minutes and then stained with Coomassie blue (Fluka, Italy) solution for 10 minutes. Samples were analysed under an optical microscope (Olympus AX 70) and the ratio between the number of cells on the polymeric structures and the total area of the polymer substrate was calculated as an index of cell density [21].

We used an image processing software routine developed in Matlab and based on segmentation which allows calculation of the area occupied by cells on the two-dimensional scaffold and the overall substrate area. For each sample, three areas were randomly chosen and photographed with 10X magnification. From each photo, 5 areas of the two-dimensional scaffold were selected. The area of the scaffold and the areas covered by the cells were calculated and averaged for all five zones, and then combined with the data from the remaining two photos of the sample to give an overall average percentage of cell coverage. This procedure was repeated three times for each sample type at each culture time, as to obtain an average value of cell density.

Both cells of a reference sample and cells seeded on two-dimensional scaffolds (cast films and microstructures fabricated by micro-moulding) were examined for comparison to evaluate the effect of substrate topography on cell growth. Type A gelatin was selected as a control.

3. Results and Discussion

3.1. Scanning Electron Microscopy (SEM)

Morphological analysis was performed on the surfaces and the fractured sections of pure components and blends. All the surfaces and the cross sections of the analysed samples were homogeneous and not-porous. Phase separation between blend components was not detected. Figure 1 shows exemplary SEM images of GL, GE and GL/GE 75/25 blend, after CT + DHT. In particular, fractured sections of pure gellan film samples showed a parallel thin layer morphology, which was also detected for GL/GE 75/25 blend due to its high GL content. Blends with a gelatin content higher than 25 wt.% did not show a similar stratified lamellar structure, probably due to the occurrence of blend phase inversion at a gelatin concentration ranging between 25 wt.% and 50 wt.%. As a conclusion, SEM observations suggested that GE was the blend matrix of GL/GE 50/50 and 25/75 samples, whereas GL was the continuous phase of GL/GE 75/25 blend.

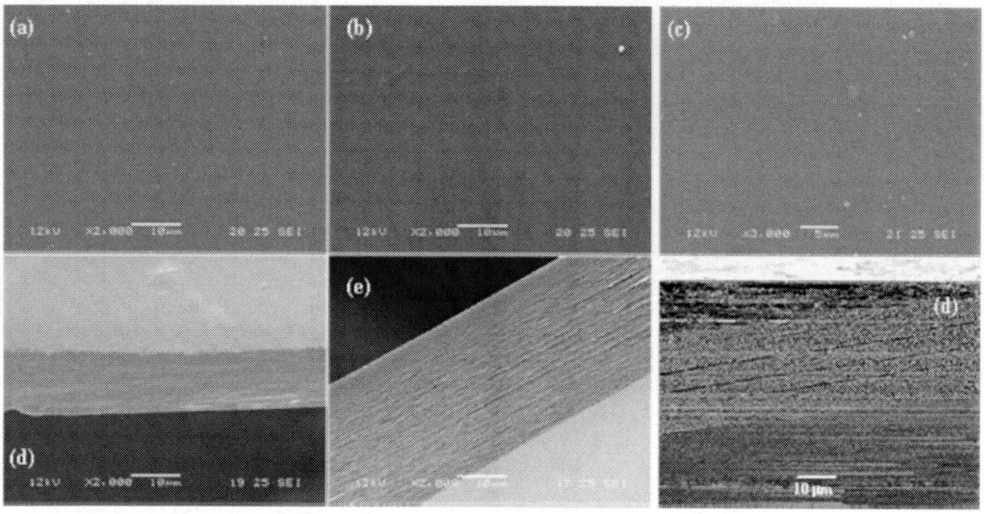

Figure 1. SEM images of the surface (a, b, c) and section (d, e, f) of cast films after CT+DHT: gelatin (a, d); gellan (b, e); GL/GE 75/25 (c, f).

3.2. Infrared Spectroscopy (FT-IR)

FTIR analysis has been frequently used to identify intermolecular interactions between the components of polymer blends [27].

Gellan FTIR-ATR spectrum showed the typical bands of polysaccharides: OH stretching vibrations at 3300 cm^{-1}, C-O-C stretching at 1040 cm^{-1} and COO^- stretching at 1600 cm^{-1}. The band associated with CH stretching vibrations was detected at 2900 cm^{-1}. Gelatin FTIR spectrum showed the carbonyl stretching vibrations of amide groups at 1650 cm^{-1} (amide I) and N-H bending and C-N stretching of amide and amino groups at 1550 cm^{-1} (amide II).

The FTIR-ATR spectra of GL/GE blends displayed all the characteristic bands of both components, except for the COO^- stretching band of GL (1600 cm^{-1}), which overlapped with gelatin amide I band.

Figure 2 shows the wavenumber variation of amide II band as a function of blend composition and the type of treatment. For each blend series (untreated, treated by CT, treated by CT + DHT), the wavenumber of amide II band shifted to higher values with increasing the gellan amount. On the other hand, for each composition, CT and more strongly CT+DHT further decreased amide II wavenumber. These results suggested that H-bonding interactions involving gelatin amide groups became stronger with decreasing the gellan amount and/or by appropriate treatments (CT and CT+DHT). A new band at 1730 cm^{-1} (C=O stretching of ester groups) was detected in the FTIR-ATR spectra of GL/GE blends after CT+DHT, being particularly evident for GL/GE 75/25 (figure 3) and pure gellan (data not shown). The appearance of this band after CT+DHT was attributed to the formation of ester groups by reaction of GL and/or GE hydroxyl groups with GL and/or GE carboxyl moieties. On the other hand, the possible formation of amide bonds during DHT [18] by reaction of GE and/or GL carboxyl groups with GE amine groups was not confirmed by FTIR-ATR analysis, due to the overlapping of the characteristic FTIR adsorption bands of the involved moieties (carboxyl, amine and amide groups) with those originally present in the gelatin and gellan spectra as well as to the low expected degree of amidation due to the low content of reactive amino groups in gelatin [27].

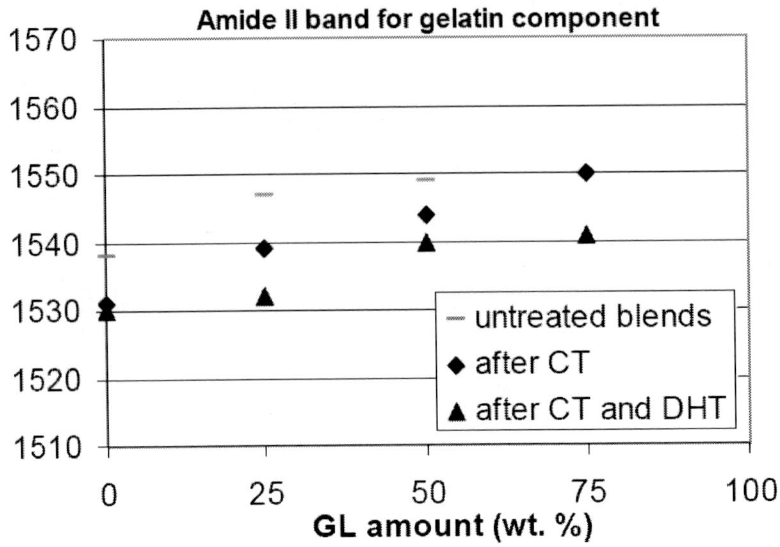

Figure 2. Wavenumber value of gelatine amide II as a function of blend composition and performed treatment.

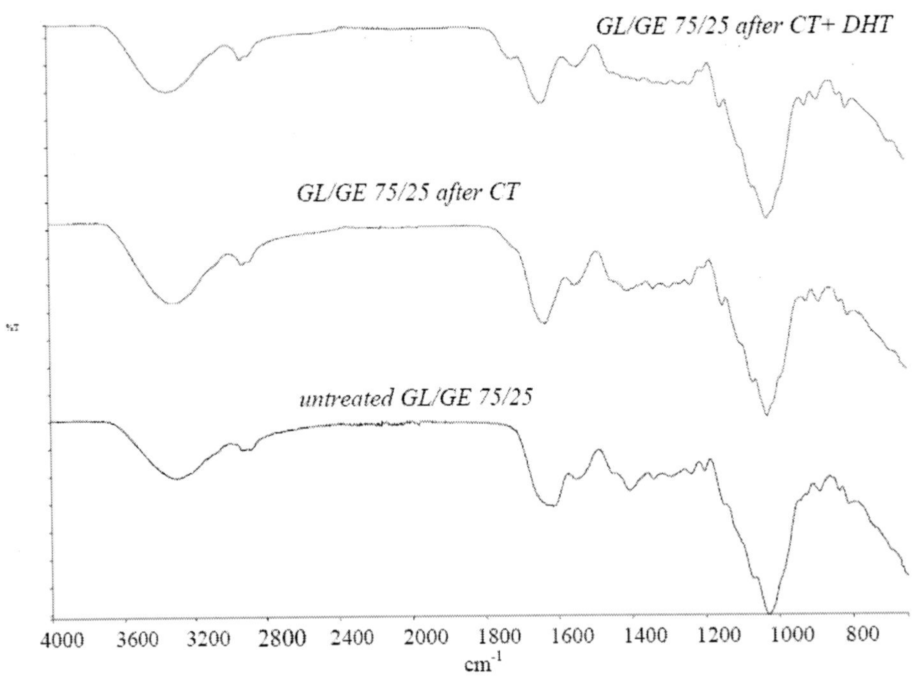

Figure 3. FTIR-ATR spectra of GL/GE 75/25 blends: untreated, after CT and after CT+DHT. A new adsorption band was detected at 1730 cm^{-1} (C=O stretching of ester groups) after CT+DHT.

3.3. Release Tests

Release tests were carried out to evaluate the amount of gelatin and gellan released from GL/GE samples, after CT and CT+DHT, using the spectrophotometric methods described in the experimental section. The weight percentage of released GE and GL, as well as the final composition of the blends after release tests were reported in table 1.

Table 1. Amount of released biopolymer expressed as weight percentage of its initial content, for blends with different composition after the chemical (CT) and chemical plus dehydro-thermal treatment (CT+DHT). The compositions of blends after releases are also collected. Measurements were performed in triplicate

Sample	Chemical treatment(CT)			Chemical treatment + dehydro-thermal treatment (CT+DHT)		
GL/GE (wt/wt)	GL released (wt.%)	GE released (wt.%)	Composition after releases (wt/wt)	GL released (wt.%)	GE released (wt.%)	Composition after releases (wt/wt)
25/75	3.5	70	33.7/66.3	0	16	28.4/71.6
50/50	2.4	67	74.7/25.3	0	5	51.3/48.7
75/25	2.5	72	84.9/15.1	0	2	75.4/24.6

CT and CT+DHT were performed with the aim to stabilise GL/GE behaviour in aqueous media, mainly decreasing their dissolution degree. At a suitable pH, gellan may form polyelectrolyte complexes with various polybases. In the case of GL/GE solutions, ionic interactions involving the carboxylic groups of GL (and/or GE) in the form of $-COO^-$ and the amino groups of GE in the form of $-NH_3^+$ may lead to the precipitation of a polycomplex. A similar behaviour has previously been studied for poly-(acrylic acid)/collagen blends [28]. The phenomenon of complex formation does not allow the preparation of homogeneous films by solution casting, unless the pH of the starting solution is higher than the isoelectric point of type A gelatine. Type A gelatin is obtained from collagen by acidic hydrolysis, and has an isoelectric point between 7 and 9 [29]. At a higher pH than GE isoelectric point, GE and GL macromolecules are both negatively charged, so that the repulsive forces prevent the formation of an insoluble complex and a homogeneous film may be obtained by casting. After solution casting, the ionic interactions between the macromolecules may be restored by dipping the dried cast film in a medium with a lower pH than gelatin isoelectric point. At pH=3 the carboxylic groups of GL were mostly (about 68.4 mol.%) in the protonated form (COOH), as the pK_a of glucuronic acid is about 3.5 [30]. On the other hand, the α-carboxylic groups of gelatin residues were mostly in their ionized form (p$K_{a(\alpha\text{-COOH})}$=1.7-2.6 [31]), whereas the GE α-amino groups as well as the $-(CH_2)_4-NH_2$ groups of lysine residues were almost completely protonated. Lateral carboxyl groups of glutammic acid ($-CH_2-CH_2-COOH$; p$K_{a(\text{glutammic acid})}$ = 4.3) and aspartic acid ($-CH_2-COOH$; p$K_{a(\text{aspartic acid})}$ = 3.9) residues were only partially in their ionized form at pH 3 [31]. The treatment at pH 3 induced ionic interactions between the dissociated carboxylic groups of GL and/or GE and the protonated amino groups of GE, thus increasing sample stability in water solutions. However, the numerical data collected in table 1 showed that the amount of released gelatin after CT was high (about 70 wt.%). Therefore, the induced ionic interactions were not sufficient to insolubilize the materials in aqueous media and a final thermal treatment was necessary to further increase sample resistance to water dissolution. After CT+DHT, gellan and gelatin releases decreased strongly, due to the formation of a crosslinked network between the macromolecules. As a consequence of CT+DHT, the composition of GL/GE blends did not change significantly upon permanence in a water bath.

3.4. Mechanical Tests

Young's modulus (*E*) is a measure of the stiffness for a given material. Its value can be experimentally determined from the slope of the stress-strain curve obtained from tensile tests.

As reported in literature [32], gelatin and gellan differ significantly in terms of mechanical properties: gelatin is able to form soft and elastic gels whereas the main features of gellan gels are hardness and brittleness. Our aim was that to evaluate the influence of gellan/gelatin weight ratio on the mechanical characteristics (Young's Modulus) of the blend film samples.

E values were found to decrease with increasing the gelatin content of blends (figure 4), as gelatin probably behaved as a plasticizer into the gellan network, reducing blend brittleness. This finding was in accordance with what reported by Lee et al. [33], which showed that for gellan/gelatin blends tensile strenght increased whereas elongation at break decreased with increasing the gellan content.

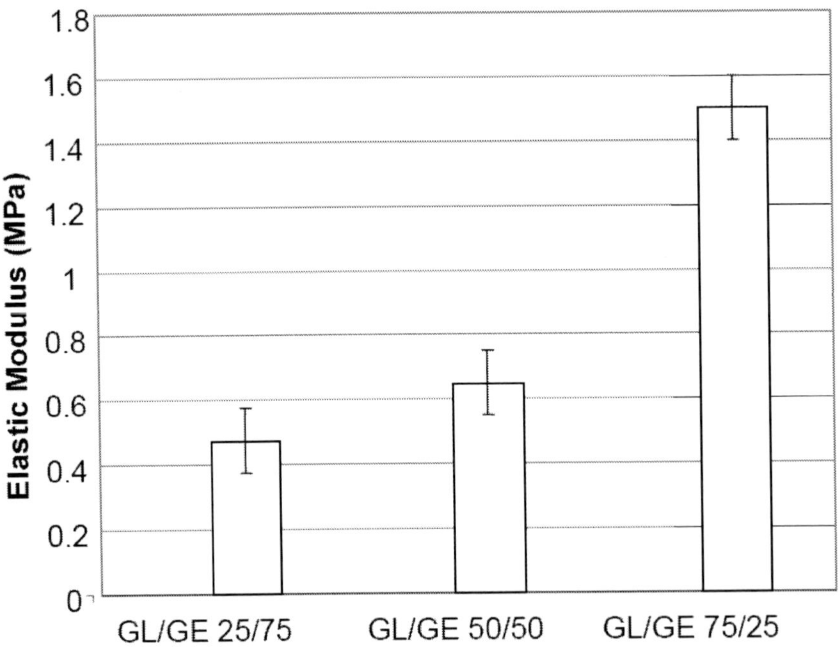

Figure 4. Young's modulus (*E*) of GL/GE blends as a function of blend composition. The columns indicate the average values whereas the bars indicate the standard deviations (n=3).

The mechanical behaviour of polymer blends is generally determined by the nature of blend matrix, the interactions between blend components and the dispersion degree of the minor component [34]. *E* values for GL/GE 25/75 and 50/50 blends were found to be similar and much lower than Young's Modulus of GL/GE 75/25 blend, suggesting that phase inversion probably occurred between 50 and 75 wt.% GL, as suggested by SEM analysis: GE was the dispersed phase for a GL content ≥ 75 wt.%, whereas it was the continuous phase for a GL amount ≤ 50 wt.%.

In conclusion, the mechanical properties of GL/GE blends showed the potential to be modulated by simply varying GL/GE weight ratio and to be tailored to those of the damaged tissue, for which regeneration of the scaffold is designed.

3.5. Cell Adhesion Tests

Figure 5 shows cell density on GL/GE two-dimensional scaffolds obtained by micro-moulding technique. Samples were used as prepared or covered with a gelatin layer before cell seeding. Results of cell culture tests on the various substrates were then compared.

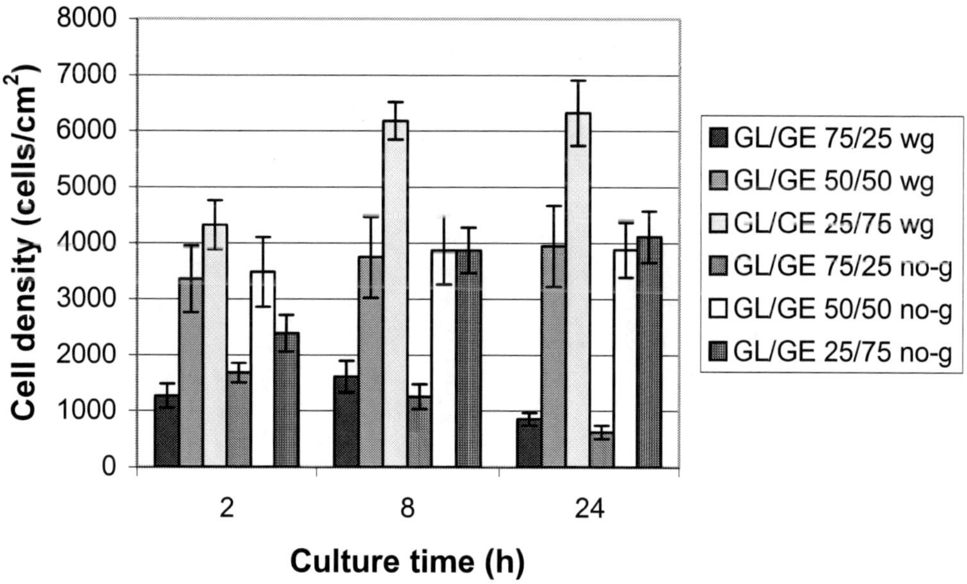

Figure 5. Primary endothelial cells density on GL/GE scaffolds as a function of blend composition and culture time. Blends were treated (wg) or not (no-g) with a gelatin solution before cell seeding. Columns indicate the average values, whereas the bars indicate the standard deviations (*n*=3).

At each culture time, endothelial cells density was found to increase with increasing the gelatin content of blends. Cell density also increased with time for all blend samples, except for the blend with 75 wt.% GL content, which conversely poorly supported cell attachment. The increase of cell density with increasing time became more pronounced for the gelatin-richest samples. Cell behaviour towards samples with different composition could be due to differences in blend mechanical properties and/or surface chemistry. As suggested by the results of the mechanical characterisation, samples containing 75 wt.% GL were stiffer and GL was their continuous phase: a mismatch in the mechanical properties between adhered cells and the substrate as well as a poor affinity between GL and primary endothelial cells could explain the low cell density found for this composition.

Pre-treatment of GL/GE cast films with a 1% (w/v) gelatin solution before cell seeding caused an increase of cell density only on GL/GE 25/75 blend, probably due to a significant absorption of gelatin by this blend as compared to samples with a lower gelatin content.

In conclusion, endothelial cells adhesion was generally enhanced by the presence of gelatin either as a blend component or as a surface modifier.

Figure 6 shows pictures of primary endothelial cells adhered on GL/GE 25/75 two-dimensional scaffolds at different times (2 h (A), 8 h (B), 24 h (C)) as compared to cells adhered on gelatin control layer (D) after 24 h. Cell morphology on these substrates was not altered with respect to the control layer.

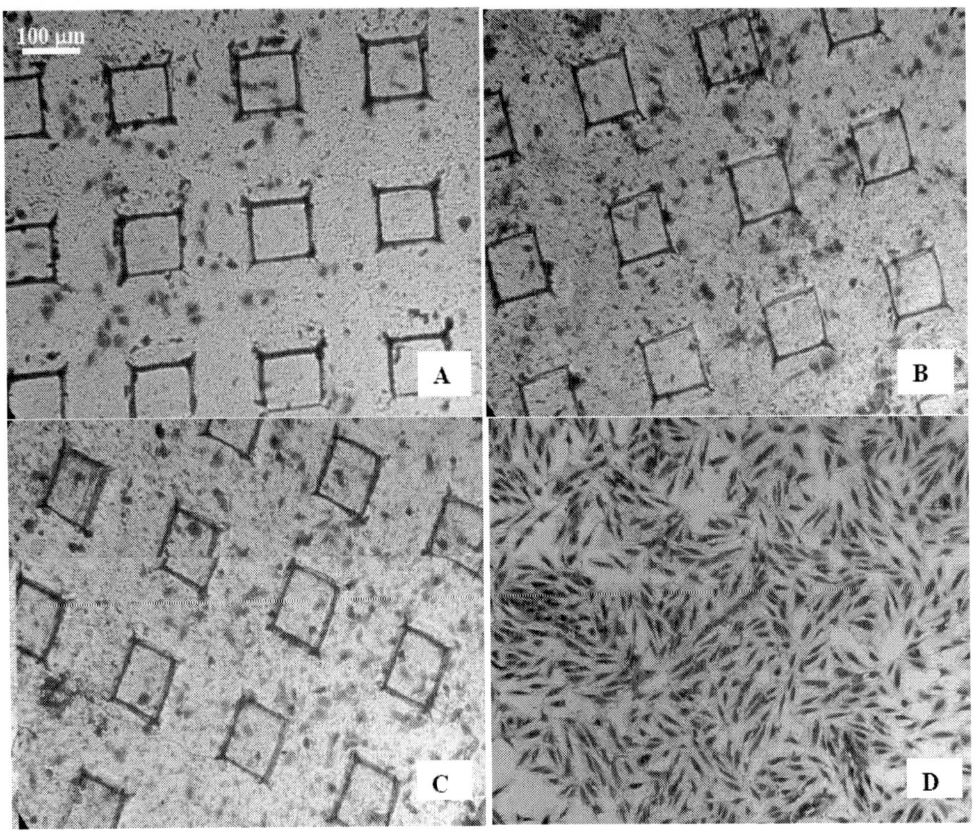

Figure 6. Optical microscopy images of primary endothelial cells adhered on GL/GE 25/75 scaffolds (A, B, C) at various times (2h (A), 4h (B), 24 h (C)) as compared to cell morphology on gelatin control layer after 24 h culture time (D). A same magnification was used.

CONCLUSION

In the last years, our research group focused the attention on the main themes related to tissue engineering and in particular on the realisation of scaffolds as templates to guide cell growth and tissue development [35]. Biocompatibility and biodegradability are two of the main properties that an ideal scaffold should have. Although many synthetic polymers, either biodegradable or not, have been used for the preparation of scaffolds, their cell adhesion is generally poor as a consequence of their low hydrophilicity and lack of surface cell recognition sites [36]. Our investigations was focussed on two natural polymers (gellan and gelatin), which biochemistry resembles that of natural tissues. GL/GE blends were produced with the aim to obtain biocompatible materials for tissue engineering applications.

Physico-chemical and mechanical properties of GL/GE blends were found to be easily modulated by varying the weight ratio of the components. GL/GE blends were compatible due to interactions between blend components by hydrogen bonding, evidenced by FTIR analysis. Ionic interactions between macromolecules were restored after film production by solution casting, using a chemical treatment in a coagulation bath with a lower pH than gelatin isoelectric point. A further dehydro-thermal treatment promoted the formation of ester and/or amide bonds between macromolecules, strongly reducing their dissolution in water media. Primary endothelial cells from human umbilical cords were cultured on micro-moulded scaffolds to evaluate the ability of GL/GE blends to support cells attachment. Cell adhesion was high for GL/GE blends with a high GE content (> 25 wt.%), showing their suitability for the production of tissue engineered constructs, in particular for cardiovascular applications. On the other hand, GL/GE blends having GL as the continuous phase did not support endothelial cells attachment.

Another important aspect of scaffold production is the control of scaffold architecture, due to the well-defined organisation found in most tissues in vivo. Different processing techniques have been studied to realize a scaffold. The technique utilized in this work, micro-moulding, which is based on the "soft lithography" approaches and makes use of a poly(dimethylsiloxane) mould, allows the formation of controlled micropatterned substrates. In the future, cells tests will be carried out at high times to investigate the influence of these structures on the organization, migration and differentiation of tissues.

REFERENCES

[1] Whang, K; Thomas, CH; Healy, KE; Nuber, G. A novel method to fabricate bioabsorbable scaffolds. *Polymer*, 1995, 36, 837-842.

[2] Ma, PX. Scaffolds for tissue fabrication. *Materials Today*, 2004, 7, 30-40.

[3] Xia, Y; George, M; Whitesides, M. Soft lithography. *Angew Chem Int Ed*, 1998, 37, 550–575.

[4] Widden, TK; Ferry, DK; Kozicki, MN; Kim, E; Kumar, A; Wilbur, J; Whitesides, GM. Pattern transfer to silicon by microcontact printing and RIE. *Nanotechnology*, 1996, 7, 447–451.

[5] Xia, Y; Kim, E; Zhao, XM; Rogers, JA; Prentiss, M; Whitesides, GM. Complex optical surfaces formed by replica molding against elastomeric masters. Science, 1996, 273, 347–349.

[6] Zhao, B; Moore, JS; Beebe, DJ. Surface-directed liquid flow inside microchannels. *Science*, 2001, 291, 1023-1026.

[7] Kane, RS; Takayama, S; Ostuni, E; Ingber, DE; Whitesides, GM. Patterning proteins and cells using soft lithography. Biomaterials, 1999, 20, 2363-2376.

[8] Mao, JS; Zhao, LG; Yin, YJ; Yao, KD. Structure and properties of bilayer chitosan–gelatin scaffolds. Biomaterials, 2003, 24, 1067-1074.

[9] Kang, HW; Tabata, Y; Ikada, Y. Fabrication of porous gelatin scaffolds for tissue engineering. *Biomaterials*, 1999, 20, 1339-1344.

[10] Vendruscolo, CT; Pereira, JL; Scamparini. AR. *Food hydrocolloids: structure, properties and function.* New York: Plenum; 1993. pp.91-5

[11] Crescenzi, V; Dea, ICM; Paletti, S; Stivala, S; Sutherland, IW. Recent development in industrial polysaccharides. *Biomedical and biotechnological advances in industrial polysaccharides.* NewYork: Gordon and Breach; 1989.

[12] Word, AG; Courts A. The science and technology of gelatin. London: Academic Press; 1977.

[13] Vandelli, MA; Romagnoli, M; Monti, A; Gozzi, M; Guerra, P; Rivasi, F; Forni, F. Microwave-treated gelatin microspheres as drug delivery system. *J. Control Release*, 2004, 96, 67-84

[14] Yannas, V; Tobolsky, AV. Crosslinking of gelatine by dehydration. *Nature*, 1967, 215, 509-510.

[15] Welz, MM; Ofner III, CM. Examination of self-crosslinked gelatin as a hydrogel for controlled release. *J. Pharm. Sci*, 1992, 81, 85-90.

[16] Ueda, H; Nakamura, T; Yamamoto, M; Nagata, N; Fukuda, S; Tabata, Y; Shimizu, Y. Repairing of rabbit skin defect by dehydrothernally crosslinked sponges incorporating transforming growth factor β1. *J. Control Release*, 2000, 88, 55-64. *

[17] Wess, TJ; Orgel, JP. Changes in collagen structure: drying, dehydrothermal treatment and relation to long term deterioration. *Thermochimica Acta*. 2000, 365, 119-128.*

[18] Pulieri, E; Chiono, V; Ciardelli, G; Vozzi, G; Ahluwalia, A; Domenica, C; Vozzi, F; Giusti, P. Chitosan/gelatin blends for biomedical applications. *J. Biomed. Mater Res. Part A*, 2007 in press.

[19] Vaz, CM; De Graaf, LA; Reis, RL; Cunha, AM. Effect of crosslinking, thermal treatment and UV irradiation on the mechancal properties and in vitro degradation behavior of several natural proteins aimed to be used in the biomedical field. *J Mater Sci Mater Med*, 2003, 14, 789-796.

[20] Vozzi, G; Flaim, C; Ahlwalia, A; Bathia, S. Fabrication of PLGA scaffolds using soft lithography and microsyringe deposition. *Biomaterials*, 2003, 24, 2533-2540.

[21] Vozzi, G; Flaim, C; Bianchi, F; Ahlwalia, A; Bathia, S. Microfabricated PLGA scaffolds: a comparative study for application to tissue engineering. *Biomaterials*, 2002, 20, 43-47.

[22] Taylor, KA; Buchanan-Smith, JG. A colorimetric method for the quantitation of uronic acids and a specific assay for galacturonic acid. *Anal. Biochem*, 1992, 201, 190-196.

[23] Taylor, KA. A colorimetric method for the quantitation of galacturonic acid. *Appl. Biochem. Biotechnol*, 1993, 43, 51-52.

[24] Stoscheck, CM. Quantitation of Protein. *Methods in Enzymology*, 1990, 182, 50-69.

[25] Ciardelli, G; Chiono, V; Vozzi, G; Pracella, M; Ahluwalia, A; Barbani, N; Cristallini, C; Giusti, P. Blends of poly-(ε-caprolactone) and polysaccharides in tissue engineering applications. *Biomacromolecules*, 2005, 6, 1961-1976.

[26] Daniliuc, L; De Kesel, C; David, C. Intermolecular interactions in blends of poly(vinyl alcohol) with poly(acrylic acid)-1.FTIR and DSC studies. *Eur. Polym. J*, 1992, 28, 1365-1371.

[27] Bertoni, F; Barbani, N; Giusti, P; Ciardelli, G. Transglutaminase reactivity with gelatine: perspective applications in tissue engineering. *Biotechnol. Lett*, 2006; 28, 697-702.

[28] Barbani, N; Lazzeri, L; Cristallini, C; Cascone, MG; Polacco, G; Pizzirani, G. Bioartificial materials based on blends of collagen and poly(acrylic acid). *J. Appl. Polym. Sci*, 1999, 72, 971-976.

[29] Wade, A; Weller, PJ. Editors, Handbook of Pharmaceutical Excipients. Second edition. London: American Pharmaceutical Association (1994).

[30] Mao, R; Tang, J; Swanson, BG. Gelling temperatures of gellan solutions as affected by citrate buffers. *J. Food Science*, 1999, 64, 648-652.

[31] Graham Solomons, TW. Capitolo 24: Aminoacidi e proteine. In: Zanichelli, editor. *Chimica Organica* (Seconda Edizione Italiana). Bologna: John Wiley and Sons Inc.; 1998; 982-983.

[32] Kwang, YL; Shim, J; Lee, HG. Mechanical properties of gellan and gelatin composite films. *Carbohyd. Polym*, 2004, 56, 251-254.

[33] Lee, KY; Shim, J; Lee, HG. Mechanical properties of gellan and gelatin composite films. *Carbohydrate Polymers*, 2004, 56, 251-254.

[34] Utracki, LA; Zaccaria, VK. Polymer Blends Handbook. Volumes 1-2. Springer – Verlag; 2002.

[35] Cai, Q; Wan, Y; Bei, J; Wang, S. Synthesis and characterization of biodegradable polylactide-grafted dextran and its application as compatilizer. *Biomaterials*, 2003, 24, 3555-3562.

[36] Cai, Q; Wan, Y; Bei, J; Wang, S. A novel porous cells scaffold made of polylactide–dextran blend by combining phase-separation and particle-leaching techniques. *Biomaterials*, 2002, 23, 4483-4492.

In: Tissue Engineering: Roles, Materials and Applications ISBN: 978-1-60456-293-4
Editors: S. J. Barnes, L. P. Harris, pp. 259-274

Chapter 10

SURFACE MODIFICATION USING RF OXYGEN PLASMA IN TISSUE ENGINEERING

K.Y. Lee[a], J.J. Cooper-White[b], I. Keen[a,b] and L.Grøndahl[a,*]

[a] School of Molecular and Microbial Sciences
[b] Australian Institute of Bioengineering and Nanotechnology,
The University of Queensland, Brisbane QLD 4072, Australia

ABSTRACT

Scaffolds intended for use in tissue engineering applications often invoke a foreign body response upon implantation leading to undesirable effects, such as encapsulation or rejection of the scaffold. Therefore to develop their full potential, biodegradable tissue engineering scaffold surfaces must be modified to overcome this foreign body response. Current research has suggested that negatively charged and hydrophilic surfaces avoid this response by inducing apoptosis in the responsible immune cells. One approach which has shown potential in functionalising biodegradable polymer surfaces with a negatively charged surface is oxygen plasma treatment. It has the advantages of controllability, versatility, and simultaneous sterilisation of the implant. Examples from our studies into the use of oxygen plasma treatment of the commonly used polymer poly(lactic-*co*-glycolic acid) (PLGA) will illustrate a number of important issues. In these studies PLGA was thinly and uniformly coated on glass coverslips to mimic the internal pore surfaces of tissue engineering scaffolds produced by thermally induced phase separation (TIPS). It was found that oxygen plasma, at pressures within the limits of current testing ($4x10^{-4}$ to 10^{-4} mbar; 10-60 s), significantly reduces the advancing and receding water contact angle. Contact angles obtained from samples washed once in ethanol after treatment consistently yielded contact angle values up to 10° higher than unwashed samples, supporting theories that oxygen plasma treatment cleaves PLGA chains, liberating soluble oligomeric products. XPS data indicates an increase in atomic oxygen content after plasma treatment by 1% in washed samples and by 2% in unwashed samples. AFM was used to visualise the extent of etching resulting from the oxygen plasma treatment, and surface roughness was found to change from < 1 nm (untreated sample) to 14 ± 3 nm (10 s oxygen plasma treatment at 150 mTorr). The findings of this study highlight the dual effect of oxygen plasma treatment and cautions the use of materials subjected to oxygen plasma in biological assays that have not been washed prior to their use.

INTRODUCTION

In the almost two decades spanning 1990 to present, there has been an explosion of interest in tissue engineering, responding to increasingly complex demands expected from the clinical practice. The principle behind tissue engineering involves implantation of a scaffold to guide autologous growth of a specific tissue in order to replace tissue lost through, for example, surgical excision or trauma, or to augment pre-existing tissue [1-3]. The ideal scaffold biodegrades over time to leave only the patient's newly formed tissue, with appropriate structure and function. Numerous fabrication techniques have been developed to produce scaffolds for tissue engineering applications. These include textile processing, fibre bonding, electrospinning, solvent casting and particulate leaching, melt moulding, thermally induced phase separation (TIPS), supercritical gas foaming, and rapid prototyping [4-5]. The TIPS method of creating scaffolds (figure 1) is particularly attractive due to the architectural control and applicability to a vast range of polymers [6]. Highly porous and interconnected structures can be formed, scaffolds of complex shapes and large sizes can be made, and bioactive molecules can potentially be incorporated.

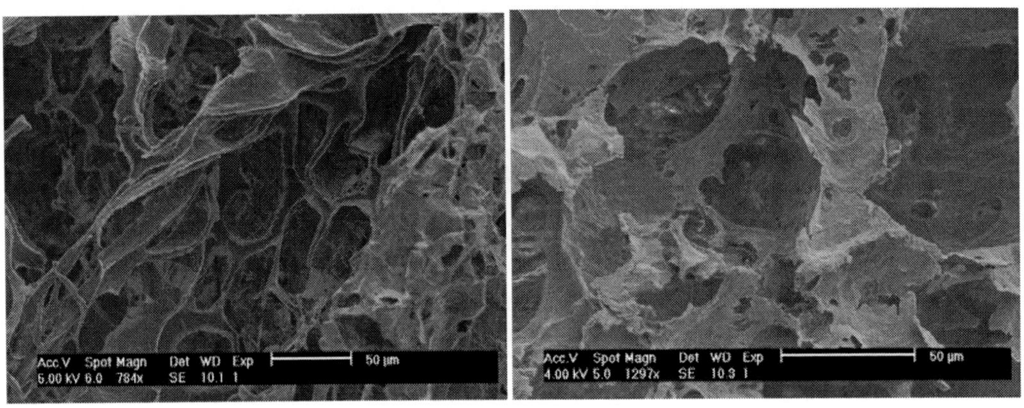

Figure 1. Examples of TIPS scaffolds prepared by previously published methods [6].

A challenge presented in the clinical application of biodegradable scaffolds is the triggering of an immune reaction upon implantation into the recipient's body. An acute response, such as inflammation in the immediate implantation site is expected, as with any surgical technique, however, it is the subsequent chronic inflammatory response which is highly undesirable. Medical devices, upon implantation, are invariably coated by blood and serum proteins which are commonly called "sticky" because they adsorb ubiquitously to all artificial surfaces exposed to the interior of the body [7-9]. An accumulative effect of this protein adsorption triggers a signalling pathway, stimulating monocyte migration from the bloodstream into the cavity surrounding the newly-coated foreign body. Further chemical signalling then initiates differentiation of these monocytes into terminal macrophages, the drivers of immune reactions. The foreign body response is characterised by amalgamation of macrophages to produce multinucleate giant cells, called foreign body giant cells (FBGCs) [10]. These FBGCs then proceed to encapsulate the foreign object and progressively wall it off from interaction with surrounding tissues. Consequent enhancement of this barricade is initiated through recruitment of fibroblasts, which lay down fibrous scar tissue. In the case of medical devices, this may result in operational and pathological complications. Biodegradable implants are particularly susceptible to this problem and are in fact prone to complications that significantly disrupt their functioning. The formation of a capsule has detrimental implications for neovascularisation in newly implanted scaffolds, which is crucial in stimulation of tissue growth and biodegradation product removal. Without proper regulation of these processes, the rate of biodegradation often becomes uncontrollable and can result in failure of the implant.

Surfaces in direct contact with serum can be modified so that the typical binding of proteins and subsequent macrophage transformation is not completely circumvented but attenuated to controllable levels. This effect can be achieved by introducing specific characteristics into implant surfaces. Anderson *et al* have demonstrated the efficacy of anionic and hydrophilic surfaces in achieving this goal [7,11]. While the specific chemical interactions involved have not been completely elucidated, evidence for correlation is compelling and suggests that surfaces with finely tuned characteristics may overcome the foreign body response altogether. Surface modification of biomaterials is a concept that has been extensively examined through wet chemical treatment [12], gamma radiation [13,14] and plasma modification [15,16]. The latter method has the advantages of simplicity, diversity, and specificity.

PLASMA

Plasma is by definition an ionized gas; it is formed when incident electrons from background radiation collide with excited particles. Primary excitation of these particles can be in the form of electromagnetic radiation, potential difference or high temperatures: the two resultant types of plasma that can be formed, depending on conditions employed, are equilibrium plasmas (where all species are at the same temperature) or non-equilibrium plasma (where electrons have a much higher temperature than other species in the plasma). For biomaterial surface modification purposes, non-equilibrium plasma is preferred because of the moderate conditions which are employed for generation of such plasma, as well as the controllability of parameters during treatment. Glow discharge is a characteristic non-equilibrium plasma – electrons in a typical glow discharge exhibit temperatures in the vicinity of 23,200K, whereas protons and other species in the equivalent matter typically measure less than 500K [17]. Despite the species being at greatly varying temperatures, all species within non-equilibrium plasma are considered reactive. The stark difference in species energy means that electron dissociation is the predominant trigger for atomic or molecular breakdown. The free nature of electrons endows electrical properties on the excited gas, usually materialised in the form of accelerated ions and zones of charge separation.

Specifically, the reactivity of plasma species is endowed by three primary forms of reactive species. Of chief importance is the effect of accelerated ions, produced by spatial charge separation in RF plasma and represents the most intense effect. The second reactive effect is that of radicals. Excited molecular gas, for example diatomic oxygen, undergoes homolytic fission in the presence of high-powered radio frequencies to yield reactive atomic species or radicals. These are not accelerated due to their lack of charge but are able to initiate fission of polymer chains by propagating homolytic fission on the polymer surface. This leads to creeping polymer chain scission and cross-linking and is long-lived, often continuing long after RF power has been disconnected. Finally, vacuum near-UV (VUV) radiation, usually in the range of 110 to 400 nm has been implicated in initiation and promotion of radical formation [18].

The plasma reactor used in this work, figure 2, was built in-house [19]. RF glow discharges, unlike DC glow discharges, do not have defined boundaries. A simplified structure breakdown separates the two active portions of this glow into bulk plasma and sheath regions [20]. Region C (figure 2) most closely resembles the negative glow seen in DC glow discharges. The DC glow discharge is generated between two plates of opposite charges, resulting in high electromagnetic frequencies which cause a breakdown of gas into the plasma state between the plates. The negative glow in these circumstances occurs near the cathode region (not in direct contact with the cathode) and is the brightest region within the entire glow. The physical composition of the negative glow is usually highly excited particles that, for the most part, do not propagate ionisation by collision. The RF energy (or potential difference in the case of DC discharges) is responsible for exciting valence electrons and causing secondary electrons to shift quanta in order to stabilise the ionised atoms. True plasma self-propagates the ionisation through collision only, thus the negative glow region is not truly in the plasma state even though it should generate the equivalent effects.

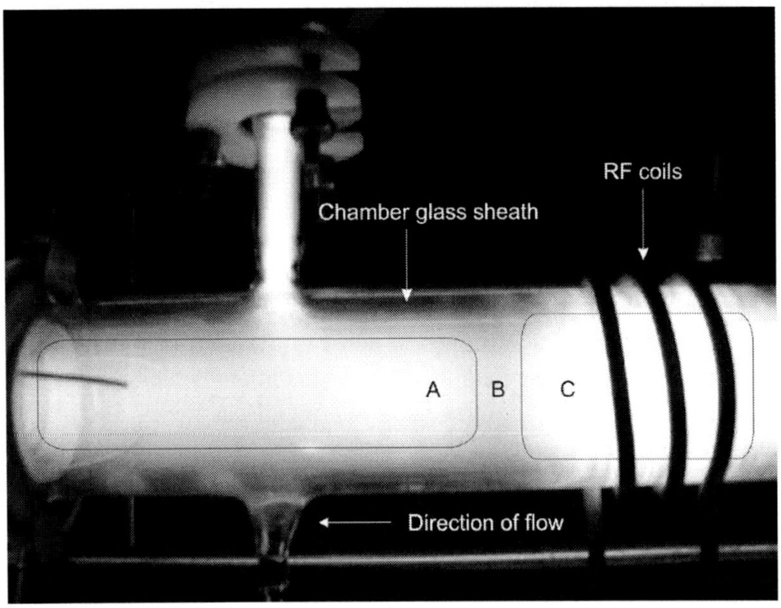

Figure 2. RF chamber used in the current study. Copper RF coils (right) energising oxygen in the chamber (evacuated to 150 mTorr). The ø45 mm internal diameter treatment chamber measured approximately 50 cm in length, giving an approximate plasma volume of 80 cm3. The tubular glass chamber had an external copper electrode which was connected to a radio frequency (RF, 27 MHz) generator which supplied the plasma power.

Region B (figure 2) is an interfacial zone between the negative glow of region C and region A, where electrons emerging from the region of most intense RF power around RF coils have not yet reached equilibrium with the oscillating RF field. This renders them unable to create secondary ions on collision with gas particles in the region. Region A (figure 2) resembles the positive column of DC discharges, having the classic characteristics of plasma, in which reactive species have reached equilibrium with the RF field and propagate the plasma state without interference from the main RF source, displaying no net electrical field. In the current study, samples were loaded into the area of region A and therefore will have encountered true plasma effects.

PLASMA MODIFICATION OF POLYMERS

The properties of RF plasma have been used to functionalise surfaces of materials in many industries, with interest in the modification of biomaterials emerging in the mid 1990s with the modification of fluoropolymers [21-24]. The potential to introduce very specific moieties into the surfaces of biopolymers has since been exploited by many groups with the intention of elucidating a connection between surface characteristics and improvements in the physical, chemical and biological properties of polymers (table 1).

Table 1. Examples of gas plasma treatment of polyesters for use in tissue engineering

Gas	Polymer	Surface Characterisation	Biological assays	Reference
Oxygen	PLGA	Contact angle, XPS, SEM, AFM	In vitro 3T3 fibroblast	[16]
		XPS, AFM, SEM	In vitro mineralisation (SBF)	[25]
	PLA	Contact angle, SEM	In vitro human embroyonic palatal mesenchyme cells	[15]
	PHBV	Content, water contact angle, XPS, SEM	none	[26]
		none	In vitro stromal osteoblastic cells	[27]
Ammonia	PLGA, PLA	Contact angle, XPS	In vitro human skin fibroblast	[28]
	PHB	Contact angle, XPS, AFM, FTIR	none	[29]
	PHBV	Contact angle, XPS, Raman	none	[30]
Corona discharge	PLGA	Contact angle, XPS	In vitro 3T3 fibroblast	[31]
Argon	PLA	Contact angle, XPS	none	[32]

PHB: Poly(3-hydroxybutyrate), PHBV: Poly(3-hydroxybutyrate-*co*-3-hydroxyvalerate), PLA: Poly(DL-lactic acid), Poly(L-lactic acid), PLGA: Poly(lactic-*co*-glycolic acid), XPS: x-ray photoelectron spectroscopy, SEM: scanning electron microscopy, AFM: atomic force microscopy, FTIR: Fourier transform infrared.

There are many parameters that affect the surface chemistry generated by plasma treatment of polymers including plasma variables (e.g. time, power, gas pressure), the gas type and the nature of the polymer [33]. The use of an inert gas (such as argon) to generate a plasma is known to cause mainly etching and crosslinking with no new functionalities introduced into the substrate, since the plasma state of this inert gas is predominantly in dynamic equilibrium between polarised and radicalised argon atoms. Because of the lack of charged particles there is no ionising effect, instead resulting in the polymer being homolytically attacked by the large excess of radicals, causing ablation and cross-linking of surface chains. On the other hand, the use of other gases such as oxygen, ammonia or nitrogen, will generally lead to the introduction of functional groups into the substrate surface. It should be noted, however, that a mixture of various functional groups are generally introduced using one simple gas, and the ratio between the different functionalities can be varied by varying the plasma parameters. In general, polymers rich in carbon and hydrogen are more susceptible to plasma treatment that polymers which contain carbon-oxygen or carbon-nitrogen moieties. This is due to the relative reactivity of carbon atom types, the most reactive are those bonded only to other carbon and hydrogen atoms.

Griesser *et al* modified commercially available fluoropolymers using oxygen, water vapour, argon and air plasmas in order to improve their biocompatibility [21]. Although they noted effective treatment by all of these plasmas, the effects were attenuated with time, until reaching a complete loss of treatment characteristics in most cases. Such effects of ion-treating of fluorinated polymer surfaces [21] have been explained in terms of surface rearrangement occurring in order to lower the net surface energy of the polymer. This effectively shields newly incorporated ionic groups, again returning the surface to the *status quo*. The effect has also been observed for polyesters with amine groups introduced using ammonia plasmas [30]. Inagake *et al.* [34] compared long term (30 days) aging effects in ammonia and oxygen plasma treated PET samples and found that ammonia treated samples reverted to the original surface composition whereas the oxygen plasma treated samples - although they did change- were still significantly different from the parent polymer after ageing.

Although a large body of work exists in the literature concerning plasma treatment effects on polymers, almost all of these have focussed on the net effects of plasma treatment for a single time point. In one case, this time point was suggested by the manufacturer of a biomaterial as being ideal for sterilisation without degradation of the substrate [35]. In other cases long term time points appear to have been selected, ranging from 1 to 40 minutes [15, 16, 25, 36] while extreme examples assess plasma treatment effects over periods of up to 10 hours [37]. Single time points below 1 min have been investigated [38, 39] and a few studies on PET have examined multiple time points below 1 min [34, 40].

Experimental Rationale

A detailed investigation into the effects of oxygen plasma surface modification is presented in this chapter. The effect of time is of primary importance if an understanding of true plasma effects is to be gained. Plasma treatment was thus limited to time points shorter than or equal to 60 seconds. Three parameters were assessed in this study: the effects of time, oxygen pressure, and post-treatment rinsing (accounting for possible interference from by-products that are produced over the period of treatment).

The current study investigated the modification of the commonly used FDA-approved, polymer PLGA (Birmingham Polymer (Chicago, IL)). In order to fully characterise the effects of varying plasma treatments, it is necessary to use a flat, uniform model substrate rather than a 3D scaffold. Solvent cast films have been employed to mimic the internal structures of PLGA scaffolds, however it has been suggested that these films -with thicknesses typically in the hundreds of micrometers- have physical characteristics of their own and do not mimic the internal pore structures of 3D TIPS scaffolds [12]. A more suitable method of dip-coating glass coverslips in PLGA solution to obtain uniformly thin (1-2μm) layers has been demonstrated in preparation for characterisation of wet chemical surface treatments [12, 41]. This 2D technique was employed in the current study and provided uniformity (and hence reproducibility) and used smaller amounts of PLGA in a more controllable form. Using data from treatment on these 2D surfaces, optimised treatments can be applied to 3D scaffolds made by the TIPS method.

ETCHING EFFECTS OF OXYGEN PLASMA

The etching effect of argon plasma was demonstrated by Inagaki *et al* in their treatment of polylactic acid (PLA) films [32]. On first examination post-treatment, the contact angle on a treated film was found to decrease significantly, indicating an increase in hydrophilicity. However upon washing sequentially in ethanol and acetone, it was found that the contact angle reverted to the original value, in effect returning the surface to the *status quo*. Nonetheless, the surface topography was greatly changed by inert gas plasma treatment, creating a uniform roughness, a surface feature that is known to affect cellular attachment [42]. In the plasma treatment of PET [34] it was observed that oxygen plasma displayed a significantly higher etching rate than hydrogen, nitrogen, argon and ammonium plasmas. However, the induced surface roughness was least pronounced for the oxygen plasma treated samples compared to other gas plasmas. The etching effect of oxygen plasma is well established and was recently studied by Qu *et al.* [25] who found that the surface roughness of PLGA (70/30) films increased with increasing treatment times (1 – 30 min).

One advantage of the PLGA thin film substrate used in the current study is the very smooth surfaces obtained. Thus, images of untreated PLGA surfaces obtained by AFM show smooth undulatory patterns which do not exceed (R_{max}) 5 nm in height (figure 3). With such smooth starting surfaces even very slight alterations to surface roughness that are undetectable by SEM can still be evaluated. Indeed, treated samples demonstrate significant pitting of the surface after 10 s of treatment at 150 mTorr of oxygen plasma. The mean amplitude of surface features relative to the centre plane, R_a, over at least 4 separate 15 μm^2 areas increased from <1 nm to 11±3 nm after treatment. This evidence of drastic topographical modification suggests significant bombardment of ions and radical attack of species within plasma and subsequent removal of degraded substances by rinsing with ethanol.

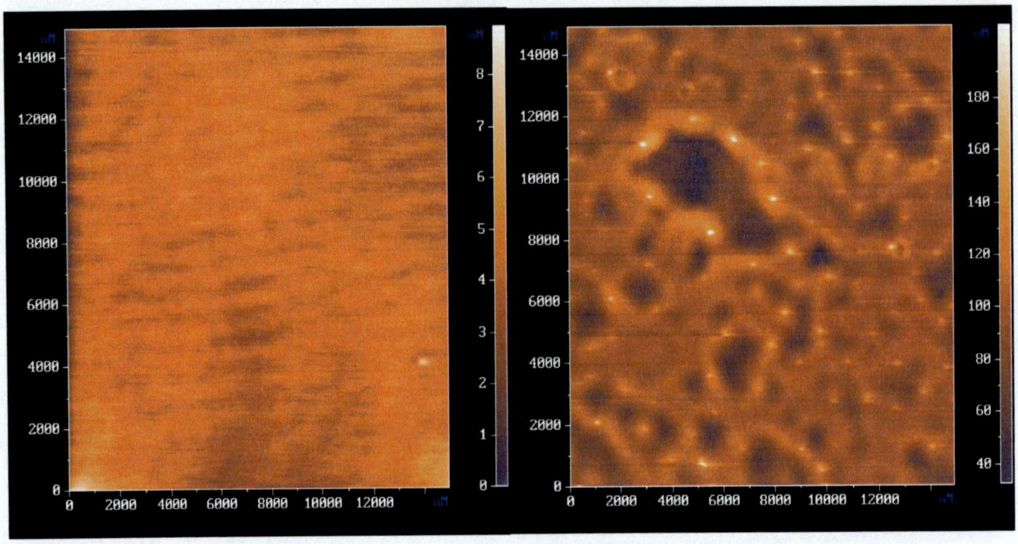

Figure 3. AFM images of (A) untreated PLGA thin film, R_a <1nm, and (B) PLGA thin film, treated with 10s of 150 mTorr O_2 plasma and rinsed, R_a=11 ± 3nm. AFM was recorded on an NT-MDT Solver P47 scanning probe microscope (NT-MDT Co., Zelenograd, Russia). Samples were scanned in semi-contact (tapping) mode with an NT-MDT NSG-11B "Golden" boron-doped silicon cantilever. A force constant of 5.5 N/m was derived and scanning was performed at a resonant frequency of 150kHz with an effective frequency of 116 kHz, over a nominal area of 15 x 15 μm. A minimum of four areas per surface were scanned, with images and R_a values generated from overall representative areas.

RF Oxygen Plasma Treatment of PLGA – Effect of Rinsing

Although etching effects of oxygen plasma have been observed by a number of research groups, no mention of washing procedures prior to substrate analysis is generally given and seems to be done only in the work by Inagaki [32, 34]. As demonstrated by Inagaki, the soluble oligomeric by-products of plasma treatment significantly decrease reliability of contact angle and XPS characterisations [32], which were the most commonly used techniques for surface characterisation in this field (table 1). In the present study, characterisation of plasma-treated polymer films was performed on both un-rinsed and rinsed (three times immersion in fresh ethanol) samples. It was found that the oxygen content did not change significantly for the 10 second treatment time compared to untreated samples and the O/C ratio was similar for 20 and 60 seconds of treatment. Samples treated for more than 10 seconds that were rinsed prior to surface characterisation by XPS demonstrated only a slight increase in the O/C ratio, whereas similar unrinsed samples displayed significant increases in O/C ratio (table 2). This indicates that although a net increase in oxygen content is observed in unrinsed samples, much of this is attributed to soluble oligomers which are subsequently removed by the washing steps.

High resolution C 1s spectra for an untreated sample and twp samples treated for 60 seconds (one rinsed and one unrinsed after treatment) are displayed in figure 4. The untreated sample displayed the three peaks characteristic for PLGA corresponding to the C-C, O-C-C(O)-O and O-C(O) chemical moieties. In addition, a small peak at 284.4 eV is evident in this sample and correlates to C-Si, attributed to the underlying silanised glass substrate (based on binding energy[43] and confirmed by a Si 2p peak in the XPS survey spectrum). The C 1s spectrum of the plasma treated but unrinsed sample displays two new peaks at 285.4 and 286.7 eV which are assigned to the terminal ester, C-C(O)-O, and the hydroxyl carbon, C-OH, respectively [43]. After rinsing, the C 1s spectrum of the plasma treated sample displays only one of these additional peaks; that of the terminal ester group.

Table 2. Characterisation of PLGA (50:50) samples exposed to O_2 plasma

Sample	Treatment time (s)	O/C atomic ratio ± 0.01	θ_A (°)	θ_R (°) ± 3	θ_A - θ_R (°)
Untreated	0	0.71	68± 2	64	4
Unrinsed (15 mTorr)	10	0.70	47± 4	31	16
	15	-	46± 4	36	10
	20	0.76	45± 4	37	8
	60	0.75	45± 4	36	9
Rinsed (15 mTorr)	10	0.70	60± 2	45	15
	15	-	60± 2	48	12
	20	0.73	59± 2	45	13
	60	0.72	58± 2	42	16

Contact angles were measured on a DataPhysics OCA-20 (Filderstadt, Germany) contact angle goniometer by the sessile drop method. Three drops were measured on each side, giving six readings per sample using the method outlined in Croll *et al* [12].

The atomic % obtained from curve fits to the C1s region is displayed in table 3. The untreated and the rinsed samples display the same relative ratio of the three main peaks. The value for the C-C moiety is slightly higher than the theoretical value which is attributed to fortuitous carbon (i.e. hydrocarbon impurities). In contrast, the un-rinsed sample displayed an increased atomic % of the O-C(O) peak relative to that of the O-C-C(O) peak. Sample wettability was found to increase significantly for samples that were left unrinsed whereas samples that were rinsed after treatment displayed no significant change in the advancing angle (table 2) indicating that the hydrophilic groups, e.g. C-OH, were situated mainly in the soluble oligomers in agreement with the XPS data. The receding angles, however, were significantly lower in all samples whether rinsing was performed or not.

Table 3. Atomic% of carbon species based on high resolution XPS data

Sample	C-Si (284.5 eV)	C-C (285.0 eV)	C-C(O)-O (285.6 eV)	C-OH (286.8 eV)	O-C-C(O) (287.1 eV)	O-C(O) (289.2 eV)
Untreated	6.1	24.0	-	-	34.4	35.5
60s unrinsed	-	23.4	3.5	8.9	27.4	36.9
60s rinsed	-	27.3	2.7	-	35.1	34.9
Theoretical	-	20	-	-	40	40

Rinsed samples demonstrated a large degree of hysteresis of 12 to 16° for all samples. This plasma-induced hysteresis is attributed to the surface etching observed for these samples. In contrast, the corresponding unrinsed samples displayed a difference between advancing and receding angles of no more than 10° at all time points except for 10 s. Lower advancing angles as well as lower hysteresis in un-rinsed samples may be attributed, as suggested by XPS data, to free oligomeric PLGA chains which become soluble in the water drop used to measure contact angles. These oligomers interfere with the surface tension of the liquid phase and affect the true wettability of the sample, as well as the fidelity of the reading itself. A greater degree of error was observed in the contact angles of unrinsed samples, confirming the latter point. Approximation of surface energy based on contact angles taken on un-rinsed samples is therefore unreliable and should be avoided.

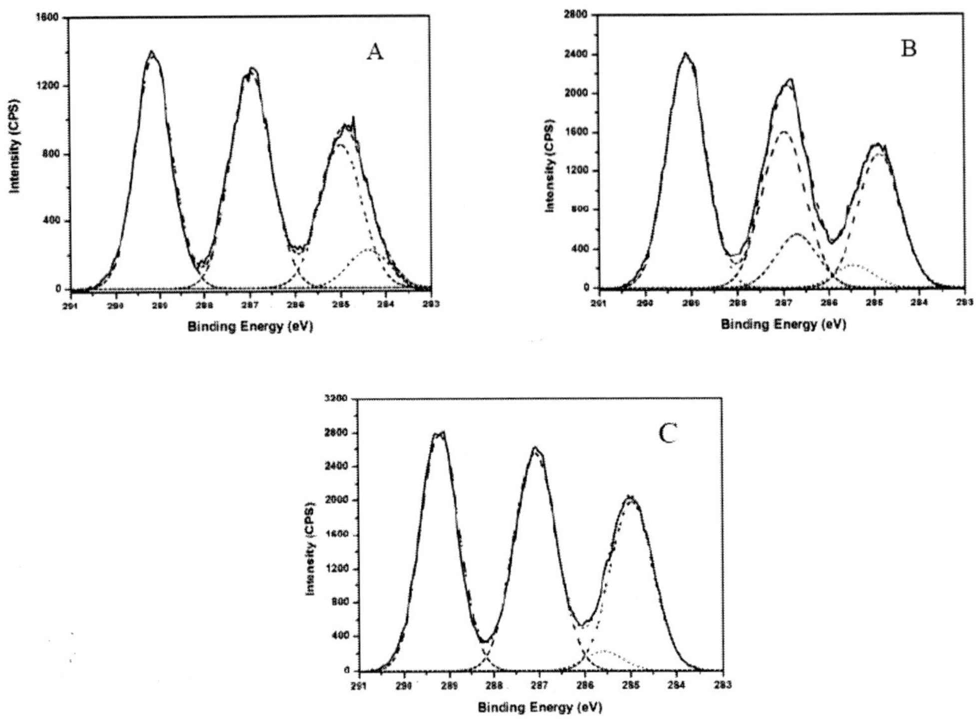

Figure 4. High resolution XPS spectra of the C 1s region of 50/50 PLGA on glass coverslip (A) untreated; (B) 60 s O_2 plasma at 15 mTorr, unrinsed, (C) 60 s O_2 plasma at 15 mTorr, rinsed. XPS was performed on an Axis Ultra XPS spectrometer (Kratos Analytical, UK). Single pass survey scans were conducted with 160 eV analyser pass energy on three randomly chosen spots on each coated coverslip sampled, using a monochromated Al Kα X-ray source operating at 15 kV and 10 mA. Binding energies were charge corrected against aliphatic C 1s at 285.0 eV, while peak widths were fixed at 1.0 for C 1s.

RF Oxygen Plasma Treatment – Effect of Gas Pressure

Generation of a glow discharge is dependent on the RF frequency endowing sufficient energy on gas molecules so as to overcome the first ionisation energy. This produces free electrons and ions. The electrons accelerate rapidly in the RF field, colliding elastically and inelastically with other particles. Inelastic collisions with other atoms may ionise those if the kinetic energy of the incident electron is greater than the ionisation energy of the atom. Otherwise, energy imparted on the atom causes excitation of valence electrons, producing radicals. While de-excitation of radicals occurs rapidly in most cases (releasing electromagnetic energy in wavelengths that produce characteristic glows), some radicals continue to react due to further excitation of the RF field. This effect is propagative and when the frequency of ionisation is low (due to fewer collisions in lower pressure plasmas), excited, rather than ionised species dominate. Thus, a direct correlation exists between pressure and the type of species which will dominate in the plasma. The current study compared the outcome at two different oxygen pressures; 15 mTorr (with a corresponding gas flow rate of 7.5 sccm) or 150 mTorr (corresponding gas flow rate of 75.0 sccm).

Comparing the data in tables 2 and 4 for two different PLGA substrates treated in an oxygen plasma at 15 mTorr and subjected to rinsing, the same trends in O/C ratios and contact angles are observed. Thus, no significant change in the O/C ratio or advancing contact angle are observed at this gas pressure, however, a significant reduction in the receding contact angle suggests surface etching as discussed above.

Table 4. Characterisation of PLGA (75:25) samples exposed to O_2 plasma

Sample	Treatment time (s)	O/C atomic ratio ± 0.01	θ_A (°) ± 3	θ_R (°) ± 3	$\theta_A - \theta_R$ (°) ± 3
Untreated	0	0.64	70	59	11
Rinsed (15 mTorr)	10	0.63	68	36	32
	15	0.64	66	38	28
	20	0.63	65	34	31
	60	0.63	68	40	28
Rinsed (150 mTorr)	10	0.65	70	44	26
	15	0.67	69	49	20
	20	0.66	70	50	20
	60	0.63	68	48	20

The effect of oxygen pressure was studied for the (75/25) PLGA polymer. At 150 mTorr, a small net increase in O/C ratio was observed at time points up to 20s (inclusive). Such enhanced oxygen levels were not seen at the lower gas pressure, neither was it observed at the 60 second time point (table 4). Clearly, subtle differences in dominating reactions, e.g. incorporation of oxygen-containing functionalities into the polymer backbone or crosslinking/chain scission, are achieved based on small changes in plasma parameters.

Although the advancing contact angles were unchanged after oxygen plasma treatment and subsequent rinsing, large changes were observed in the receding angles for all treatment times and pressures (table 4). Contact angle hysteresis was found to differ significantly between samples treated at the two pressures. The greater degree of hysteresis in samples treated at 15 mTorr suggests a connection between treatment pressure and an increase in surface roughening. This could be due to the inverse relationship between pressure and the path length of bombarding ions, which gain more momentum prior to colliding with the polymer surface. This has the potential to penetrate more deeply into the polymer surface than at lower pressures, allowing further chain scission by oxygen radicals.

Mechanistic Aspects of Oxygen Plasma Treatment

The plasma state generated in an RF oxygen plasma is very complex. It contains a variety of species including excited molecules, radicals, and ions. The different species all have the potential to interact with an exposed surface causing a large array of chemical reactions to take place simultaneously. While the oxygen plasma is largely ionising, there is a fair degree of radical formation in the plasma state [44] and reactions occurring in an argon plasma are likely to also occur in the oxygen plasma. Inagaki *et al* [32] suggest a degradation scheme for PLLA treatment with argon plasma, which has been modified for PLGA in Scheme 1.

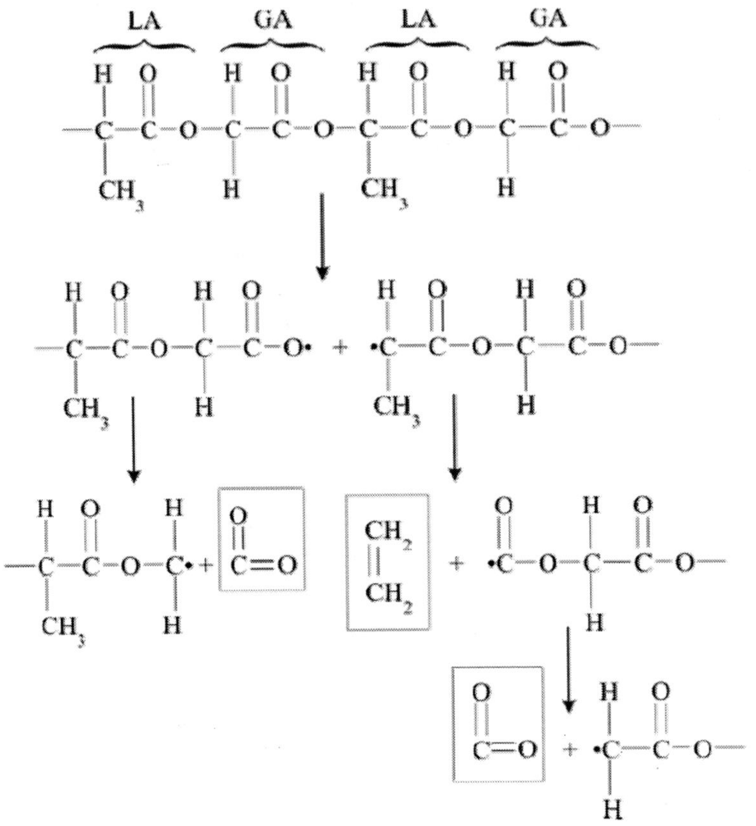

Scheme 1. Degradation of PLGA by radical propagation, producing carbon dioxide and ethylene as primary by-products.

The initial radiation induced scission of the C-O bonds in the ester group was proposed by Inagaki based on work on gamma irradiation induced chain scission. This scheme demonstrates the propensity of –COO· and –·CH–CH$_3$ (but not –·CH$_2$ as would be formed by glycolic acid in PLGA) to form good leaving groups (CO_2 and $H_2C{=}CH_2$). The resulting radical end groups have been observed by EPR in gamma irradiation treated PGA and PLGA [45]. Subsequently, these radical end groups can terminate through crosslinking or disproportionation to form carbon-based end groups such as the terminal ester observed in the C 1s high resolution XPS spectra.

Not all reactions of excited oxygen species in plasma cause chain scission. Hydrogen abstraction from the PLGA backbone or side group of LA can be facilitated by initiating oxygen free radicals (Scheme 2A) [46]. Subsequent addition of oxygen can take place by reaction with either oxygen free radicals (Scheme 2B) or molecular oxygen. In this manner, C-OH groups, observed in the C 1s high resolution spectra, may be introduced.

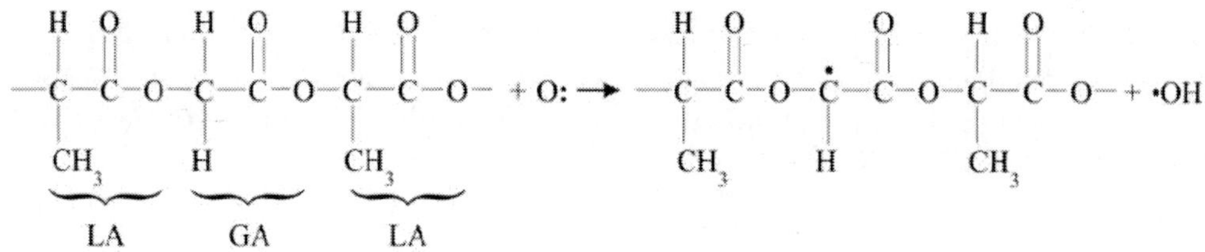

Scheme 2A. Non-scissional activation (hydrogen abstraction) of PLGA by reaction with an atomic oxygen free radical.

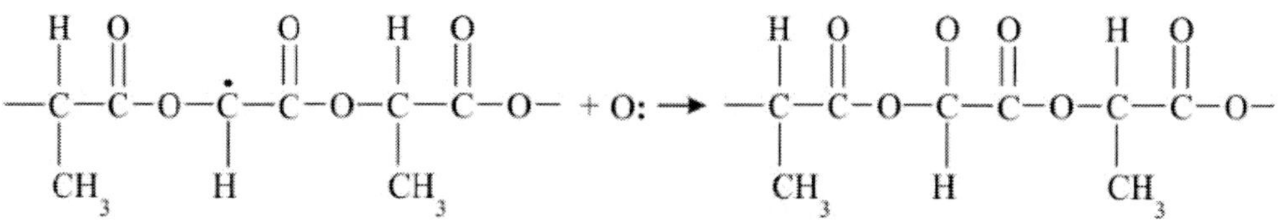

Scheme 2B. Single step addition of oxygen to activated PLGA by reaction with with an atomic oxygen free radical.

IMPLICATIONS OF USING OXYGEN PLASMA FOR TISSUE ENGINEERING APPLICATIONS

Surface modification of materials for tissue engineering applications is an active research field. It has been recognised that it is difficult to change a single surface property (e.g. wettability, chemistry, or topography) without simultaneously changing others. Identification of the specific surface characteristics responsible for influencing biological response therefore becomes challenging. In addition, a synergistic effect of surface morphology and chemistry on bone cells grown has been shown in a study by Chesmel *et al.* [47], thus increasing this complexity. The dual effect of oxygen plasma treatment of PLGA (etching as well as introducing new chemistry) is an example of the complexity encountered in surface modification of materials. In reports of improved biological response to oxygen plasma treated polymers it is thus not possible to assess its cause; it can be due to changed surface chemistry, surface topography or due to a combination of both these effects. An added level of complexity for oxygen plasma treated samples used in biological assays is the issue of rinsing away soluble oligomers prior to biological testing. The lack of reporting of any rinsing processes of oxygen plasma treated materials used in biological assays makes it very difficult to interpret results obtained, as unpredictable side effects may arise from the presence of soluble oligomers.

A key issue when modifying the surface properties of 3D tissue engineering scaffolds is to achieve modification throughout all the pore surfaces within a scaffold, and not only on the outermost exposed surface. Few studies exist in the literature where convincing evidence is presented to demonstrate successful surface modification within a porous scaffold. One such study involves hydrolysis and aminolysis of PLGA TIPS scaffolds [41]. A crucial aspect of using wet chemical treatments is to be able to modify only the surface without altering the bulk properties which has been successfully demonstrated for these PLGA scaffolds. The use of ammonia gas plasma to modify small (thickness 4.4 mm) solvent cast/particulate leached scaffolds has been demonstrated to be successful throughout the scaffolds using a simple dye assay [48]. A recent study used plasma polymerisation of allylamine in PLGA gas-foamed scaffolds (thickness of a few mm) [49] and, using XPS, modification was shown to be throughout. However, the outmost surface was modified more effectively than the internal pores by this procedure. The lack of easily detectable changes in the surface properties after oxygen plasma treatment makes it more difficult to assess the penetration depth of this plasma into the pores of 3D scaffolds. Improved cell penetration into the interior of oxygen plasma treated scaffolds has been demonstrated (scaffold dimensions not given) [15, 27] and can be viewed as indirect proof that indeed oxygen plasma treatment is a viable method for surface modifying 3D scaffolds.

Conclusion

This chapter has demonstrated a number of important issues associated with the use of oxygen plasma treatment for tissue engineering applications. Oxygen plasma offers great potential with respect to increasing surface hydrophilicity and surface topography modification, though much work is needed to fine-tune parameters yielding controllable outcomes and the ability to achieve uniform modification throughout 3D scaffolds. An important aspect of producing scaffolds for tissue engineering applications is that of supplying fully characterised samples. In that regard, it is of utmost importance that a rinsing step be introduced as part of the fabrication process in order to eliminate undesirable side effects. Data presented in this chapter represent the first detailed study into the effect of rinsing oxygen plasma-treated samples; this study is therefore also the only thorough study into changes in surface properties caused by oxygen plasma of PLGA.

Acknowledgements

The authors would likc to thank the Dr Barry Wood (Brisbane Surface Analysis Centre, UQ) for assistance with obtaining XPS data; and Dr Thor Bostrom (Analytical Electron Microscopy Facility, Queensland University of Technology, Brisbane) for AFM measurements.

References

[1] Lysaght, M.J.; Hazlehurst, A.L. *Tissue Engineering* 2004, 10, 309-320.

[2] Holy, C.E.; Shoichet, M.S.; Davies, J.E. *J. Biomed. Mater. Res.* 2000, 51, 376-382.

[3] Papadaki, M. IEEE engineering in medicine and biology magazine 2004, 95-98.

[4] Hutmacher, D. W. *Biomaterials* 2000, 21, 2529-2543.

[5] Pavlov, M. P.; Mano, J. F.; Neves, N. M.; Reis, R. L. *Macromol. Bioscience* 2004, 4, 776-784.

[6] Cao, Y.; Croll, T. I.; Lees, J. G.; Tuch, B. E.; Cooper-White, J. J. *Aust. J. Chem.* 2005, 58, 691-703.

[7] Anderson, J.M.; Defife, K.; Mcnally, A.; Collier, T.; Jenney, C. *J. Mater. Sci. Mater. Med.* 1999, 10, 579-588.

[8] Azeez, A.; Yun, J.; Defife, K.; Colton, E.; Cahallan, L.; Verhoeven, M.; Cahallan, P.; Anderson, J.M.; Hiltner, A. *J. Appl. Polym. Sci.* 1995, 58, 1741-1749.

[9] Tang, L.P.; Jennings, T.A.; Eaton, J.W. *Proc. Nat. Acad. Sci.* 1998, 95, 8841-8846.

[10] Anderson, J. *Biomaterials science: an introduction to materials in medicine*, Ratner, B.D.; Ed.; 2004, Elsevier Academic Press: Amsterdam ; Boston. pp 851.

[11] Brodbeck, W.G.; Voskerician, G.; Ziats, N.P.; Nakayama, Y.; Matsuda, T.; Anderson, J.M. *J. Biom. Mater. Res.* 2003. 64A, 320-329.

[12] Croll, T.I.; O'Connor, A.J.; Stevens, G.W; Cooper-White, J.J. *Biomacromolecules* 2004, 5, 463-473.

[13] Martinez-Sancho, C.; Herrero-Vanrell, R.; Negro, S. *J. Cont. Rel.* 2004, 99, 41-52.

[14] Grøndahl, L.; Chandler-Temple, A.; Trau, M. *Biomacromolecules* 2005, 6, 2197-2203.

[15] Chim, H.; Ong, J.L.; Schantz, J.T.; Hutmacher, D.W.; Agrawal, C.M. *J. Biomed. Mater. Res.* 2003, 65A, 327-335.

[16] Wan, Y.; Qu, X.; Lu, J.; Zhu, C.; Wan, L.; Yang, J.; Bei, J.; Wang, S *Biomaterials* 2004, 25, 4777-4783.

[17] Chapman, B.N. *Glow discharge processes: sputtering and plasma etching*. 1980, New York; Chichester; Brisbane: Wiley. XV, pp 406.

[18] Friedrich, J.; Unger, W.; Lippitz, A.; Koprinarov, I.; Ghode, A.; Geng, S.; Kuhn, G. *Comp. Interfaces* 2003, 10, 139-171.

[19] George G.A.; Cash, G.A.; Le T.L.; Goss B.G.S.; Wood B.J.; Brown J.R.; St. John N.A. *Polym. Adv. Tech.* 1996, 7, 343-355.

[20] Makabe, T.; Nakano, N.; Yamaguchi, Y. *Physical Rev. A* 1992, 45, 2520-2531.

[21] Griesser, H.J.; Chatelier, R.C.; Gengenbach, T.R.; Vasic, Z.R.; Johnson, G.; Steele, J.G. *Polym. Int.* 1992, 27, 109-117.

[22] Kang, E.T.; Tan, K.L.; Kato, K.; Uyama, Y.; Ikada, Y. *Macromolecules* 1996, 29, 6872-6879.

[23] Xie, X.M.; Gengenbach, T.R.; Griesser, H.J. *J. Adhesion Sci. Tech.* 1992, 6, 1411-1431.

[24] Yun, J.K.; Defife, K.; Colton, E.; Stack, S.; Azeez, A.; Cahalan, L.; Verhoeven, M.; Cahalan, P.; Anderson, J.M. *J. Biomed. Mater. Res.* 1995. 29, 257-268.

[25] Qu, X.; Cui, W.; Yang, F.; Min, C.; Shen, H.; Bei, J.; Wang, S. *Biomaterials* 2007, 28, 9-18.

[26] Hasirci, V.; Tezcaner, A.; Hasirci, N.; Suzer, S. *J. Appl. Polym. Sci.* 2003, 87, 1285-1289.

[27] Kose, G.T.; Korkusuz, F.; Korkusuz, P.; Purali, N.; Ozkul, A.; Hasirci, V. *Biomaterials* 2003, 24, 4999.

[28] Yang, J.; Shi, G.X.; Bei, J.Z.; Wang, S.G.; Cao, Y.L.; Shang, Q.X.; Yang, G.G.; Wang, W.J. *J. Biomed. Mater. Res.* 2002, 62, 438-446.

[29] Nitschke, M.; Schmack, G.; Janke, A.; Simon, F.; Pleul, D.; Werner, C. *J. Biomed. Mater. Res.* 2002, 59, 632-638.

[30] Keen, I.; Broota, P.; Rintoul, L.; Fredericks, P.; Trau, M.; Grøndahl, L *Biomacromolecules* 2006, 7, 427-434.

[31] Khang, G.; Rhee, J.M.; Lee, J.H.; Lee, I.; Lee, H.B. *Korea Polym. J.* 2000, 8, 276-284.

[32] Inagaki, N.; Narushima, K.; Tsutsui, Y.; Ohyama, Y. *J. Adhesion Sci. Tech.* 2002, 16, 1041-1054
[33] Grace, J.M.; Gerenser, L.J. *J. Dispersion Sci. Tech.* 2003, 24, 305-341.
[34] Inagaki, N.; Narushim, K.; Tuchida, N.; Miyazaki, K. *J. Appl. Polym. Sci Part B: Polym Phys* 2004, 42, 3727-3740.
[35] Holy, C.E.; Cheng, C.; Davies, J.E.; Shoichet, M.S. *Biomaterials* 2001, 22, 25-31.
[36] Yang, J.; Bei, J.Z.; Wang, S.G. *Biomaterials* 2002, 23, 2607-2614.
[37] Friedrich, J.; Loeschcke, I.; Frommelt, H.; Reiner, H.D.; Zimmermann, H.; Lutgen, P. *Polym. Degrad. Stab.* 1991, 31, 97-114.
[38] Kang, I.K.; Choi, S.H.; Shin, D.S.; Yoon, S.C. *Int. J. Biol. Macromolecules* 2001; 28, 205-212.
[39] Khang, G.; Jeon, J.H.; Lee, J.W.; Cho, S.C.; Lee, H.B. *Bio-Med. Mater. Engineering* 1997, 7, 357-368.
[40] Cioffi, M.O.H.; Voorwals, H.J.C.; Mota, R.P. *Mater. Charac.* 2003, 50, 209-215.
[41] Cao, Y.; Croll, T.I.; Cooper-White, J.J.; O'Connor, A.J.; Stevens, G.W. *Production and Surface Modification of Polylactide-Based Polymeric Scaffolds for Soft-Tissue Engineering*, in *Methods in Molecular Biology*, A.P. Hollander and P.V. Hatton, Editors. 2004, Humana Press Inc.: Totowa, NJ. pp 87-111.
[42] Thissen, H.,; Hayes, J.P.; Kingshott, P.; Johnson, G.; Harvey, E.C.; Griesser, H.J. *Smart Materials and Structures* 2002, 11, 792-799.
[43] Beamson G.; Briggs D. *High resolution XPS of organic polymers*; The Scienta ESCA300 Database, Wiley, 1992.
[44] Shohet, J.L. *The plasma state*. 1971, New York: Academic Press. pp 333.
[45] Babanalbandi, A.; Hill, D.J.T.; O'Donnell, J.H.; Pomery, P.J. *Polym. Degrad. Stab.* 1996, 52, 59-66
[46] Egitto, F.D. *Pure Appl. Chem.* 1990, 62, 1699-1708.
[47] Chesmel, K.D.; Clark, C.C.; Brighton, C.T.; Black. J. *J. Biomed. Mater. Res.* 1995, 29, 1101-1110.
[48] Yang, J.; Wan, Y.Q.; Yang, J.L.; Bei, J.Z.; Wang, S.G. *J. Biomed. Mater. Res.* 2003, 67A, 1139-1147
[49] Barry, J.J.A.; Silva, M.M.C.G.; Shakensheff, K.M.; Howdle, S.M.; Alexander, M.R. *Adv. Func. Mater.* 2005, 15, 1134-1140.

In: Tissue Engineering: Roles, Materials and Applications ISBN: 978-1-60456-293-4
Editors: S. J. Barnes, L. P. Harris, pp. 275-285

Chapter 11

ALTERNATIVE METHODS IN CARDIOVASCULAR TISSUE ENGINEERING

Geofrey De Visscher and Willem Flameng
Laboratory for Experimental Cardiac Surgery,
Katholieke Universiteit Leuven
Minderbroedersstraat 17, 3000 Leuven, Belgium

ABSTRACT

The current review will address alternative methods for the tissue engineering of blood vessels and heart valves. This is in contrast with most current reports on cardiovascular tissue engineering (CVTE), which are generally limited to one specific paradigm or certain aspects thereof. The paradigm in question described the creating of autologous tissue, by in vitro seeding of mainly synthetic scaffolds followed by a phase of *in vitro* maturation, finally resulting in an implantable construct suitable for only one specific patient. Although this approach is very promising it has suffered from several draw-backs, such as costs, instigating several groups to search for alternative approaches. Mainly aiming at the creation if living tissue by *in vivo* approaches and thereby reducing the costs and risks associated to the *ex vivo* treatment of cells, these approaches have already proven the ability to create living blood vessels and blood vessels by exploiting and guiding naturally occurring processes, such as foreign body reaction and wound healing. The techniques to obtain such living prostheses can be based on intraperitoneal implantation and seeding, as was shown by the group of Campbell for blood vessels [1] and by our group for heart valves [2]. In both cases the methods have been developed and evaluated *in vivo*. A second alternative are the smart matrices. These scaffolds are designed to enhance *in situ* recellularisation [3,4] but have some implications in the *in vitro* tissue engineering as well.

INTRODUCTION

Tissue engineering in general and cardiovascular tissue engineering in particular has been appointed the most promising solution to current shortcomings of medical replacements of tissues and organs. The concept tissue engineering itself, coined in 1985 by Fung Y.C. [5], was later defined as "an interdisciplinary field that applies the principles of engineering and life sciences toward the development of biological substitutes that restore, maintain, or improve tissue function or a whole organ" [6].

A literature search on cardiovascular tissue engineering showed a total of 1218 papers, reviews included, from 1989 till 10 October 2007. Following the initial phase of a Gartner's Hype Cycle [7], the first publication was slowly followed with an expansion of the field which seems to be topping of or even declining (figure 1A). Blood vessel tissue engineering clearly shows the same pattern (figure 1B). Similarly, an initial report in 1995 by Shinoka and co-workers [8] has spurred several groups to participate in the field of heart valve engineering. Entering the term heart valve tissue engineering in the literature search yielded 299 papers published between 1995 and 10 October 2007. These results (figure 1C), clearly illustrate the massive growth of the field after this first report in 1995.

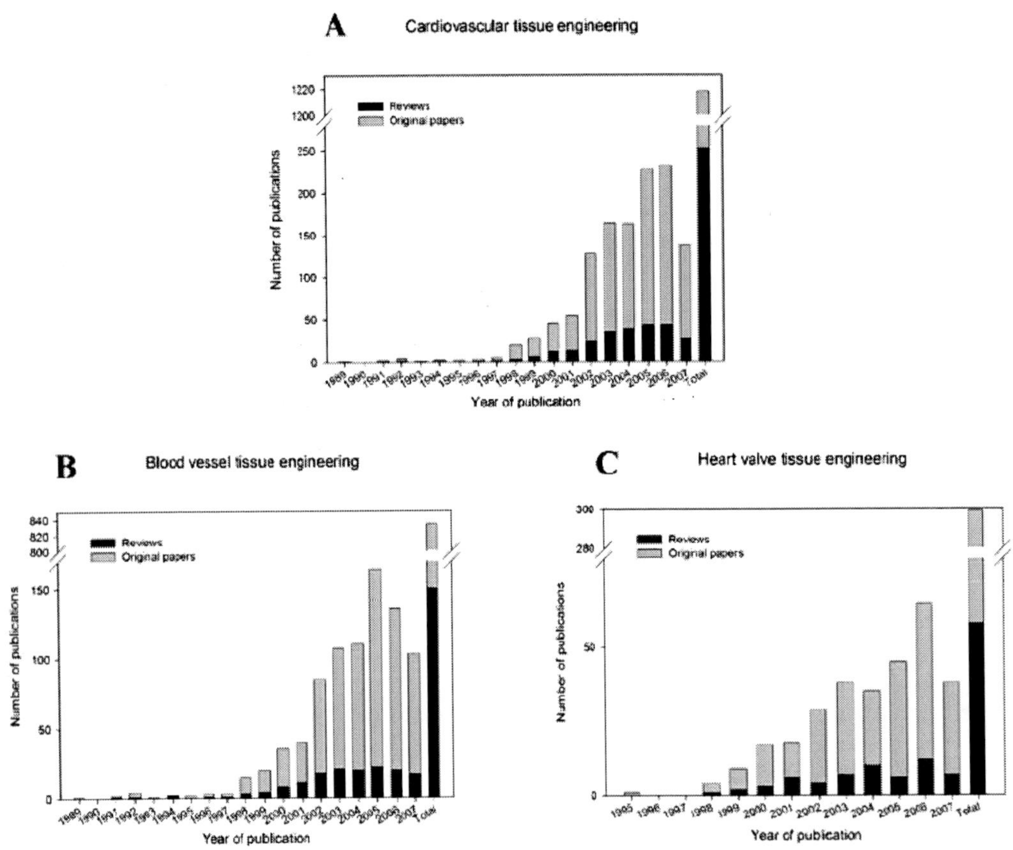

Figure 1. Publications track record on cardiovascular (A), blood vessel (B) and heart valve (C) tissue engineering from the first paper till 10 October 2007.

When looking at the tissue engineering concept some general principles apply and they have already been nicely summarized [9]. However for the case of clarity some of these need to be recapitulated. As described in the review, tissue engineering efforts can be classified into groups according to the paradigm they use. The "complete paradigm", used by most tissue engineering groups is comprised out of 4 components: (1) cells seeded on (2) scaffolds are (3) matured *in vitro* in a bioreactor and the (4) matured construct is implanted as a prosthesis and subjected to *in vivo* remodeling. The major example for this type of tissue engineering is the work of Hoerstrup and coworkers [10]. From the general description 3 applicable "incomplete paradigms" could be derived. Closest to the complete paradigm is the cell seeded scaffold approach, where a matrix is seeded with cells and immediately implanted avoiding maturation in a bioreactor. Using this method Dohmen and coworkers have first studied en endothelial cell seeded heart valve in sheep and subsequently [11] applied the method in the clinic [12]. Another incomplete paradigm is the cell transplant method. Only using cells this method is often used in *in situ* myocardium tissue engineering, where preharvested cells, e.g. stem cells, are injected either in the infarcted myocardium itself [13] or in the coronary artery [14]. Similar approaches have also been reported for heart valve prostheses [15]. The final paradigm is the scaffold that attracts endogenous cells in situ. It is exactly this field where amongst others our group is active [16].

Although remaining as promising as initially stated, the current heart valve and blood vessel tissue engineering solutions according to the full paradigm suffer from their own shortcomings as well. However, this is beyond the scope of this review, but briefly it is extremely expensive, no real applicable solution has been obtained yet and recent results have shown that even cultured syngenic engineered heart tissue can be completely absorbed by the recipients' immune response [17]

History of Alternative Cardiovascular Tissue Engineering

Despite that the origin of tissue engineering is often placed in the 80's, several attempts predate this. Particularly in the field in blood vessels, tissue engineering has a long history. Going back in time, one often discovers the pioneering work of Alexis Carrel, a French surgeon and Nobel prize winner. The research he has done in the mid-30's together with Charles A. Lindbergh resulting in the book "The Culture of Organs" is seen as a precursor of tissue engineering [5]. More critically, however, one could conclude that the book describes the development of a bioreactor rather than the development of engineered tissue [18].

From the 50's onwards synthetic tubes as a replacement of arterial prostheses emerge [19], but it is only at the beginning of the next decade that the first real tissue engineered grafts are reported. Early pioneering work by Eiken [20] and Schilling and coworkers [21,22] in dogs, all based in connective tissue tubes, never resulted in clinical trials or application.

In 1969, however, Dr. Charles H. Sparks reported the evaluation of a "Die-grown reinforced arterial graft" in 48 dogs and 2 patients [23]. The graft was the first to intentionally combine a matrix, a non biodegradable knitted Dacron® tube, with the dog's or patient's own tissue. In order to achieve this combination the tube was suspended between an internal mandril and a perforated ingrowth tube 0.034 to 0.036 inch apart. These construct were implanted, e.g. subcutaneously [24], to induce a foreign body reaction resulting in the deposition of granulation tissue in between the mandril and the ingrowth tube. This process took five to eight weeks and after removal of the mandril and ingrowth tube, the resulting Dacron reinforced granulation tissue tube was used as a vascular graft: aorta or iliac arteries in dogs and a femoro-popliteral artery or femoro-femoro bypass graft in the patients. Soon thereafter more clinical results showed the technique to be sufficiently encouraging [25], however, the long term follow up proved to be devastating for this technique. After Dr. Sparks death in 1973, two of his co-workers Hallin and Sweetman followed all of his and his associates patients for up to seven years [26]. In total seventy three grafts were made of which thirteen could not be used because of inappropriate maturation or inflammation. The remaining sixty were implanted in fifty four patients of whom eight showed initial failure. Overall, the complications observed were (1) difficulties with the maturation, (2) initial thrombosis, (3) late thrombosis, (4) bleeding, (5) aneurism formation, (6) anastomotic aneurism and (7) infection. These findings were confirmed the next year, when Roberts and Hopkinson reported a 2 year follow up in which only two grafts remained patent [27].

Although the Sparks mandril technique in itself was given up, it is exactly this work that lies at the origin of a very typical or alternative branch of cardiovascular tissue engineering.

Current Alternative Cardiovascular Tissue Engineering

In this section we will review several cardiovascular tissue engineering solutions based on the aforementioned incomplete paradigm. All these efforts have one common denominator, that is, they all use no or rather the recipients' body as a bioreactor.

An immediate derivative of the Sparks mandril technique is the work done by the group of Prof. Julie H. Campbell. After a hiatus of 22 years, a first report on an adapted granulation tissue based vascular graft was published [1]. Instead of using skin granulation tissue of the original Sparks mandril, they implanted silastic tubing into the peritoneal cavity of rats and rabbits. This resulted in significant adaptations of the original technique: (1) intraperitoneal granulation tissue is thicker [28], (2) no Dacron mesh was used for support and (3) a mesothelial layer somewhat similar to the endothelium of normal vessels lined these vessels. After removing the silastic tubing these grafts were everted to internalize the mesothelial layer and successfully implanted in the rats' abdominal aorta or the rabbits' carotid artery. The obtained patency rates were 67% and 70% respectively. Clear evidence was provided that the mesothelial lining showed some protection against thrombosis in these small caliber grafts (3 and 5 mm inner ∅, respectively) and the grafts became vasoactive as well. Further evaluating these grafts, tubing were implanted in the peritoneal cavity of the dog, but also in the pleural cavity a more appropriate site if one wants to make coronary artery grafts [29]. The obtained connective tissue tubes had bursting strengths comparable to the dog's femoral artery and also had sufficient suture holding strength. Of the eleven implanted tubes eight (73%) remained patent at six months. With respect to the cell phenotypes, the grafts developed blood vessel properties. However 1 out of 10 cells still stained positive for the panleucocytic marker CD45. Overall, the obtained results were positive enough to start commercialisation and at present clinical trials are being prepared (www.vascam.com).

Using the body as its own bioreactor in exactly the same way as described above, Hayashida and co-workers made a heart valve prosthesis [29]. This group constructed a three dimensional structure out of silicone and polyurethane that was implanted subcutaneously in rabbits for 4 weeks. Upon retrieval the silicone parts of the mold are removed, but the microporous polyurethane remains in place. Generally, the aspects of these valves and the leaflets, in particular, were very good. Nevertheless, it remains to be determined how they behave in the circulation.

Very soon after their initial paper, Campbell's group recognized the haematopoietic origin of the myofibroblast population in the deposited granulation tissue [30,31], and concluded that the most probable origin of these cells were the peritoneal macrophages. This was in accordance with their findings in neointima deposition after coronary artery injury [32]. At this time it has already been recognized that macrophages could transdifferentiate into myofibroblasts [30,33,34,35], but it has also been established that the bone marrow derived stromal cell can acquire the same phenotype [36,37].

This is where our group recognized another possibility. Although haematopoietic or bone marrow derived cells clearly played a role in the process of foreign body reaction, their role has not been elucidated yet. However, the early phases of this reaction had not been studied, although they provide a very interesting natural defence system able to recruit primitive cells to a very specific site. Initially it was proven that short, i.e. 2 to 3 days, implantation of scaffolds into the peritoneal cavity were optimal for obtaining primitive cells [38]. These cells could also be extracted from the general population and could be differentiated along the osteo-, adipo- and myogenic differentiation pathways [38]. Therefore, this seeding protocol was applied for heart valve prostheses [39,2]. Patches of acellular matrix material were seeded by implanting them in the abdominal cavity of sheep for three days. The matrix seeded with cells comprising amongst the immune cells a seven to ten percent primitive cell fraction were converted into valves and immediately implanted as a pulmonary valve in the same sheep. This protocol succeeded in full recellularisation of the valve but an excess in contractile cells was also observed. While the valves were all functional, with nice leaflet motion at 5 months, they also showed an increase in the proportion of regurgitation, albeit very mild. As a matter of fact the contractile cells had remodeled the valves thus far that they mildly disturbed the coaptation of the leaflets. Longer leaflets would have solved the problem, but it is our opinion that altering the composition of the cell population is a better and safer solution [39].

Another interesting procedure using the body as a bioreactor is the development of cardiac tissue by the group of Morritt [40]. In this set up neonatal rat ventricular myocytes were harvested according to standard tissue engineering procedures, but the cells, suspended in matrigel, were then seeded in a vascularized bioreactor chamber implanted in the groin of a rat. Ten weeks later the matrigel-cell mixture developed into a vascularised tissue showing spontaneous contractile function at a frequency much slower than the heart rate.

Considering the aforementioned alternative approaches to tissue engineering of cardiovascular structures, we see that they are all based on ectopic tissue generation within the same organism. Consequently they require an additional surgical procedure to produce a perishable not of the shelf product. Despite this, these solutions, the blood vessels in particular, seem to be close to clinical application.

More patient friendly and real of the shelf products, which can also be used in emergencies, are also being developed. These are the solutions that fit the "scaffold that attract endogenous cell", as did some of the previous examples, however the following examples do so in orthotopic location. This means that upon implantation they perform their function immediately, and become cellularised and thus living in time.

Originally discovered as a cell binding moiety of fibronectin in 1984 by Pierschbacher and Ruoslahti [41], RGD (arginine-glycine-aspartic acid) was used as one of the common recellularisation enhancing coatings of matrices, such as poly(urethane) [42], poly(lactic acid) [43], nanocomposite material [44] and even collagen[45]. Besides found in fibronectin, RGD is also part of other naturally occurring proteins such as fibrillin-1 [46] osteopontin [47] and collagen [41]. Matrices providing similar cues have been developed by several groups, for example hyaluronic based matrices [45,48]. All these examples provide adhesion points for cells. As such they are very relevant for the complete tissue engineering paradigm, because the cells are seeded in vitro and therefore the cell population is well defined. However this approach lacks specificity if one wants to attract a specific cell population to a prosthetic device.

Other groups, including our own, have been searching for more specific cues which can be incorporated in prostheses. Figure 2 illustrates the method of action of such prostheses. The first two examples are scaffolds for heart valve tissue engineering that are primed for *in situ* recellularisation. Recently, the group of Ota in cooperation with Medtronic, one of the companies commercialising heart valve prostheses, reported on in situ reendothelialization of decellularized porcine aortic roots [4]. These roots were coated with a fusion protein of fibronectin and hepatocyte growth factor. The rationale behind this was that the fibronectin works as an adhesive, binding the fusion protein to the graft. The hepatocyte growth factor then attracts endothelial cells, a previously described property [49], thus inducing the reendothelialization of the cell-free matrix. The coated acellular porcine aortic roots were implanted in beagle dogs up till one month. Histological examination showed the lining of the graft with a Factor VIII positive endothelial layer and some layer underneath that were vimentin positive, a marker for mesenchymal cells [50]. Overall they showed that enhancing recellularisation in situ was possible and that appropriate cell types could be obtained, albeit that the number of interstitial heart valve cells was rather low.

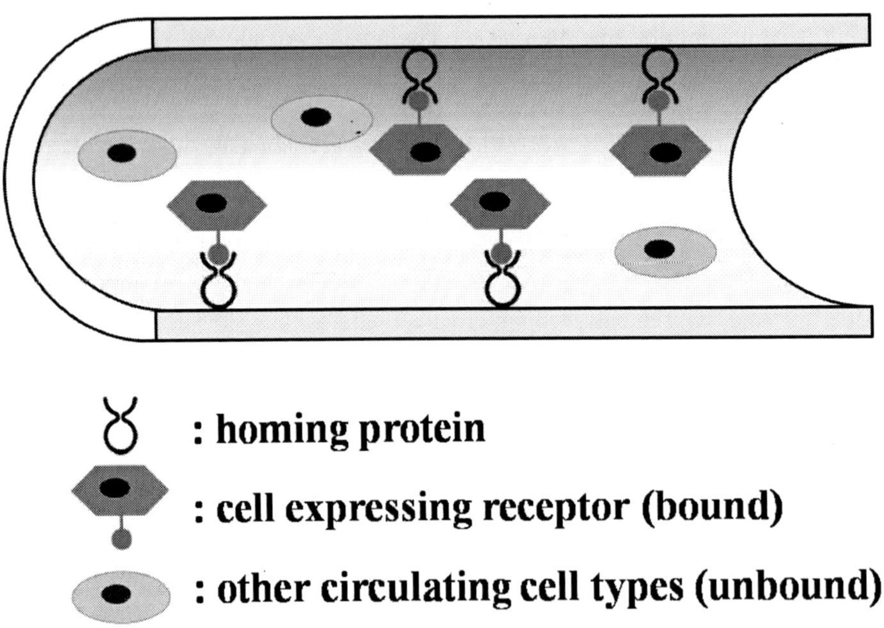

Figure 2. Schematic representation of selective self seeding prostheses.

As aforementioned we studied the foreign body reaction as a recellularisation paradigm. However, the same reaction was also considered as an adequate recruiter of primitive cells.

When analysing the gene expression during this reaction we identified several genes responsible for homing, signalling and structural proteins [51]. From this analysis we have identified 2 possible candidates with known effects in stem cells: stem cell factor (SCF) and stromal cell derived factor -1 (SDF-1). Because both were significantly upregulated we hypothesized that they might be involved in homing of primitive cells in the foreign body reaction. Supplementary, granulation tissue does comprise fibroblasts and myofibroblast, which are in effect the interstitial cells of heart valves [52]. An initial rat study provided evidence of differential homing, but the main effects were obtained in a sheep study very similar to the one described above [16]. We found that valves coated with SDF-1 in combination with fibronectin showed properties very similar to native valves, this in contrast to valves coated with fibronectin alone or intraperitoneally seeded valves. The valves coated with fibronectin and SDF-1 had cell numbers comparable to native valves. More importantly the cells expressing a contractile phenotype were significantly lowered close to the 2 to 5% level already observed in native valves [53]. Surprisingly, because the coating aims at recellularisation with interstitial cells, we found that the valves were endothelialized as well. The coating resulted, at least in the pulmonary valve in sheep, in an almost natural recellularisation of the matrix with preserved biomechanical properties. There are two remarks though: a small presence of leucocytes was still present, as was the case in blood vessels of the Campbell group, and we do not know if the endothelialization will occur spontaneously in other species. Accordingly, it is worth mentioning, that Askari and co-workers when implanting cardiac fibroblasts stably transfected to express SDF-1 in infracted myocardium also obtained homing of CD117-positive stem cells [54].

Conclusion

As become apparent from the studies reviewed here, alternatives or rather incomplete paradigms of tissue engineering do exist, but they are scarce. On the other hand one has to acknowledge that most if not all these approaches are feasible and simple solutions. Therefore it is almost certain that the application of these solutions into real clinical products is realistic.

References

[1] Campbell, JH; Efendy, JL; Campbell, GR. Novel vascular graft grown within recipient's own peritoneal cavity. *Circ. Res.* 1999 85, 1173-1178

[2] De Visscher, G; Vranken, I; Lebacq, A; Van Kerrebroeck, C; Ganame, J; Verbeken, E; Flameng, W. In vivo cellularisation of a cross-linked matrix by intraperitoneal implantation: a new tool in heart valve tissue engineering. *Eur. Heart J.* 2007 28, 1389-1396

[3] Flameng, W. A new approach to heart valve tissue engineering. *Int. J. Cardiol.* 2004 95 Suppl. 1, S55-S56

[4] Ota, T; Sawa, Y; Iwai, S; Kitajima, T; Ueda, Y; Coppin, C; Matsuda, H; Okita, Y. Fibronectin-hepatocyte growth factor enhances reendothelialization in tissue-engineered heart valve. *Ann. Thorac. Surg.* 2005 80, 1794-1801

[5] Lal, B; Viola, J; Hicks, D; Grad O. NSF: Abt report on *"The Emergence of Tissue Engineering as a Research Field"*. 2003

[6] Langer, R; Vacanti, JP. Tissue engineering. *Science.* 1993 260, 920-926

[7] Fenn, J. Understanding Gartner's Hype Cycle, 2007. 2007 G00144727,

[8] Shinoka, T; Breuer, CK; Tanel, RE; Zund, G; Miura, T; Ma, PX; Langer, R; Vacanti, JP; Mayer, JE, Jr. Tissue engineering heart valves: valve leaflet replacement study in a lamb model. *Ann. Thorac. Surg.* 1995 60, S513-S516

[9] Rabkin, E; Schoen, FJ. Cardiovascular tissue engineering. *Cardiovasc. Pathol.* 2002 11, 305-317

[10] Hoerstrup, SP; Sodian, R; Daebritz, S; Wang, J; Bacha, EA; Martin, DP; Moran, AM; Guleserian, KJ; Sperling, JS; Kaushal, S; Vacanti, JP; Schoen, FJ; Mayer, JE, Jr. Functional living trileaflet heart valves grown in vitro. *Circulation.* 2000 102, III44-III49

[11] Dohmen, PM; Ozaki, S; Verbeken, E; Yperman, J; Flameng, W; Konertz, WF. Tissue engineering of an auto-xenograft pulmonary heart valve. *Asian Cardiovasc. Thorac. Ann.* 2002 10, 25-30

[12] Gabbieri, D; Dohmen, PM; Linneweber, J; Lembcke, A; Braun, JP; Konertz, W. Ross procedure with a tissue-engineered heart valve in complex congenital aortic valve disease. *J. Thorac. Cardiovasc. Surg.* 2007 133, 1088-1089

[13] Orlic, D; Kajstura, J; Chimenti, S; Limana, F; Jakoniuk, I; Quaini, F; Nadal-Ginard, B; Bodine, DM; Leri, A; Anversa, P. Mobilized bone marrow cells repair the infarcted heart, improving function and survival. *Proc. Natl. Acad. Sci. U.S.A.* 2001 98, 10344-10349

[14] Janssens, S; Dubois, C; Bogaert, J; Theunissen, K; Deroose, C; Desmet, W; Kalantzi, M; Herbots, L; Sinnaeve, P; Dens, J; Maertens, J; Rademakers, F; Dymarkowski, S; Gheysens, O; Van Cleemput, J; Bormans, G; Nuyts, J; Belmans, A; Mortelmans, L; Boogaerts, M; Van de, WF. Autologous bone marrow-derived stem-cell transfer in patients with ST-segment elevation myocardial infarction: double-blind, randomised controlled trial. *Lancet.* 2006 367, 113-121

[15] Vincentelli, A; Wautot, F; Juthier, F; Fouquet, O; Corseaux, D; Marechaux, S; Le Tourneau, T; Fabre, O; Susen, S; Van Belle, E; Mouquet, F; Decoene, C; Prat, A; Jude, B. In vivo autologous recellularization of a tissue-engineered heart valve: are bone marrow mesenchymal stem cells the best candidates? *J. Thorac. Cardiovasc. Surg.* 2007 134, 424-432

[16] De Visscher, G; Blockx H; Vranken, I; Verbeken, E; Van Oosterwyck H; Herregods, MC; Meuris, B; Flameng, W. Enhanced in situ recellularization by fibronectin and stromal cell derived factor-1alpha coating of heart valves acquires a quasi-native valve composition of cell phenotypes. *Circulation.* 2007 116 (suppl), II-446

[17] Zimmermann, WH; Didie, M; Wasmeier, GH; Nixdorff, U; Hess, A; Melnychenko, I; Boy, O; Neuhuber, WL; Weyand, M; Eschenhagen, T. Cardiac grafting of engineered heart tissue in syngenic rats. *Circulation.* 2002 106, I151-I157

[18] The Culture of Organs. *Can. Med. Assoc. J.* 1938 39, 416-

[19] Voorhees, AB. How it all began. In: Sawyer, PN and Kaplitt, M. J. editors. *Vascular grafts.* New York: Appleton-Century-Crofts; 1978; 3-4.

[20] Eiken, O. Autogenous connective tissue tubes for replacement of small artery defects. A preliminary report of an experimental study in dogs. *Acta Chir. Scand.* 1960 120, 47-50

[21] Schilling, JA; Shurley, HM; Joel, W; Richter, KM; White, BN. Fibrocollagenous tubes structured in vivo. Morphology and biological characteristics. *Arch. Pathol.* 1961 71, 548-553

[22] Schilling, JA; Shurley, HM; Joel, W; White, BN; Bradford, RH. Abdominal aortic grafts: Use of in vivo structured autologous and homologous fibrocollagenous tubes. *Ann. Surg.* 1964 159, 819-828

[23] Sparks, CH. Die-grown reinforced arterial grafts: observations on long-term animal grafts and clinical experience. *Ann. Surg.* 1970 172, 787-794

[24] Sparks, CH. Silicone mandril method for growing reinforced autogenous femoro-popliteal artery grafts in situ. *Ann. Surg.* 1973 177, 293-300

[25] Sparks, CH. Silicone mandril method of femoropopliteal artery bypass. Clinical experience and surgical technics. *Am. J. Surg.* 1972 124, 244-249

[26] Hallin, RW; Sweetman, WR. The Sparks' mandril graft. A seven year follow-up of mandril grafts placed by Charles H. Sparks and his associates. *Am. J. Surg.* 1976 132, 221-223

[27] Roberts, PN; Hopkinson, BR. The Sparks mandril in femoropopliteal bypass. *Br. Med. J.* 1977 2, 1190-1191

[28] Rosengren, A; Danielsen, N; Bjursten, LM. Inflammatory reaction dependence on implant localization in rat soft tissue models. *Biomaterials.* 1997 18, 979-987

[29] Chue, WL; Campbell, GR; Caplice, N; Muhammed, A; Berry, CL; Thomas, AC; Bennett, MB; Campbell, JH. Dog peritoneal and pleural cavities as bioreactors to grow autologous vascular grafts. *J. Vasc. Surg.* 2004 39, 859-867

[30] Campbell, JH; Efendy, JL; Han, C; Girjes, AA; Campbell, GR. Haemopoietic origin of myofibroblasts formed in the peritoneal cavity in response to a foreign body. *J. Vasc. Res.* 2000 37, 364-371

[31] Campbell, JH; Efendy, JL; Han, CL; Campbell, GR. Blood vessels from bone marrow. *Ann. N.Y. Acad. Sci.* 2000 902, 224-229

[32] Bayes-Genis, A; Campbell, JH; Carlson, PJ; Holmes, DR, Jr.; Schwartz, RS. Macrophages, myofibroblasts and neointimal hyperplasia after coronary artery injury and repair. *Atherosclerosis.* 2002 163, 89-98

[33] Jabs, A; Moncada, GA; Nichols, CE; Waller, EK; Wilcox, JN. Peripheral blood mononuclear cells acquire myofibroblast characteristics in granulation tissue. *J. Vasc. Res.* 2005 42, 174-180

[34] Ninomiya, K; Takahashi, A; Fujioka, Y; Ishikawa, Y; Yokoyama, M. Transforming growth factor-beta signaling enhances transdifferentiation of macrophages into smooth muscle-like cells. *Hypertens. Res.* 2006 29, 269-276

[35] Schmitt-Graff, A; Desmouliere, A; Gabbiani, G. Heterogeneity of myofibroblast phenotypic features: an example of fibroblastic cell plasticity. *Virchows Arch.* 1994 425, 3-24

[36] Galmiche, MC; Koteliansky, VE; Briere, J; Herve, P; Charbord, P. Stromal cells from human long-term marrow cultures are mesenchymal cells that differentiate following a vascular smooth muscle differentiation pathway. *Blood.* 1993 82, 66-76

[37] Bonanno, E; Ercoli, L; Missori, P; Rocchi, G; Spagnoli, LG. Homogeneous stromal cell population from normal human adult bone marrow expressing alpha-smooth muscle actin filaments. *Lab. Invest.* 1994 71, 308-315

[38] Vranken, I; De Visscher, G; Lebacq, A; Verbeken, E; Flameng, W. The recruitment of primitive Lin- Sca-1+, CD34+, C-kit+ and CD271+ cells during the early intraperitoneal foreign body reaction. *Biomaterials.* 2008 29(7), 797-808

[39] De Visscher, G; Blockx H; Meuris, B; Van Oosterwyck H; Verbeken, E; Herregods, MC; Flameng, W. Functional and biomechanical evaluation of a completely recellularized stentless pulmonary bioprosthesis in sheep. *J. Thorac. Cardiovasc. Surg.* 2008 135(2), 395-404

[40] Morritt, AN; Bortolotto, SK; Dilley, RJ; Han, X; Kompa, AR; McCombe, D; Wright, CE; Itescu, S; Angus, JA; Morrison, WA. Cardiac tissue engineering in an in vivo vascularized chamber. *Circulation.* 2007 115, 353-360

[41] Pierschbacher, MD; Ruoslahti, E. Variants of the cell recognition site of fibronectin that retain attachment-promoting activity. *Proc. Natl. Acad. Sci. U.S.A.* 1984 81, 5985-5988

[42] Tiwari, A; Salacinski, HJ; Punshon, G; Hamilton, G; Seifalian, AM. Development of a hybrid cardiovascular graft using a tissue engineering approach. *FASEB J.* 2002 16, 791-796

[43] Quirk, RA; Chan, WC; Davies, MC; Tendler, SJ; Shakesheff, KM. Poly(L-lysine)-GRGDS as a biomimetic surface modifier for poly(lactic acid). *Biomaterials.* 2001 22, 865-872

[44] Alobaid, N; Salacinski, HJ; Sales, KM; Ramesh, B; Kannan, RY; Hamilton, G; Seifalian, AM. Nanocomposite containing bioactive peptides promote endothelialisation by circulating progenitor cells: an in vitro evaluation. *Eur. J. Vasc. Endovasc. Surg.* 2006 32, 76-83

[45] Burgess, BT; Myles, JL; Dickinson, RB. Quantitative analysis of adhesion-mediated cell migration in three-dimensional gels of RGD-grafted collagen. *Ann. Biomed. Eng.* 2000 28, 110-118

[46] Stephan, S; Ball, SG; Williamson, M; Bax, DV; Lomas, A; Shuttleworth, CA; Kielty, CM. Cell-matrix biology in vascular tissue engineering. *J. Anat.* 2006 209, 495-502

[47] Denhardt, DT; Guo, X. Osteopontin: a protein with diverse functions. *FASEB J.* 1993 7, 1475-1482

[48] Turner, NJ; Kielty, CM; Walker, MG; Canfield, AE. A novel hyaluronan-based biomaterial (Hyaff-11) as a scaffold for endothelial cells in tissue engineered vascular grafts. *Biomaterials.* 2004 25, 5955-5964

[49] Bussolino, F; Di Renzo, MF; Ziche, M; Bocchietto, E; Olivero, M; Naldini, L; Gaudino, G; Tamagnone, L; Coffer, A; Comoglio, PM. Hepatocyte growth factor is a potent angiogenic factor which stimulates endothelial cell motility and growth. *J. Cell Biol.* 1992 119, 629-641

[50] Guarita-Souza, LC; Carvalho, KA; Woitowicz, V; Rebelatto, C; Senegaglia, A; Hansen, P; Miyague, N; Francisco, JC; Olandoski, M; Faria-Neto, JR; Brofman, P. Simultaneous autologous transplantation of cocultured mesenchymal stem cells and skeletal myoblasts improves ventricular function in a murine model of Chagas disease. *Circulation.* 2006 114, I120-I124

[51] Lebacq, A; De Visscher, G; Vranken, I; Flameng, W. Homing, signaling, and structural molecule gene xpression profiling in the early-phase foreign body reaction. Int.J.Lower Extremity Wounds. 2007 6, 198

[52] Della Rocca, F; Sartore, S; Guidolin, D; Bertiplaglia, B; Gerosa, G; Casarotto, D; Pauletto, P. Cell composition of the human pulmonary valve: a comparative study with

the aortic valve--the VESALIO Project. Vitalitate Exornatum Succedaneum Aorticum labore Ingegnoso Obtinebitur. *Ann. Thorac. Surg.* 2000 70, 1594-1600

[53] Mendelson, K; Schoen, FJ. Heart valve tissue engineering: concepts, approaches, progress, and challenges. *Ann. Biomed. Eng.* 2006 34, 1799-1819

[54] Askari, AT; Unzek, S; Popovic, ZB; Goldman, CK; Forudi, F; Kiedrowski, M; Rovner, A; Ellis, SG; Thomas, JD; DiCorleto, PE; Topol, EJ; Penn, MS. Effect of stromal-cell-derived factor 1 on stem-cell homing and tissue regeneration in ischaemic cardiomyopathy. *Lancet.* 2003 362, 697-703

In: Tissue Engineering: Roles, Materials and Applications ISBN: 978-1-60456-293-4
Editors: S. J. Barnes, L. P. Harris, pp. 287-301

Chapter 12

TISSUE ENGINEERING USING ADIPOSE TISSUE-DERIVED ADULT STEM CELLS FOR SKELETAL MUSCLE REPAIR

MiJung Kim** *and Yu Suk Choi

Department of Laboratory Medicine, University of Ulsan College of Medicine/ Cell and Molecular Biology Laboratory, Asan Institute for Life Sciences, Asan Medical Center, University of Ulsan, Seoul 138736, Korea.

ABSTRACT

The use of adult stem cells for cell-based tissue engineering and regeneration represents a promising approach for repairing damaged tissue. This emerging field of regenerative medicine will require a reliable source of stem cells in addition to biomaterial scaffolds and growth factors. Recent studies have shown that stem cells can be isolated from a wide variety of tissues including bone marrow, muscle, and adipose tissue. The cellular component of this regenerative approach will play a key role in bringing tissue-engineered constructs from the laboratory bench to the clinical bedside. An ideal source of stem cells still remains undefined, however, and may differ depending upon the required application.

Muscle tissue engineering approaches could revolutionize current therapies for irreversible skeletal muscle damage and muscular dystrophies, and may significantly improve the quality of life of millions of patients. Skeletal muscle tissue engineering approaches rely mainly on myoblasts and biomaterials; however, myoblasts lack expansion capability. Cells from other sources, including adult stem cells, are being investigated to overcome this problem. Several cell sources have been proposed for creation of engineered skeletal tissue, but the literature to date unfortunately does not assist in the identification of an ideal cell source. We describe here the advantages and limitations of candidate cell sources for skeletal muscle tissue engineering.

As with bone marrow, adipose tissue is a mesoderm-derived organ that contains a stromal population including microvascular endothelial cells, smooth muscle cells, and stem cells. This population shares many characteristics with its bone marrow counterpart

* Correspondence : Associate Professor of Dept. of Laboratory Medicine, University of Ulsan College of Medicine / Cell & Molecular Biology Lab., AILS, Seoul, Korea. e-mail: mjkimhsc@korea.com, Fax: 82-2-3010-4182.

including extensive proliferative potential and an ability to undergo multilinear differentiation. Adipose tissue also has the ability to dynamically expand and shrink throughout the life of an adult. This capacity is mediated by the presence of vascular and non-vascular cells that provide a pool of stem and progenitor cells with unique regenerative capacities. Human adipose tissue-derived adult stem cells (ASCs) therefore meet many of the requirements of "ideal" cells for stem cell therapy and tissue engineering. ASCs have been used in cell-based tissue engineering of hard tissues such as bone and cartilage, but their capacity to assist in regeneration of damaged muscle has not yet been investigated in detail.

Here we review recent progress in the muscle tissue engineering field and describe one of our experiments as an example of successful skeletal muscle tissue engineering. We engineered skeletal muscle tissue using a combination of ASCs and injectable PLGA spheres. These findings suggest that mixtures of ASCs and injectable scaffolds may provide excellent tools for muscle tissue engineering and that such combinations may find applications in clinical settings requiring muscle tissue regeneration therapy.

1. Introduction

Loss of voluntary power of skeletal muscle may result from interruption of innervation or may occur because of deficiency in or loss of muscle mass. Muscle loss or deficiency may be primary in congenital anomalies such as Prune-belly syndrome and advanced forms of anal atresia, or may be secondary after radical resection of malignant tumors, trauma, or surgical interventions. Intrinsic diseases such as muscular dystrophies (Duchenne type or Steinert's disease), periodic diseases (Myasthenia gravis), developmental defects (hypoplasia or aplasia), and endocrine or metabolic disturbances, are also responsible for functional losses of skeletal muscle [1].

Skeletal muscles are composed of bundles of highly oriented and dense muscle fibers, each being a multinucleated cell derived from a myoblast. The muscle fibers in native skeletal muscle are closely packed together in an extracellular three-dimensional matrix to form an organized tissue with high cell density and an advanced degree of cellular orientation, to generate longitudinal contraction. After muscle injuries, myofibers become necrotic and are removed by macrophages [2]. A specialized myoblast subpopulation termed satellite cells, scattered below the basal lamina of myofibers, is capable of muscle regeneration [3]. The incidence of satellite cells in skeletal muscle is very low (1–5%) and depends on age and muscle fiber composition [4]. These cells usually remain in a quiescent and undifferentiated state but can enter the mitotic cycle in response to specific local factors [5]. Such factors include proliferation and fusion of myoblasts to form multinucleated and elongated myotubes, which self-assemble to form a more organized structure, namely the muscle fiber [3]. Satellite cells migrate to and proliferate in injured areas and can also form a connective tissue network ("muscle fibrosis"). This process is termed "scar tissue formation" and leads to loss of muscle functionality [5,6].

Tissue engineering is a fast-developing field that aims to create or regenerate tissues and organs by using combinations of cells, scaffolds, and/or biomolecules. Tissue engineering strategies promise to revolutionize current therapies for irreversible skeletal muscle damage, such as are caused by the muscular dystrophies, and to significantly improve the quality of life of millions of patients. Skeletal muscle tissue engineering approaches rely mainly on the use of combinations of myoblasts and biomaterials, which can deliver cells safely to defective sites and then provide appropriate environments for tissue assembly and neovascularization. Because myoblasts lack expansion capacity, cells from other sources, including adult stem cells, are being investigated for use in muscle tissue engineering [7].

In addition to cells, the second element necessary in cell-based tissue engineering is the scaffold, which must be biocompatible, biodegradable, and possess considerable stability, thus allowing for successful integration between newly formed tissue and host tissue. Unlike hard tissue such as bone and cartilage, muscle tissue engineering may require rather soft and flexible biomaterial scaffolds in therapeutic applications. Injectable biodegradable scaffolds such as PLGA spheres may fulfill these requirements.

2. Cells for Skeletal Muscle Tissue Engineering

The optimal cell source for engineered skeletal tissue must meet several requirements. Cells should be easy to harvest, capable of proliferating to fill a void, nonimmunogenic, and either show myoblast functions or have the capacity to differentiate into functional muscle cells. Unfortunately, a cell source with all these desirable features remains undescribed in the literature. Several cell sources have been proposed for skeletal tissue engineering, as shown in table 1, which describes the advantages and limitations of seven cell sources [8–15].

Table 1. Advantages and disadvantages of various sources of stem cells

Cells	Easy Accessibility	High Expandability	Myogenic Capacity	Autologous Origin
Skeletal myoblast	Yes	Depends on age	Yes	Yes
MDSCs	Yes	No	Yes	Yes
ESCs	No	Yes	Yes	No
BMSCs	Yes	Depends on age	Yes	Yes
UCBCs	Yes	Yes	Yes	No
HSCs	Yes	Yes	Debated	Yes
ASCs	Yes	Yes	Yes	Yes

* MDSC; muscle-derived stem cell, ESC; embryonic stem cell, BMSC; bone marrow-derived stem cell, UCBC; umbilical cord blood stem cell, HSC; hematopoietic stem cell, ASC; adipose tissue-derived stem cell.

2.1. Muscle-derived cells (MDCs)

Different populations of muscle-derived progenitor cells appear to exhibit varying degrees of pluripotency. The best-characterized muscle progenitor cells are satellite cells [16]. In addition to participating in the formation of myofibers, muscle satellite cells can also differentiate into other lineages, such as osteogenic, chondrogenic, and adipogenic cells [17,18]. Other populations of muscle-derived cells are mostly distinct from satellite cells [19] and can be divided into many different subpopulations by examining surface marker profiles and the abilities of the cells to undergo myogenic differentiation. Muscle-derived stem cells (MDSCs) can be separated into two main subpopulations: CD45+ cells and CD45- cells, based on their relationships to hematopoietic cells. CD45+ cells isolated from normal muscle have a very limited myogenic potential *in vitro* and *in vivo*, but they possess a high degree of hematopoietic potential [20]. In contrast, CD45- cells readily differentiate into myogenic lineages both *in vitro* and *in vivo* [20,21] and can also differentiate into other lineages, such as blood cells [22].

The different populations of muscle-derived cells are closely related. For example, CD45+ cells, which normally possess low myogenic potential, can undergo active myogenesis during muscle injury by responding to the combined signaling of Wnt 5a, Wnt 5b, and Wnt 7a [23]. Thus, the differentiation potential of muscle-derived cells may be modified by environmental cues. This published study also demonstrated that CD 45+ muscle-derived cells are the resident muscle stem cells responsible for regeneration of injured muscle when the pool of satellite cells has been depleted because of cardiotoxin insult [23]. More research is, however, necessary to determine the relative contributions of different populations of muscle progenitor cells, including CD45+ cells, CD45- cells, and satellite cells, to muscle regeneration after common injuries that do not lead to the depletion of a particular group of cells. Nevertheless, it has been shown that under the influence of environmental factors (such as Wnt signaling), CD45+ muscle-derived cells can contribute to muscle regeneration [8].

2.2. Embryonic Stem Cells

Embryonic stem (ES) cells, which are multipotent cells derived from the inner cell mass of the pre-implantation blastocyst, have the potential to be useful as a new approach for cell transplantation therapy [24,25]. When transferred to suspension cultures, mouse ES cells begin to differentiate into multicellular aggregates of differentiated and undifferentiated cells, termed embryoid bodies (EBs) which resemble early post-implantation embryos. Cell differentiation in EBs is disorganized and frequently varies from one EB to another, even in the same culture [24]. Although EBs develop skeletal muscle cells and other cells *in vitro*, transplantation of EBs, without any induction to direct development along a specific pathway, leads to a failure of integration into recipient tissues and often results in teratoma formation in transplanted tissues [26]. Growth factors which selectively promote myogenic differentiation have not yet been identified.

ES cells may have a developmental advantage over myogenic precursor cells derived from muscle or bone marrow in their capacity to form differentiated skeletal muscle tissue *in vivo*. Compared to other forms of cell transplantation [27–31], in which occasional dystrophin-positive myonuclei were seen within multinucleate myotubes, ES cell transplantation results in the development of normally organized and vascularized skeletal muscle tissue *in vivo* [10].

Many results indicate that ES cells can contribute to the formation of new viable muscle tissue [24–26]. There are, however, disadvantages to the use of human ES cells. 1) ES cells are mainly derived from IVF surplus embryos, which are of allogenic origin, and this may result in rejection by the recipient's immune system; 2) the isolation and use of ES cells are still topics of intense ethical debate, and; 3) ES cells are by definition tumorigenic when undifferentiated. This emphasizes the need for a thorough understanding of the pathways of differentiation of ES cells into myoblasts. Only then can ES cells be used for regenerating damaged muscle tissue. The alternative adult stem cells may have fewer restrictions on use.

2.3. Bone Marrow-Derived Stem Cells (BMSCs)

Ferrari et al. first described transdifferentiation of BMSCs into skeletal muscle [30]. Unfractionated BMSCs from C57M/*LacZ* transgenic mice were injected into the chemically injured tibialis anterior muscle of immunodeficient mice. Two to five weeks later, four of six mice that received unfractionated BMSCs exhibited β-gal$^+$ nuclei within the damaged muscle [30]. After bone marrow transplantation, intravenously injected BMSCs also homed to damaged muscle and differentiated to show a skeletal muscle phenotype [30]. Subsequent studies with mouse models of Duchenne muscular dystrophy demonstrated that *mdx* transgenic mice transplanted with wild-type bone marrow contained donor-derived cells that were positive for muscle-specific markers (myogenin and Myf-5), and expressed dystrophin [31]. Similarly, when irradiated mice were transplanted with EGFP+ BMSCs, BMSC-derived satellite cells and multinucleated muscle fibers could be identified in recipient mice. Although these studies indicate that BMSCs can differentiate into a skeletal muscle phenotype, the specific type of BMSC that exhibits this plastic behavior remains to be identified.

2.4. Adipose Tissue-Derived Stem Cells (ASCs)

Adipose tissue has recently been identified as an alternative, uniquely abundant, and accessible source of pluripotent cells. In adult mammals, adipose tissue and an association of adipocytes are held together in a framework of collagen fibers [11]. In addition to mature adipocytes, adipose tissue contains the stromal vascular fraction (SVF), which is composed of a heterogeneous cell population including adipose precursor cells of varying degrees of differentiation, vascular cells, and a supportive stroma. As with BMSCs, adipose tissue-derived stem cells (ASCs) can differentiate *in vitro* into adipogenic, chondrogenic, osteogenic and myogenic cells [32], as well as into other non-mesenchymal lineages such as neurons and hepatocytes [33,34] in the presence of lineage-specific inductive media. Also, heterogeneous population of ASCs has been shown to include progenitors of endothelial cells. Recently, short-term cultures of ASCs have been shown to develop into regenerating muscle fibers with low efficiency after transplantation into damaged muscles. The full myogenic potential of freshly isolated ASCs and their ability to respond to myogenic cues producing terminally differentiated myotubes are largely unknown, however.

2.5. Summary

Stem cell-based skeletal muscle tissue engineering is potentially a powerful application for skeletal muscle diseases. As seen above, we have considered four stem cell candidates for skeletal muscle tissue engineering. ES cells and MDCs have been studied in much more detail than have ASCs, however, but the use of ES cells is associated with ethical and immunogenic rejection problems. MDSCs and MSCs do not replicate to any great extent *in vitro*, and are restricted in donor expandability. ASCs, which can be obtained by minimally invasive procedures from subcutaneous fat, are highly expandable in culture and can be readily induced to differentiate into muscle tissue-forming cells by exposure to well-established myogenic media. Adipose tissue is routinely available in liter quantities from patients undergoing liposuction.

3. Scaffolds for Skeletal Muscle Tissue Engineering

In addition to cells, the second element necessary for cell-based tissue engineering is the scaffold, which must be biocompatible, biodegradable, and possess considerable stability, thus allowing for successful integration between the newly formed tissue and the host tissue. Biomaterials provide mechanical stability to the construct in the short term and serve as templates for the three-dimensional organization of the developing tissue. The effective repair and regeneration of injured tissues and organs depend on early re-establishment of the blood flow needed for metabolic support. An implantable biomaterial should be three-dimensional and highly porous to enable cell growth and flow transport of nutrients and metabolic waste. The scaffold should be biodegradable and should have a surface chemistry suitable for cell attachment and proliferation. Furthermore, the scaffold should be capable of easy laboratory processing to form a variety of shapes and sizes [35].

Many attempts have been made to reconstruct skeletal muscle tissue *in vitro* [36–40]; the results of these studies taken together suggest it is indeed feasible to engineer bioartificial muscles. With current technology, however, tissue-engineered skeletal muscle analogs are far from clinical reality. Morphologically, these materials fall short of actual skeletal muscle in many respects, including small diameters of myofibers, low myofiber organization, and excessive extracellular matrix content. Because myofibers need to be packed parallel to each other to generate sufficient force for muscle contraction [39,41], the lack of structural organization in engineered constructs results in too little active force capacity for utility in clinical applications.

Injectable biodegradable scaffolds, such as poly(lactic-co-glycolic acid) (PLGA) spheres, provide appropriate rigidity, self-aggregate, and function similarly to porous scaffolds *in vivo*. In addition, we have shown that adipose tissue can be generated from PLGA spheres combined with BMSCs or ASCs. PLGA is a highly biocompatible material approved by the United States Food and Drug Administration (FDA) for certain human clinical applications, including bioresorbable surgical sutures and screws. Moreover, these scaffolds have not been observed to result in any post-operation scars [42–45].

4. Engineering Skeletal Muscle Tissue

Much work has been done using the cells and scaffolds mentioned above, but skeletal muscle tissue engineering still remains challenging. We will introduce one of our previous experiments as an example of successful skeletal muscle tissue engineering using PLGA [44].

4.1. Isolation and Culture of AdiposeTissue-Derived Stem Cells

A variety of terms have been used to describe the plastic-adherent cell population isolated from collagenase digests of adipose tissue, such as adipose-derived stem cells (ADCs), adipose-derived adult stem (ADAS) cells, and processed lipoaspirate (PLA) cells. Fortunately, the International Fat Applied Technology Society has recently clarified the situation and adopted the standard terminology “adipose tissue-derived stem cells (ASCs)”. ASCs are one of several populations in the stromal vascular fraction, which is a heterogeneous population of stem cells, microvascular endothelial cells, fibroblasts, and smooth muscle cells. ASCs can be isolated by liposuction using a very simple and minimally invasive method. Figure 1 is a graphical representation of our study. Our concept was to cure skeletal muscle failure in patients using a combination of tissue engineering techniques with autologous adult stem cells. Briefly, we collected lipoaspirate samples from young, healthy plastic surgery patients by liposuction, digested the samples with collagenase for 1 h, lysed red blood cells, washed the remaining cells, and plated them on dishes. Each raw lipoaspirate (~ 300ml) was washed extensively with an equal volume of phosphate buffered saline (PBS) and the extracellular matrix was digested at 37°C for 30 min with 0.075% (w/v) collagenase (Sigma; St. Louis, MO). Enzyme activity was neutralized with Dulbecco’s modified Eagle’s medium (DMEM; Gibco; Grand Island, NY) containing 10% (v/v) fetal bovine serum (FBS), and the suspension was then centrifuged at 800G for 10 min (Sorvall RT 7 Plus). The pellet, containing the high density stromal vascular fraction (SVF) was resuspended in 0.16 M NH_4Cl and incubated at room temperature for 5 min to lyse any remaining red blood cells. The SVF was collected by centrifugation at 800G for 5 min, filtered through a 100 μm nylon mesh to remove cellular debris, and incubated overnight at 37°C in a humidified atmosphere containing 5% (v/v) CO_2 in control medium (DMEM, 10% [v/v] FBS, 1% [w/v] antibiotic/antimycotic solution). Each plate was washed extensively with PBS to remove residual nonadherent red blood cells. The resulting cell population, the ASCs, was maintained at 37°C in a humidified atmosphere containing 5% (v/v) CO_2 in control medium.

Sub-passaging yielded spindle-shaped cells which readily expanded in simple growth media. We used these stem cells in previous studies for adipose tissue and muscle tissue engineering (figure 2) [43,44]. Figure 3 shows the cells, and describes expression of cell surface markers. ASCs are positive for CD 73, CD 90, and CD 105, but are negative for CD 34 and CD 45. These results are similar to those obtained using bone marrow stromal cells [43].

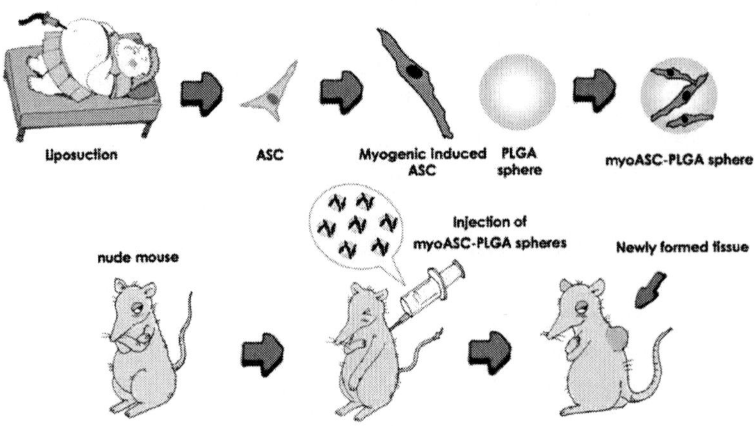

Figure 1. Schematic diagram of skeletal muscle tissue engineering using ASCs.

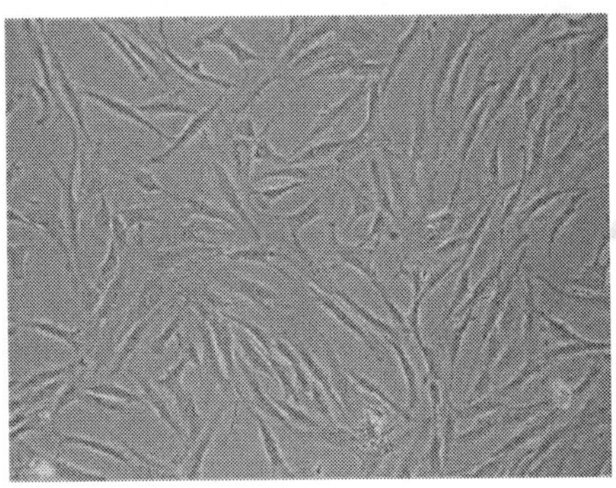

Figure 2. Morphology of adipose tissue-derived adult stem cells (ASCs). Human primary ASCs were cultured for 7 days in complete medium (DMEM, 10% [w/v] FBS, 1% [w/v] antibiotics). After the second passage, the ASCs appeared to assume a more uniform fibroblast-like shape (magnification x100).

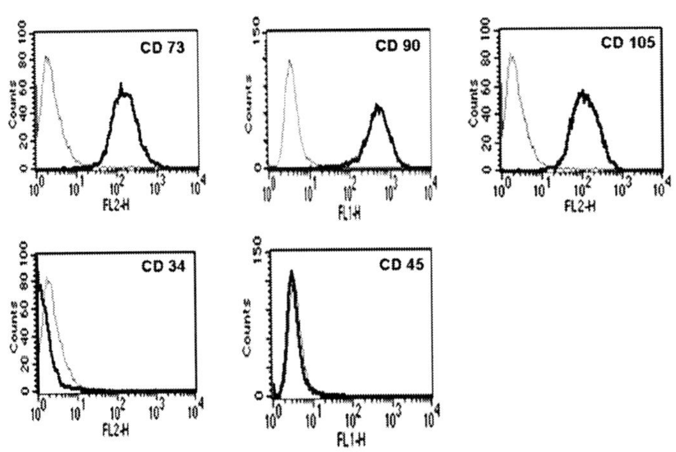

Figure 3. Characterization of ASCs using flow cytometry.

4.2.Preparation of Scaffolds and *In Vitro* Cultivation of ASC-Scaffold Mixtures

PLGA [85:15] (MW = 90,000; Purac; Park Road Blair, Nebraska) was dissolved 12:100 (w/v) in dichloromethane, added dropwise to a stirred solution of 0.27% (w/v) polyvinyl alcohol (PVA; MW = 30,000–70,000, Sigma), and stirred for 72 h, allowing for complete evaporation of the solvent. The spheres were isolated by filtration, washed with deionized water, air-dried for 2 h, and vacuum-dried for an additional 24 h. The spheres were separated into size ranges of 100–150 μm using commercially available sieves and stored desiccated until further use.

4.3. Myogenic Potency of Adipose Tissue-Derived Stem Cells

For muscle regeneration, we first verified that *in vitro* cultures of ASCs could differentiate into muscle cells. We found that the myogenic induction of ASCs was time-dependent; the length of culture time in myogenic medium was an important factor. During very early stages of myogenic differentiation in myogenic conditioned medium, ASCs started to differentiate into a myogenic lineage expressing Myo D at day 7 (figure 4).

ASCs can transdifferentiate into cells of other lineages, but, at terminal differentiation, they have the properties of mature muscle cells. We therefore myogenically induced ASCs for 3 weeks, at which stage they expressed both Myo D1 and MYS, confirming that they had differentiated into muscle cells of early to intermediate stages (figures 5 and 6). Because ASCs can differentiate into various cell types, it was possible for injected ASCs attached to PLGA spheres to differentiate into undesirable cell lineages. To ascertain whether any ASCs had differentiated into osteocytes, chondrocytes, or adipocytes, we cultured cells in myogenic medium for 42 days, and stained these cells with alizarin red S, safranin O, or oil red O. Cells did not stain with these chemicals and we found no significant differences between cells grown thereafter in myogenic media and cells grown in nondifferentiating control media. Once induced in myogenic medium, the ASCs were committed to the myogenic lineage and differentiated into muscle cells, suggesting that injected ASCs would not differentiate *in vivo* into cells of other lineages.

4.4. *In Vivo* Study

PLGA spheres, either alone or attached to ASCs, were cultured in myogenic media for 21 days and injected subcutaneously into the necks of six athymic nude mice for each group; six untreated mice served as controls. After 30 and 60 days, the mice were sacrificed and newly formed tissues were harvested (figure 7). We observed newly formed tissues under the skins of mice injected with ASCs attached to PLGA, but not in control groups. A gross view of incised dorsal skin showed that the largest volume of newly formed tissue was regenerated 60 days after mice were injected with ASCs attached to PLGA.

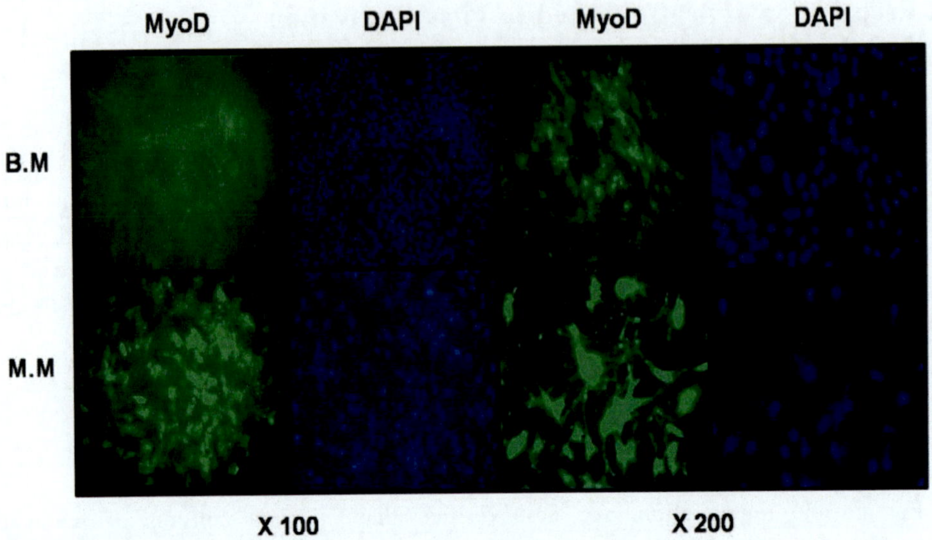

Figure 4. Myogenic differentiation of ASCs *in vitro* culture in myogenic conditioned medium at day 7. Myo D and DAPI staining were done for differentiation and cellularity, respectively. (B.M.;basal media, M.M.;myogenic media, X100, X200;each magnification designated)

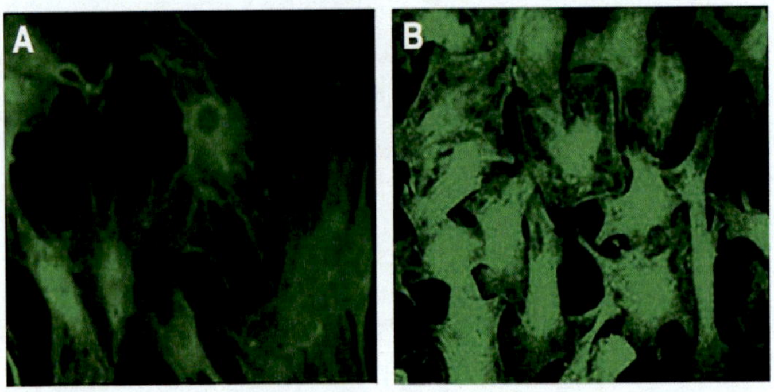

Figure 5. MyoD1 expression by ASCs cultured in myogenic media. ASCs were stained with Myo D1-FITC and visualized using a confocal microscope on day 0 (A) or day 42 (B).

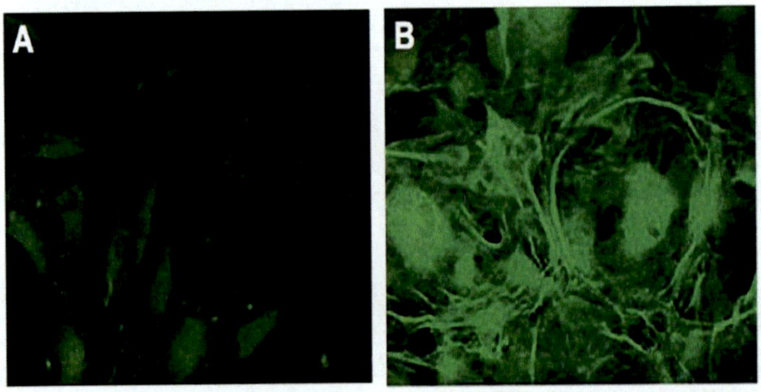

Figure 6. Myosin heavy-chain (MYS) expression by ASCs cultured in myogenic media. ASCs were stained with MYS-FITC and visualized using a confocal microscope on day 0 (A) or day 42 (B).

H and E staining showed that muscle tissues were regenerated in mice injected with ASCs attached to PLGA at 60 day post-injection, and the regenerated muscle tissue was morphologically similar to native muscle (figure 8). When we stained regenerated tissues with FITC-Myo D1, we found morphologies similar to that of native muscle, and myogenic markers were expressed at both 30 and 60 day post-injection (figure 9). This finding was confirmed by MYS staining and by RT-PCR, in that messages encoding Myo D1 and MYS were observed at both 30 and 60 day post-injection; at 60 days, the band intensities were similar to those observed in native muscle tissue (data not shown).

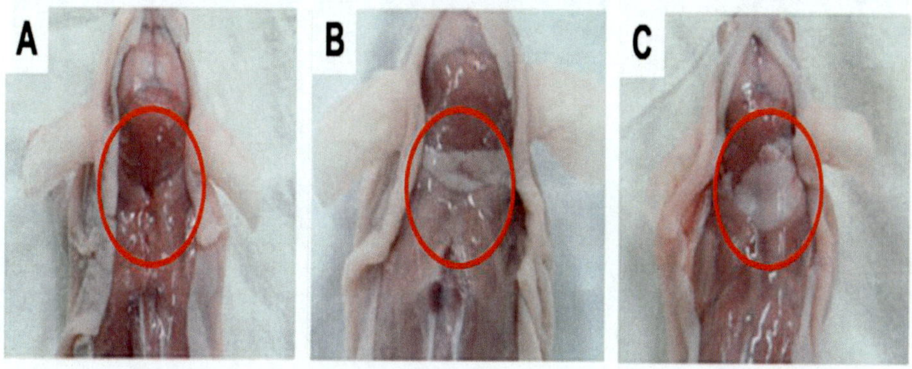

Figure 7. Gross images of newly formed tissue in untreated nude mice (A), and in nude mice injected with ASCs attached to PLGA, after 30 days (B) and 60 days (C).

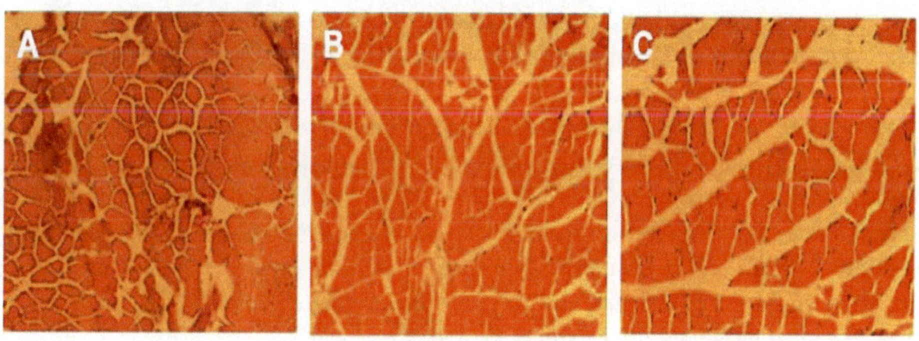

Figure 8. H&E staining of engineered skeletal muscle of nude mice injected with ASCs attached to PLGA after 30 days (A) or 60 days (B), and of native skeletal muscle (C).

4.5. SUMMARY

We have shown that ASCs attached to PLGA spheres generated skeletal muscle tissue when injected subcutaneously into nude mice. These findings suggest that ASCs and injectable spheres are useful materials for skeletal muscle tissue engineering, and indicate that this combination can be used in a clinical setting for such engineering.

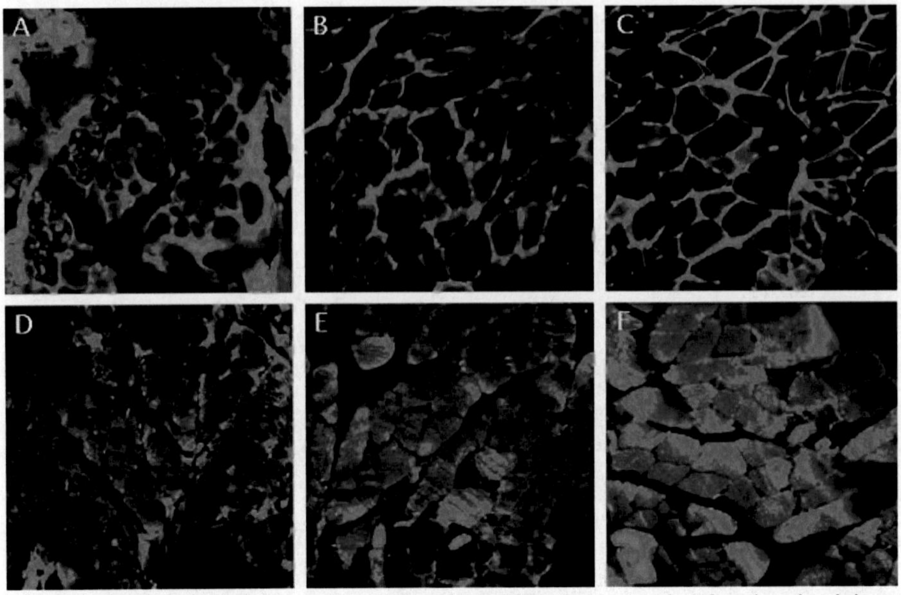

Figure 9. Myo D1 and Mys expression in engineered skeletal muscle of nude mice injected with ASCs attached to PLGA after 30 (A,D) or 60 days (B,E), and in native skeletal muscle (C,F), stained with FITC-antibodies to MyoD1(A-C) and MYS(D-F).

5. DISCUSSION

Our ASCs have several important advantages compared to previously reported stem cell systems. ASCs can easily be isolated from patients by liposuction and can be efficiently expanded *in vitro* (the doubling time is about 24 hours). Furthermore, injection of ASCs should encounter few ethical problems, because the employment of these cells avoids controversies surrounding the use of ES cells. In addition, PLGA, the scaffolding material, has been clinically approved by the U.S. FDA [45]. Autologous transplantation of ASCs should minimize the risks of immune rejection. Although our results indicate that the combination of ASCs and injectable PLGA spheres can participate in the process of skeletal muscle tissue engineering and give rise to fully differentiated muscle tissue, there are still problems that must be solved prior to clinical use.

First, questions that remain to be answered by researchers studying ASCs are similar to those encountered by others working in the stem cell field, and involve such issues as 1) the efficiency of myogenic differentiation, 2) the mechanisms underlying development from stem cells to mature muscle cells, 3) the cell viability and proliferation capacity of differentiated cells, and, 4) contractile functions of differentiated cells.

Second, the major obstacle in the engineering of thick complex tissues such as skeletal muscle is that of vascularization, which is necessary to maintain cell viability during tissue growth and for the induction of structural organization. Most approaches to engineering new tissue have relied on the host for vascularization. Although this approach has been useful in many tissues, it has not been very successful in thick highly vascularized tissue such as muscle, which must have an abundant blood vessel supply. Also, researchers need to evaluate the contractile function of their engineered tissue.

ACKNOWLEDGMENTS

This research was supported by a grant to Dr. MiJung Kim (#SC3280) from the Stem Cell Research Center of the 21st Century Frontier Research Program, funded by the Ministry of Science and Technology, Republic of Korea.

REFERENCES

[1] Saxena, AK; Marler, J; Benvenuto, M; Willital, GH; Vacanti, JP. Skeletal muscle tissue engineering using isolated myoblasts on synthetic biodegradable polymers: preliminary studies. *Tissue Eng*, 1999, 5, 525-532.

[2] Hurme, T; Kalimo, H; Lehto, M; Jarvinen, M. Healing of skeletal muscle injury: and ultrastructural and immunohistochemical study. *Med. Sci. Sports Exerc*, 1991, 23, 801-810.

[3] Campion, DR. The muscle satellite cell: a review. *Int. Rev. Cytol*, 1984, 87, 25-251.

[4] Allen, RE; Temm-Grove, CJ; Sheehan, SM; Rice, G. Skeletal muscle satellite cell culture. *Methods Cell Biol*, 1997, 52, 155-176.

[5] Hill, M; Wernig. A; Goldspink, G. Muscle satellite (stem) cell activation during local tissue injury and repair. *J. Anat*, 2003, 203, 89-99.

[6] Li, Y; Huard, J. Differentiation of muscle-derived cells into myofibroblasts in injured skeletal muscle. *Am. J. Pathol*, 2002, 161, 895-907.

[7] Bach, AD; Beier, JP; Stern-Staeter, J; Horch, RE. Skeletal muscle tissue engineering. *J. Cell Mol. Med*, 2004, 8, 413-422.

[8] Peng, H; Huard, J. Muscle-derived stem cells for musculoskeletal tissue regeneration and repair. *Transpl. Immunol*, 2004, 12, 311-319.

[9] Dawn, B; Bolli, R. Adult bone marrow-derived cells: Regenerative potential, plasticity, and tissue commitment. *Basic Res. Cardiol*, 2005, 100, 494-503.

[10] Bhagavati, S; Xu, W. Generation of skeletal muscle from transplanted embryonic stem cells in dystrophic mice. *Biochem. Biophys. Res. Commun*, 2005, 333, 644-649.

[11] Di Rocco, G; Iachininoto, MG; Tritarelli, A; Straino, S; Zacheo, A; Germani, A; Crea, F; Capogrossi, MC. Myogenic potential of adipose-tissue-derived cells. *J. Cell Sci*, 2006, 119, 2945-2952.

[12] Sinanan, AC; Buxton, PG; Lewis, MP. Muscling in on stem cells. *Biol. Cell*, 2006, 98, 203-214.

[13] Dellavalle, A; Sampaolesi, M; Tonlorenzi, R; Tagliafico, E; Sacchetti, B; Perani, L; Innocenzi, A; Galvez, BG; Messina, G; Morosetti, R; Li, S; Belicchi, M; Peretti, G; Chamberlain, JS; Wright, WE; Torrente, Y; Ferrari, S; Bianco, P; Cossu, G. Pericytes of human skeletal muscle are myogenic precursors distinct from satellite cells. *Nat. Cell Biol*, 2007, 9, 255-267.

[14] Levenberg, S; Rouwkema, J; Macdonald, M; Garfein, ES; Kohane, DS; Darland, DC; Marini, R; van Blitterswijk, CA; Mulligan, RC; D'Amore, PA; Langer, R. Engineering vascularized skeletal muscle tissue. *Nat. Biotechnol*, 2005, 23, 879-884.

[15] Dezawa, M; Ishikawa, H; Itokazu, Y; Yoshihara, T; Hoshino, M; Takeda, S; Ide, C; Nabeshima, Y. Bone marrow stromal cells generate muscle cells and repair muscle degeneration. *Science*, 2005, 309, 314-317.

[16] Partridge, TA. Cells that participate in regeneration of skeletal muscle. *Gene Ther*, 2002, 9, 752-753.

[17] Asakura, A; Komaki, M; Rudnicki, M. Muscle satellite cells are multipotential stem cells that exhibit myogenic, osteogenic, and adipogenic differentiation. *Differentiation*, 2001, 68, 245-253.

[18] Wada, MR; Inagawa-Ogashiwa, M; Shimizu, S; Yasumoto, S; Hashimoto, N. Generation of different fates from multipotent muscle stem cells. *Development*, 2002, 129, 2987-2995.

[19] Asakura, A; Seale, P; Girgis-Gabardo, A; Rudnicki, MA. Myogenic specification of side population cells in skeletal muscle. *J. Cell Biol*, 2002, 159, 123-134.

[20] McKirmey-Freeman,SL; Jackson, KA; Camargo, FD; Ferrari, G; Mavilio, F; Goodell, MA. Muscle-derived hematopoietic stem cells are hematopoietic in origin. *Proc. Natl. Acad. Sci. USA*, 2002, 99, 1341-1346.

[21] Qu-Petersen, Z; Deasy, B; Jankowski, R; Ikezawa, M; Cummins, J; Pruchnic, R; Mytinger, J; Cao, B; Gates, C; Wernig, A; Huard, J. Identification of a novel population of muscle stem cells in mice: potential for muscle regeneration. *J. Cell Biol*, 2002, 157, 851-864.

[22] Cao B, Zheng B, Jankowski RJ, Kimura S, Ikezawa M, Deasy B, Cummins J, Epperly M, Qu-Petersen Z, Huard J. Muscle stem cells differentiate into haematopoietic lineages but remain myogenic potential. *Nat. Cell Biol*, 2003, 5, 640-646.

[23] Polesskaya, A; Seale, P; Rudnicki, MA. Wnt signalling induces the myogenic specification of resident CD45+ adult stem cells during muscle regeneration. *Cell*, 2003, 113, 841-852.

[24] Odorico JS, Kaufman DS, Thomson JA. Multilineage differentiation from human embryonic stem cell lines. *Stem cells*, 2001, 19, 193-204.

[25] Thomson JA, Itskovitz-Eldor J, Shapiro SS, Waknitz MA, Swiergiel JJ, Marshall VS, Jones JM. Embryonic stem cells lines derived from human blastocysts. *Science*, 1998, 282, 1145-1147.

[26] Smith, AG. Embryo-derived stem cells: of mice and men. *Annu. Rev. Cell Dev. Biol*, 2001, 17, 435-462.

[27] Chamberlain, JS. Gene therapy of muscular dystrophy. *Hum. Mol. Genet*, 2002, 11, 2355-2362.

[28] Smythe, GM; Hodgetts, SI; Grounds, MD. Problems and solutions in myoblast transfer therapy. *J. Cell. Mol. Med*, 2001, 5, 33-47.

[29] Gibson, AJ; Karasinski, J; Relvas, J; Moss, J; Sheratt, TG; Strong, PN; Watt, DJ. Dermal fibroblasts convert to a myogenic lineage in mdx mouse muscle. *J. Cell Sci*, 1995, 108, 207-214.

[30] Ferrari, G; Cusella-De Angelis, G; Coletta, M; Paolucci, E; Stornaiulo, A; Cossu, G; Mavilio, F. Muscle regeneration by bone marrow derived progenitors. *Science*, 1998, 279, 1528-1530.

[31] Gussoni, E; Soneoka, Y; Strickland, CD; Buzney, EA; Khan, MA; Flint, AF; Kunkel, LM; Mulligan, RC. Dystrophin expression in the mdx mouse restored by stem cell transplantation. *Nature*, 1999, 401, 390-391.

[32] Zuk, PA; Zhu, M; Mizuno, H; Huang, J; Futrell, JW; Katz, AJ; Benhaim, P; Lorenz, HP; Hedrick, MH. Multilineage cells from human adipose tissue: implications for cell-based therapies. *Tissue Eng*, 2001, 7, 211-228.

[33] Safford, KM; Hicok, KC; Safford, SD; Halvorsen, YD; Wilkison, WO; Gimble, JM; Rice, HE. Neurogenic differentiation of murine and human adipose-derived stromal cells. *Biochem. Biophys. Res. Commun*, 2002, 294, 371-379.

[34] Seo, M; Suh, SY; Bae, YC; Jung, JS. Differentiation of human adipose stromal cells into hepatic lineage in vitro and in vivo. *Biochem. Biophys. Res. Commun*, 2005, 328, 258-264.

[35] Hutmacher, DW. Scaffold design and fabrication technologies for engineering tissues-stat of the art and future perspectives. *J. Miomater. Sci. Polym. Ed*, 2001, 12, 107-124.

[36] Powell, CA; Smiley, BL; Mills, J; Vandenburgh, HH. Mechanical stimulation improves tissue-engineered human skeletal muscle. *Am. J. Physiol. Cell Physiol*, 2002, 283, C1557-C1565

[37] Kosnik Jr, P; Dennis. RG; Faulkner, JA. Functional development of engineered skeletal muscle from adult and neonatal rats. *Tissue eng*, 2001, 7, 573-584.

[38] Okano, T; Satoh, S; Oka, T; Matsuda, T. Tissue engineering of skeletal muscle: highly dense, highly oriented hybrid muscular tissue biomimicking native tissues. *ASAIO J*, 1997, 43, M749-M753.

[39] Saxena, AK; Willital, GH; Vacanti, JP. Vascularized three-dimensional skeletal muscle tissue-engineering. *Biomed. Mater Eng*, 2001, 11, 275-281.

[40] Mulder, MM; Hitchcock, RW; Tresco, PA. Skeletal myogenesis on elastomeric substrates: implications for tissue engineering. *J. Miomater Sci. Polym. Ed*, 1998, 44, 355-370.

[41] Vandenburgh, HH. Functional assessment and tissue design of skeletal muscle. *Ann. N Y Acad. Sci*, 2002, 961, 201-202.

[42] Choi, YS; Park, SN; Suh, H. Adipose tissue engineering using mesenchymal stem cells attached to injectable PLGA spheres. *Biomaterials*. 2005, 26, 5855-5863.

[43] Choi, YS; Cha, SM; Lee, YY; Kwon, SW; Park, CJ; Kim, M. Adipogenic differentiation of adipose tissue derived adult stem cells in nude mouse. *Biochem. Biophys. Res. Commun*, 2006, 345, 631-637.

[44] Kim, M; Choi, YS; Yang, SH; Hong, HN; Cho, SW; Cha, SM; Pak, JH; Kim, CW; Kwon, SW; Park, CJ. Muscle regeneration by adipose tissue-derived adult stem cells attached to injectable PLGA spheres. *Biochem. Biophys. Res. Commun*, 2006, 348, 386-392.

[45] Food and Drug Administration. Web site: <www.fda.gov>

INDEX

A

B

C

D

E

F

I

J

K

L

N

O

P

Q

R

S

T

U

V

W

X

Y

Z